Springer Theses

Recognizing Outstanding Ph.D. Research

Aims and Scope

The series "Springer Theses" brings together a selection of the very best Ph.D. theses from around the world and across the physical sciences. Nominated and endorsed by two recognized specialists, each published volume has been selected for its scientific excellence and the high impact of its contents for the pertinent field of research. For greater accessibility to non-specialists, the published versions include an extended introduction, as well as a foreword by the student's supervisor explaining the special relevance of the work for the field. As a whole, the series will provide a valuable resource both for newcomers to the research fields described, and for other scientists seeking detailed background information on special questions. Finally, it provides an accredited documentation of the valuable contributions made by today's younger generation of scientists.

Theses are accepted into the series by invited nomination only and must fulfill all of the following criteria

- They must be written in good English.
- The topic should fall within the confines of Chemistry, Physics, Earth Sciences, Engineering and related interdisciplinary fields such as Materials, Nanoscience, Chemical Engineering, Complex Systems and Biophysics.
- The work reported in the thesis must represent a significant scientific advance.
- If the thesis includes previously published material, permission to reproduce this must be gained from the respective copyright holder.
- They must have been examined and passed during the 12 months prior to nomination.
- Each thesis should include a foreword by the supervisor outlining the significance of its content.
- The theses should have a clearly defined structure including an introduction accessible to scientists not expert in that particular field.

More information about this series at http://www.springer.com/series/8790

Florian Eyben

Real-time Speech and Music Classification by Large Audio Feature Space Extraction

Doctoral Thesis accepted by
the Technische Universität München, Germany

Springer

openSMILE:)
by audEERING™

audEERING™
intelligent Audio Engineering

Author
Dr. Florian Eyben
Institute for Human-Machine Communication (MMK)
Technische Universität München
Munich, Germany

Supervisors
Prof. Björn Schuller
Department of Computing
Imperial College
London, UK

and

Chair of Complex and Intelligent Systems
University of Passau
Passau, Germany

Prof. Werner Hemmert
Bio-Inspired Information Processing
Institute of Medical Engineering (IMETUM)
Technische Universität München
Munich, Germany

ISSN 2190-5053 ISSN 2190-5061 (electronic)
Springer Theses
ISBN 978-3-319-27298-6 ISBN 978-3-319-27299-3 (eBook)
DOI 10.1007/978-3-319-27299-3

Library of Congress Control Number: 2015957094

Printed on acid-free paper

This Springer imprint is published by SpringerNature
The registered company is Springer International Publishing AG Switzerland

Parts of this thesis have been published in the following articles:

Journals

- F. Eyben, M.Wöllmer, A. Graves, B. Schuller, E. Douglas-Cowie, and R. Cowie. Online emotion recognition in a 3-D activation-valence-time continuum using acoustic and linguistic cues. *Journal on Multimodal User Interfaces (JMUI)*, 3(1–2):7–19, March 2010. doi:10.1007/s12193-009-0032-6
- F. Eyben, M.Wöllmer, and B. Schuller. A Multi-Task Approach to Continuous Five- Dimensional Affect Sensing in Natural Speech. *ACM Transactions on Interactive Intelligent Systems, Special Issue on Affective Interaction in Natural Environments*, 2(1), March 2012. Article No. 6, 29 pages
- F. Eyben, A. Batliner, and B. Schuller. Towards a standard set of acoustic features for the processing of emotion in speech. *Proceedings of Meetings on Acoustics (POMA)*, 9(1):1–12, July 2012
- F. Eyben, K. Scherer, B. Schuller, J. Sundberg, E. André, C. Busso, L. Devillers, J. Epps, P. Laukka, S. Narayanan, and K. Truong. The Geneva Minimalistic Acoustic Parameter Set (GeMAPS) for Voice Research and Affective Computing. *IEEE Transactions on Affective Computing*, 2015. 14 pages, in press
- F. Eyben, G. L. Salomão, J. Sundberg, K. R. Scherer, and B. Schuller. Emotion in The Singing Voice—A Deeper Look at Acoustic Features in the Light of Automatic Classification. EURASIP *Journal on Audio, Speech, and Music Processing, Special Issue on Scalable Audio-Content Analysis*, 2015, 14 pages, in press

Conferences

- B. Schuller, B. Vlasenko, F. Eyben, G. Rigoll, and A. Wendemuth. Acoustic emotion recognition: A benchmark comparison of performances. In *Proc. of the IEEE Workshop on Automatic Speech Recognition and Understanding (ASRU) 2009*, pages 552–557, Merano, Italy, December 2009. IEEE
- F. Eyben, M. Wöllmer, and B. Schuller. openEAR—Introducing the Munich Open-Source Emotion and Affect Recognition Toolkit. In *Proc. of the 3rd International Conference on Affective Computing and Intelligent Interaction (ACII 2009)*, volume I, pages 576–581, Amsterdam, The Netherlands, September 2009. IEEE
- F. Eyben, M. Wöllmer, and B. Schuller. openSMILE—The Munich Versatile and Fast Open-Source Audio Feature Extractor. In *Proc. of ACM Multimedia 2010*, pages 1459–1462, Florence, Italy, 2010. ACM

- F. Eyben, F.Weninger, F. Gross, and B. Schuller. Recent developments in openSMILE, the munich open-source multimedia feature extractor. In *Proc. of ACM Multimedia 2013*, pages 835–838, Barcelona, Spain, 2013. ACM
- F. Eyben, B. Schuller, and G. Rigoll. Improving Generalisation and Robustness of Acoustic Affect Recognition. In L.-P. Morency, D. Bohus, H. K. Aghajan, J. Cassell, A. Nijholt, and J. Epps, editors, *Proc. of the 14th ACM International Conference on Multimodal Interaction (ICMI) 2012*, pages 517–522, Santa Monica, CA, USA, October 2012. ACM
- F. Eyben, F. Weninger, S. Squartini, and B. Schuller. Real-life Voice Activity Detection with LSTM Recurrent Neural Networks and an Application to Hollywood Movies. In *Proc. of ICASSP 2013*, pages 483–487, Vancouver, Canada, May 2013. IEEE
- F. Eyben, F. Weninger, and B. Schuller. Affect recognition in real-life acoustic conditions—A new perspective on feature selection. In *Proc. of INTERSPEECH 2013*, pages 2044–2048, Lyon, France, August 2013. ISCA

Supervisor's Foreword

It is an honour for me to introduce Dr. Eyben's outstanding doctoral thesis work accepted for publication in Springer Theses with this Foreword. Dr. Eyben was one of the first two Ph.D. students who joined my Machine Intelligence and Signal Processing research group at the Insititute for Human-Machine Communication of the Technische Universität München (TUM) in 2008. He worked on the award-winning EU-FP7 research project SEMAINE for 3 years, where he developed fundamental methods for automatic, real-time speech emotion recognition and non-verbal vocal analysis in the context of emotional sensitive virtual agent characters. During this period he created and maintained the openSMILE software toolkit for acoustic feature extraction and paralinguistic speech analysis. This toolkit and the methods implemented by it, laid the foundation of his ground-breaking thesis work. It was awarded twice at the ACM International Conference on Multimedia (2010 and 2013) and has been used to provide comparative baseline evaluations in several international, renowned research competitions in the field.

During his time at my research group, he was a big help for supporting and advising master students and new Ph.D. students as the group quickly grew in size. He constructively advised other team members on their research and collaborated with them in many studies and publications, contributing largely to the success and the international standing of the whole group.

Besides his project work, in his spare time he volunteered to help me with lecture preparation and to hold a tutorial course for the Pattern Recognition lecture offered at our institute. For the excellent tutorial, which was packed with hands-on examples and online demonstrations for most of the topics covered, he was awarded the best lecture award from the student association of the faculty for electrical engineering.

Automatic paralinguistic speech analysis is a young research field pioneered by my group since 2010. It is the continuation and generalisation of automatic speech emotion recognition, which has been an active research field for almost two decades. The goal of speech emotion recognition is to automatically provide an estimate of a speaker's emotional state from the acoustic properties of her/his voice. Psychological studies have identified numerous acoustic properties which are correlated to emotion. However, these properties are high-level descriptions of acoustic vocal qualities and cannot be robustly extracted from speech recordings by standard signal processing algorithms. Thus, inspired by my earlier work on acoustic feature brute-forcing for speech emotion recognition, Dr. Eyben adopted and extended this approach for his thesis work. The fundamental idea of this method is to compute a very high number of acoustic descriptors from a speech signal, regardless of their theoretical significance or correlation with a specific task, and then apply machine learning methods on large data sets to create robust models for identification of paralinguistic speech information, such as a speaker's emotion, age, gender, personality, alcohol intoxication level, sleepiness, depression, or certain voice pathologies. Dr. Eyben conducted original research work on a previously unseen large amount of acoustic descriptors which he implemented in his open-source software toolkit openSMILE. This toolkit is the first of its kind which is capable of extracting all descriptors incrementally, in real-time. Further, he was the first to publish an open-source emotion recognition toolkit (openEAR), based on openSMILE. All of this have opened completely new possibilities for research teams across the world, such as integrating out-of-the-box emotion recognition into their projects, rapidly creating interactive prototypes from their research, and—most notably—investigating automatic recognition of paralinguistic information besides emotion.

Renowned research competitions which provide comparative evaluations for several automatic recognition tasks in the field were organised by my group annually at INTERSPEECH and other highly recognised venues. The baseline acoustic descriptors and baseline results were created based on Dr. Eyben's thesis work, which thereby has set a widely acknowledged standard for international comparison—a high standard even—as these baseline results were often hard to beat despite the large number of participating teams which included top research teams from all over the world.

Recently, audEERING, a spin-off company was founded by me, Dr. Eyben, and colleagues. Having bought the intellectual property rights to the software and methods developed by Dr. Eyben at TUM, audEERING is now successfully marketing the emotion recognition and non-verbal music and speech analytics technology for applications such as call and contact centre quality monitoring, marketing research, brand testing, human health and safety, as well as security and entertainment products.

Munich Prof. Björn Schuller
February 2015

Abstract

Automatic classification of speech and music has become an important topic with the advent of speech technologies in devices of our daily lives, such as smartphones and smartwatches. While for automatic speech recognition, commercial technologies with good accuracies are available to buy, the classification of music and paralinguistic information beyond the textual content, such as mood, emotion, voice quality, or personality, is a very young field, but possibly the next technological milestone for man–machine interfaces.

This thesis advances the state of the art in the area by defining standard acoustic feature sets for real-time speech and music analysis and by proposing solutions for real-world problems: a multi-condition learning approach to increase noise robustness, noise robust incremental segmentation of the input audio stream based on a novel, context-aware, and data-driven voice activity detector, and a method for fully (time and value) continuous affect regression tasks are introduced. Standard acoustic feature sets were defined and evaluated throughout a series of international research challenges. Further, a framework for incremental and real-time acoustic feature extraction is proposed, implemented, and published as an open-source toolkit (openSMILE). The toolkit implements all of the proposed baseline acoustic feature sets and has been widely accepted by the community—the publications introducing the toolkit have been cited over 400 times.

The proposed acoustic feature sets are systematically evaluated on 13 databases containing speech affect and music style classification tasks. Experiments over a wide range of conditions are performed, i. e. training instance balancing, feature value normalisation, and various classifier parameters. Also, the proposed methods for real-time, noise robust, and incremental input segmentation and noise robust multi-condition classification are extensively evaluated on several databases. Finally, fully continuous (time and value), automatic recognition of affect in five dimensions with long short-term memory recurrent neural networks is evaluated on a database of natural and spontaneous affect expressions (SEMAINE). The superior performance of the proposed large feature sets over smaller sets was shown successfully for a multitude of tasks.

All in all, this thesis is a significant contribution to the field of speech and music analysis and hopefully expedites the process of bringing real-world speech and music analysis applications, such as robust emotion or music mood recognition, a bit closer to daily use.

Acknowledgments

This thesis is a result of my research work conducted at the Institute for Human-Machine Communication at the Technische Universität München, Germany.

I would like to thank my supervisor Prof. Björn Schuller for providing the opportunity and resources to write my thesis in his Machine Intelligence and Signal Processing group at the Institute. He was an excellent mentor and supervisor, leaving me enough freedom to conduct research in the areas of my interest but giving me enough guidance to help me focus my research on important issues, learn how to write scientific papers, and publish my work successfully. I enjoyed the great collaborative atmosphere while working in his team and I am very grateful to have been introduced to and get to work with many high-level international collaborators through him and his projects. I further want to thank him for providing me the opportunity to do two internships abroad during the time of my thesis.

I am grateful to Prof. Gerhard Rigoll, who—as the head of the Institute—made my stay at the institute possible in the first place and constantly supported my research and the research of the group. I would like to thank, both him and Prof. Schuller, for their support which enabled me to attend many international conferences and project meetings.

I would further like to thank my second examiner Prof. Werner Hemmert for his time and effort dedicated to my thesis, as well as Prof. Wolfgang Utschick for hosting my defence at his institute.

Special thanks go to all my colleagues who have worked with me and who have always been available for inspiring and challenging discussions. I especially want to thank my co-authors, Felix Weninger, with whom I had long and inspiring discussions and have developed many ideas and publications together, and Martin Wöllmer, who extensively collaborated with me and contributed to the success of the SEMAINE project. Further, I want to thank my colleague Erik Marchi for his testing and coding work on the openSMILE GUI parts and for valuable in-office discussions, as well as Jürgen Geiger and Fabien Ringeval, for their support and discussions.

For his great coordination of the SEMAINE project, his integration efforts, and the wonderful coder-camps, I would like to thank Marc Schröder. For a fruitful collaboration in the SEMAINE project and for helping me find accommodation during my internship in London, I would like to thank Michel Valstar. For excellent discussions on acoustic parameters, for providing the GEMEP corpus, for his lead efforts in proposing the Geneva Minimalistic Parameter Set, and for his support for the large scale parameter set evaluations, I would like to express special thanks to Prof. Klaus Scherer.

I also want to thank my master students Christoph Kozielski, Benedikt Gollan, Marcel Knapp, and Bernd Huber, for their contributions of ideas and for testing and improving openSMILE throughout their thesis work. Also, I want to thank all of the many researchers around the world who actively use openSMILE, give feedback, and thereby contribute to the success of the standard feature sets and the toolkit.

Most of all, for their encouraging support, love, and care, I would like to thank my wife Sarah and my parents.

Munich
June 2014

Florian Eyben

The research leading to these results has received funding from the European Community's Seventh Framework Programme (FP7/2007–2013) under grant agreements No. 211486 (SEMAINE) and No. 289021 (ASC-Inclusion). The research was further supported by an ERC Advanced Grant in the European Community's Seventh Framework Programme under grant agreement 230331-PROPEREMO (Production and perception of emotion: an affective sciences approach) awarded to Prof. Klaus Scherer and hosted by the University of Geneva.

Contents

Acronyms

ABC	Airplane Behaviour Corpus
ACF	Autocorrelation Function
AR	Autoregressive
ARFF	Attribute Relation Feature Format
AR-GARCH	Autoregressive-Generalised Autoregressive Conditional Heteroskedasticity
ASR	Automatic Speech Recognition
AUC	Area Under (ROC) Curve
AVEC	Audio-Visual Emotion Challenge
AVIC	Audio-Visual Interest Corpus
BLSTM	Bidirectional Long Short-Term Memory
BLSTM-RNN	Bidirectional Long Short-Term Memory Recurrent Neural Network
BPM	Beats per Minute
BPTT	Backpropagation Through Time
BRD	BallRoom Dance-style
CC	(Pearson) Correlation Coefficient
CEC	Constant Error Carousel
CENS	CHROMA Energy-distribution Normalised Statistics
CEP	Cepstrum
CFS	Correlation-based Feature-subset Selection
ComParE	Computational Paralinguistics ChallengE
CV	Coefficient of Variation
dB	Decibel
DC	Direct current
DCT	Discrete Cosine Transformation
DCT-II	Discrete Cosine Transformation Type-II
DES	Danish Emotional Speech
DFT	Discrete Fourier Transformation
DSP	Digital Signal Processor
EER	Equal Error Rate

eGeMAPS	Extended Geneva Minimalistic Acoustic Parameter Set
EMI	Electromagnetic Interference
EMO-DB	Berlin Emotional Speech Database
EOI	End-of-Input
FFNN	Feed-Forward Neural Network
FFT	Fast Fourier Transformation
FIR	Finite Impulse Response
FNR	False Negative Rate
FPGA	Field Programmable Gate Array
FPR	False Positive Rate
FT	Fourier Transformation
GeMAPS	Geneva Minimalistic Acoustic Parameter Set
GEMEP	Geneva Multimodal Emotion Portrayals
GMM	Gaussian Mixture Model
HMM	Hidden Markov Model
HNR	Harmonics-to-Noise Ratio
HPCP	Harmonic Pitch Class Profiles
HTK	Hidden Markov Toolkit (Young et al. 2006)
Hz	Hertz
IIR	Infinite Impulse Response
IQR	Inter-Quartile Range
IS09	INTERSPEECH 2009 Emotion Challenge
IS10	INTERSPEECH 2010 Paralinguistics Challenge
IS11	INTERSPEECH 2011 Speaker State Challenge
IS12	INTERSPEECH 2012 Speaker Trait Challenge
LLD	Low-level Descriptor
LOI	Level of Interest
LOSO	Leave-One-Subject-Out
LP	Linear Prediction
LPC	Linear Predictive Coding
LR	Likelihood Ratio
LSP	Line Spectral Pair
LSF	Line Spectral Frequencies
LSTM	Long Short-Term Memory
LSTM-RNN	Long Short-Term Memory Recurrent Neural Network
LTAS	Long-Term Average Spectrum
LTI	Linear Time Invariant
MCR	Mean-Crossing Rate
MCT	Multi-condition Training
MFB	Mel-Frequency Band
MFCC	Mel-Frequency Cepstral Coefficient
MIR	Music Information Retrieval
MLP	Multi-Layer Perceptron
MPEG	Moving Picture Experts Group
MRN	Mean Range Normalisation

MVN	Mean Variance Normalisation
NMF	Non-negative Matrix Factorisation
NN	Neural Network
PCA	Principal Component Analysis
PCP	Pitch Class Profiles
PDA	Pitch Detection Algorithm
PLP	Perceptual Linear Prediction
PLP-CC	Perceptual Linear Prediction Cepstral Coefficients
PMF	Probability Mass Function
RASTA	RelAtive Spectral TrAnsform
RASTA-PLP	RelAtive Spectral TrAnsform Perceptual Linear Prediction
ReLU	Rectified Linear Unit
RIR	Room Impulse Response
RMS	Root Mean Square
RNN	Recurrent Neural Network
ROC	Receiver Operating Characteristic
RoP	Roll-off Point
rProp	Resilient Propagation
RTF	Real-time factor
SAD	Speech Activity Detection
SAL	Sensitive Artificial Listener
SD	Spectral Difference
SHS	Subharmonic Summation
SING	Geneva Singing Voice Database
SMO	Sequential Minimal Optimisation
SNR	Signal-to-Noise Ratio
SPL	Sound Pressure Level
STFT	Short-Time Fourier Transform
SUSAS	Speech Under Simulated and Actual Stress
SVM	Support Vector Machine
SVR	Support Vector Regression
TEO	Teager energy operator
TNR	True Negative Rate
TPR	True Positive Rate
UAR	Unweighted Average Recall
VAD	Voice Activity Detection
VAM	Vera-am-Mittag
WAR	Weighted Average Recall
ZCR	Zero-Crossing Rate

Symbols

General Signal Representation

t	Time (continuous)
n	Discrete time index, of a sample, or frame
T	Period
τ	Discrete time lag (e.g. for ACF)
N	Number of samples/frames/items/etc.
$a(t)$	Continuous amplitude of time continuous signal a at time t
$a(n)$	Continuous amplitude of time discrete signal a at index $n = \frac{t}{T_s}$
$x(n)$	Discrete amplitude of time discrete signal $a(n)$, or general time series of length N
T_s	Sampling period in seconds
f_s	Sampling frequency in Hertz $f_s = \frac{1}{T_s}$

Short-Time Processing/Windowing

k	Discrete frame index
$\hat{n}$	Discrete sample index, relative to a single frame
K	Total number of frames
N_f	Frame size in samples
L_f	Frame size in seconds
$N_f^{(T)}$	Frame step/period in samples
T_f	Frame step/frame period in seconds
O_f	Percentage of overlap between two adjacent frames
$w_{xx}(n)$	Window function of type xx, time domain
$W_{xx}(m)$	Discrete spectrum of windowing function $w_{xx}(n)$

Frequency Domain Signals

f_0	DFT base frequency, DFT frequency resolution
f	Frequency
$f^{(scale)}$	Frequency expressed in the unit of "scale", e.g. Hz for "lin"
Θ_{scale}	Frequency scale transformation function from linear f to frequency scale "scale"
m	Discrete frequency $m = \frac{f}{f_0} \in [0; M]$, bin index
m'	Real-valued frequency bin index
$m^{(scale)}$	Bin index for spectral scale "scale"
$F(m)$	Function F which converts the discrete frequency index m to a linear frequency f
M	Number of discrete frequency bins
$X(m)$	Discrete (complex) spectrum of signal $x(n)$ at discrete frequency m
$X_M(m)$	Spectral magnitude of signal $x(n)$ at discrete frequency m
$X_{M,norm}(m)$	(Magnitude) spectral density
$X_\phi(m)$	Spectral phase of signal $x(n)$ at discrete frequency m
$X_P(m)$	Power spectrum
$X_{P,norm}(m)$	Power spectral density

Low-Level Acoustic Descriptors

Spectral Bands and Filter Shapes

b	Spectral band index (for a band spectrum)
B	Number of spectral bands
$g(f)$	Spectral filter shape, continuous function in f
$\Phi(m)$	Spectral filter shape, discretised to bins
β	Filter bandwidth
f_c	Filter centre frequency for bandpass filters, filter cutoff frequency for high-/low-pass filters
f_l	Lower frequency bound
f_u	Upper frequency bound
m_l	Lower frequency bound, bin index
m_u	Upper frequency bound, bin index
α_l	Weighting factor of lower bound frequency bin
α_u	Weighting factor of upper bound frequency bin

Fundamental Frequency

F_0	Fundamental frequency ("pitch")
p_v	Probability of voicing

$X_H(m)$	Subharmonic sum spectrum
N_c	Number of pitch candidates

Energy/Loudness

E	Signal energy
E_n	Normalised signal energy (= signal power)
E_{rms}	Root-mean-square normalised signal energy
E_{log}	Logarithmic signal energy
E_l	Loudness, auditory-based model
$E_{l,approx}$	Loudness (narrow-band approximation)
$\Phi\{x(n)\}$	Teager energy operator

Spectral Dynamics

$SD^{(k)}$	Spectral difference between frame k and $k-1$
$SD_+^{(k)}$	Positive spectral difference between frame k and $k-1$
S_{flux}	Spectral flux

Spectral Statistics

$S_{centroid}$	Spectral centroid
$S_{entropy}$	Spectral entropy
$S_{flatness}$	Spectral flatness
$S_{kurtosis}$	Spectral kurtosis
$S_{rop(n)}$	n % spectral roll-off point
$S_{skewness}$	Spectral skewness
S_σ	Spectral standard deviation
$S_{variance}$	Spectral variance (spectral spread)
HM	Spectral harmonicity
$p_X(m)$	Spectrum $X(m)$ converted to a probability mass function
η	Hammarberg index
ρ_α	Alpha ratio

PLP/Auditory Spectra

$X_{P,aud}(b)$	Discrete auditory band spectrum
$E(b)$	Equal loudness weight for spectral band b

Cepstrum/Cepstral Features

$s(n)$	Vocal tract source signal in the time domain, discrete
$S(m)$	Vocal tract source signal in the frequency domain, excitation spectrum
$h(n)$	Impulse response in the time domain, discrete
$H(m)$	Impulse response in the frequency domain transfer function
L	In context of MFCC: Liftering coefficient
X_{floor}	Spectral band floor value
$C(i)$	*i*th cepstral coefficient

Linear Prediction

p	Linear predictor order
$\hat{x}(n)$	Approximation of signal $x(n)$, e.g. through a linear predictor
a_i	Linear predictor coefficients
k_j	Reflection coefficients
$e(n)$	Error signal (e.g. in linear prediction, residual signal)
α_j	Sum-squares error of linear predictor of order j
$r(d)$	Autocorrelation coefficient for discrete lag d
h_{inv}	Impulse response of the inverse vocal tract filter
H_{inv}	Impulse response of the inverse vocal tract filter

Formants

F_i	Speech formant frequency, if $i \geq 1$, e.g. F_1 and F_2
$F_i^{(bw)}$	Speech formant bandwidth for $i \geq 1$

Harmonic- and Formant-Amplitudes

H_i	Amplitude of *i*th spectral harmonic
A_i	Amplitude of highest spectral harmonic in *i*th formant range
H_{i-j}	Harmonic amplitude differences/ratios between *i*th and *j*th harmonic

Voice Quality

J_{pp}	Period-to-period jitter (local)
$\overline{J_{pp}}$	Average (within one frame) period-to-period jitter
J_{pp}	Period-to-period shimmer (local)
$\overline{S_{pp}}$	Average (within one frame) period-to-period shimmer

HNR_{wf}	Harmonics-to-noise ratio computed via the direct waveform matching method
$HNR_{wf,log}$	Logarithmic harmonics-to-noise ratio computed via the direct waveform matching method
HNR_{acf}	Harmonics-to-noise ratio computed via autocorrelation
$HNR_{acf,log}$	Logarithmic harmonics-to-noise ratio computed via autocorrelation

Musical

s	Semitone index
S	Total number of semitones per octave (for PCP)
O	Number of octaves

Delta/Derived

$\delta_i^W(n)$	ith order delta regression coefficient at discrete time n; window size W
$x_{sma}(n)$	$x(n)$ smoothed with a moving average filter (typically 3 frames)

Supra-Segmental Features

$\underline{x}$	Feature vector (frame)
$\underline{x}(n)$	Time series of feature vectors
$\underline{X}$	Supra-segmental feature vector
$\mathcal{F}$	Functional

Means

μ	Mean
μ_a	Arithmetic mean (amean)
$\mu_{\mid}a\mid$	Arithmetic mean of absolute values (absmean)
μ_{a+}	Arithmetic mean of positive values only (posamean)
μ_{a-}	Arithmetic mean of negative values only (negamean)
$\mu_a^{(nz)}$	Arithmetic mean of non-zero values only (nzamean)
μ_g	Geometric mean (nzgmean), here: always only of non-zero values
μ_q	Quadratic mean (qmean)
μ_{rq}	Root-quadratic mean (rqmean), i. e. square root of μ_q
μ_f	Flatness
μ_x	Arithmetic mean of $x(n)$

Moments

m_i	ith central (statistical) moment
$\overline{m}_i$	ith standardised (statistical) moment
σ	Standard deviation
σ^2	Variance
$\overline{\sigma}$	Coefficient of Variation

Extremes

R_x	Range of signal $x(n)$
x_{min}	Minimum value of signal $x(n)$
x_{max}	Maximum value of signal $x(n)$
n_{min}	Position n (frame) of minimum value in $x(n)$
n_{max}	Position n (frame) of maximum value in $x(n)$
$d_{max,\mu}$	Difference between maximum value and arithmetic mean of $x(n)$
$d_{min,\mu}$	Difference between minimum value and arithmetic mean of $x(n)$

Distributions

P_j	jth percentile
$x_{centroid}$	Centroid of time domain signal $x(n)$
x_θ	Absolute amplitude threshold on signal $x(n)$
x_θ^{rel}	Relative amplitude threshold on signal $x(n)$

Regression

m	Linear regression slope
o	Linear regression offset
a	Quadratic regression coefficient 1 (shape, quadratic)
b	Quadratic regression coefficient 2 (shape, linear)
c	Quadratic regression coefficient 3 (offset)
n_v	Temporal location (index) of parabola vertex (quadratic regression)
$\hat{x}(n_v)$	Amplitude of $x(n)$ at parabola vertex index n_v (quadratic regression)
$\overline{e}$	Normalised (by sequence length) sum-squares (quadratic) regression error
$\overline{e}_a$	Normalised (by sequence length) absolute (linear) regression error
m_{left}	Left slope of parabola estimated from quadratic regression
m_{right}	Right slope of parabola estimated from quadratic regression
CC	Pearson correlation coefficient

Peaks

$x_{\mu}^{(peaks)}$	Arithmetic mean of peak amplitudes
$d_{\mu}^{(peaks)}$	Mean distance between peaks (temporal)
$d_{\sigma}^{(peaks)}$	Variance of inter-peak distances
$x_{\mu}^{(minima)}$	Arithmetic mean of valley amplitudes
$d_{\mu}^{(minima)}$	Mean distance between valleys (temporal)
$d_{\sigma}^{(minima)}$	Variance of inter-valley distances
$m_{j,i}$	Slope (rising) of line connecting a valley i and the following (future) peak j
$m_{i,j}$	Slope (falling) of line connecting a peak j and the following (future) valley i
n_{rel}	Temporal (frame) index relative to length N of segment: $n_{rel} = n/N$

Modelling

Neural Networks

$g(x)$	Squashing function/neuron activation function
$\underline{\mathbf{w}}$	Neuron weight vector
$\underline{\mathbf{x}}$	Neuron input vector
c_t	LSTM cell state at time t
i_t	LSTM cell input gate activation at time t
o_t	LSTM cell output gate activation at time t
f_t	LSTM cell forget gate activation at time t

Incremental Processing

N_{pre}	Number of continuous speech frames required for the detection of the start of a speech segment
N_{post}	Number of continuous silence/non-speech frames required for the detection of the end of a speech segment
L_{max}	Maximum allowed speech segment length
L_{min}	Minimum allowed speech segment length
$N_{seg}^{(i)}$	Length of the ith speech segment
N_{win}	Window length (in frames) for incremental sub-segmentation
N_{step}	Window step length (in frames) for incremental sub-segmentation
L_{win}	Window length (in frames) for incremental sub-segmentation
L_{step}	Window step length (in frames) for incremental sub-segmentation
p_w	openSMILE data-memory level write pointer
$p_{r,i}$	openSMILE data-memory level read pointer of ith reader

N_{free}	Free space (frames/samples) in openSMILE data-memory level
N_{avail}	Maxmimum available data items (frames/samples) in openSMILE data-memory level
$N_{avail}^{(i)}$	Available data items (frames/samples) in openSMILE data-memory level for ith reader
$\overline{N}_{xx}$	Same as N_{free} and N_{avail} for ring buffer data-memory levels

List of Figures

List of Tables

Chapter 1
Introduction

In our daily lives we are interacting with a constantly growing number of computerised devices, such as in-car computers, electronic vending machines, automated telephone call-centres, smartphones, and wearable computing devices. In order to make the interaction with these devices as simple and intuitive as possible, more natural interaction methods are essential (Pentland 2005; Sinha et al. 2010). Currently, most man-machine interfaces rely on haptic input, i. e., keyboard, mouse, and recently touchscreens (cf. Grudin 2007; Schenk and Rigoll 2010). For every new device the user has to learn how to operate it by reading the manual, for example. In the future, computing and communication devices will be even more ubiquitous and high-functionality (Sinha et al. 2010), making it more and more important to have intuitive interfaces, which accept visual and spoken input, and with which we can communicate naturally and intuitively just like with any other person—without reading a manual first (Sinha et al. 2010).

An important ingredient for more natural man-machine interaction is to enable machines to analyse and understand human intentions and social behaviour as well as humans do. This is a very complex and ambitious goal that we are, yet, far from reaching (Pentland 2005). To reach such an ambitious goal, machines have to be able to analyse social (Schuller et al. 2008; Vinciarelli et al. 2009), natural affective (Zeng et al. 2009; Schuller et al. 2009; Batliner et al. 2011), paralinguistic (Schuller and Batliner 2013), and contextual signals (Vinciarelli et al. 2009), for example. Further, machines have to be able to deal with ambiguities (ter Maat and Heylen 2009), and robustly interpret all input and deduce the correct intention of the user from them, i. e., the machines should behave in a way that would be considered socially intelligent (Schroeder and McKeown 2010).

F. Eyben, *Real-time Speech and Music Classification by Large Audio Feature Space Extraction*, Springer Theses, DOI 10.1007/978-3-319-27299-3_1

1.1 Acoustic Analysis of Speech and Music

This thesis focuses on the analysis of human signals, limited to the acoustic channel, in particular speech and music, thereby connecting fields of automatic speech recognition (cf. e. g., Furui 1996) and Music Information Retrieval (MIR) (cf. Bello et al. 2000; Fingerhut 2004). The human voice is presumably the most important mean of direct human to human communication: Speech carries linguistic content, i. e., sentences and words, and paralinguistic content (Schuller and Batliner 2013), such as mood, affect, speaker states such as intoxication and sleepiness, and speaker traits such as age, gender, and personality (Schuller and Batliner 2013). As was shown by the author of this thesis in other studies, speech possibly also carries information about eye contact in dyadic conversations (Eyben et al. 2013b), or heart-rate and skin conductivity in some cases (Schuller et al. 2013). Further, the average subjective likeability of a voice was estimated automatically in Eyben et al. (2013a). Singing and—more generally—music allows for an even better expression of mood and emotions, in particular, through rhythm and melody.

From a technical point of view (cf. Schuller 2013), speech and music are acoustic signals, represented in the physical world by micro variations of pressure, mostly air pressure, in the range from approx. 50–8,000 Hz. When speech or music are recorded with a microphone, the air pressure modulation is converted to an electrical signal. In digital sound processing, as this thesis deals with exclusively, the analogue electrical signal is digitised by analogue digital conversion and sampling. Digital signal processing techniques (Oppenheim and Schafer 1975) are then applied to the audio samples to reduce the amount of information to the aspects of interest, and discard unnecessary or redundant information. This process is known as feature extraction (cf. e. g., Ruske 1993).

1.2 Deficiencies of the State-of-the-Art

While the automatic, computer-based recognition of the linguistic content from speech has matured over three decades of research (cf. Rabiner 1989; Furui 1996; Dahl et al. 2012), the analysis of speech paralinguistics (Schuller and Batliner 2013) and high level music attributes such as mood (e. g., Schmidt and Kim 2010) is a comparably young field (Schuller 2013).

Automatic recognition of emotions, for example, in acted speech and in acoustic laboratory conditions, i. e., on the same database, where training and test data have the same acoustic conditions, was shown to be at human performance levels (e. g., Schuller 2006). However, recognition of affect and paralinguistic information "in the wild" (Dhall et al. 2013), i. e., in noisy and reverberated acoustic conditions, on spontaneous speech, and in a continuum of non-prototypical, real-life emotions, or in music, for example, remains a challenge (cf. e. g., You et al. 2006; Schuller et al. 2006, 2007; Schmidt and Kim 2010). This is most evident, if the good performance of, e. g., emotion classification within a corpus of prototypical emotions is contrasted to the performance obtained in cross corpus classification (e. g., Schuller et al. 2010; Eyben et al. 2010a), i. e., when training a model on one database, and evaluating on another which contains similar emotions but was likely recorded with a different protocol and in a different environment. Robustness under such conditions is very important for, e. g., use in in-car environments (Eyben et al. 2010b).

Experiments conducted so far, were mostly done using small acoustic feature sets (cf. e. g., Ververidis and Kotropoulos 2006), extracted with various toolkits, such as Praat (Boersma 2001) or Mirtoolbox (Lartillot and Toiviainen 2007), limiting the comparability and reproducibility of experiments across research sites.

1.3 Aims of This Thesis

In order to advance the field of paralinguistic speech and music analysis research, this thesis addresses the challenge of analysis of speech and music beyond the textual and musical note content. This means a higher level analysis based on the acoustic mid- and long-term properties, in order to identify paralinguistic attributes such as affect, or high level music attributes such as genre or mood. A large space of acoustic parameters is discussed and evaluated in this context, and—to encourage comparability and reproducibility of results—standard acoustic parameter sets are proposed and evaluated for voice based affect recognition and music style identification. Further, a strong emphasis is put on the applicability of the proposed methods to real-world use-cases. This includes discussion of real-time processing issues related to the signal processing, introduction of an incremental processing software-framework, and the evaluation of novel methods to increase noise robustness.

In particular, the following four major aims have been defined for this thesis:

1. Development of an open-source framework and toolkit for incremental speech and music analysis with a strong focus on real-time, incremental feature extraction algorithms and a broad coverage of acoustic descriptors,
2. Definition of novel, large-scale standard acoustic feature sets for incremental speech and music analysis,
3. Design of speech and music analysis methods which can robustly handle real-life conditions,
4. Evaluation of the framework, the methods, and the standard acoustic feature sets on a wide range of music classification and speech affect recognition tasks.

These aims are elaborated in more details in the following sections.

1.3.1 Real-time Analysis Framework

Compared to video/visual feature extraction, the computational complexity of audio feature extraction is significantly lower on average. Still, not all acoustic features can be extracted in real-time, and some modifications to the overall chain of processing must be made in order to efficiently process inputs in real-time and—especially—obtain incremental classification results with a latency as low as possible. Also, no standard feature extraction toolkit which unites a large set of speech and music descriptors existed prior to this thesis. Thus, the first goal defined for this thesis in the scope of the SEMAINE EU-FP7 project[1] was to create a flexible and modular framework and toolkit for real-time, incremental audio feature extraction. The framework should be usable for both large-scale, efficient off-line batch processing for research, as well as for on-line, incremental processing in interactive speech dialogue systems, for example. Further, a large set of acoustic descriptors should be included and algorithms for real-time, incremental extraction of these features should be implemented, described, and made publicly available, in order to allow researchers to use a common implementation for many tasks, instead of a multitude of different libraries.

1.3.2 Baseline Feature Sets

Next, from the large set of acoustic descriptors implemented in the framework, high-dimensional feature sets based on feature brute-forcing should be proposed and published in order to set baseline standards for international evaluation campaigns and comparison of audio analysis and classification algorithms across the world. The feature sets should be shared freely with the research community, and should be extractable with an open-source toolkit (see previous aim, Sect. 1.3.1), in order to encourage reproducible research basing on standard acoustic features and standard feature extraction algorithms.

1.3.3 Real-World Robustness

As current methods mostly deal with laboratory conditions and show results on clean speech corpora, real-life conditions are mainly neglected. This thesis aims to propose (and evaluate) solutions to improve the analysis performance under real-world conditions, such as high levels of background noise, reverberation, or speaker variability. Further, another important issue to be addressed in this respect was the segmentation of continuous input audio streams, especially for real-time interactive systems. In contrast to textual speech recognition, in acoustic analysis of speech larger units of speech (Zeng et al. 2009), e. g., sentences or phrases (Batliner et al. 2010) are commonly used, and the performance of the whole approach is highly dependent on the segmentation chosen (Batliner et al. 2010). In most research the units are pre-segmented by hand and therefore all results obtained so far have the precondition of

[1] http://www.semaine-project.eu/.

perfect segmentation. In real-time, real-world settings, perfect segmentation is not possible due to limited context and often limited resources. Moreover, long segments such as sentences increase the latency of the analysis algorithm unnecessarily. Thus, for this thesis an alternate input segmentation method should be developed which is suitable for incremental processing with low latency.

1.3.4 Large-Scale Evaluation

The final, fourth goal, which links to the three previous goals, was to evaluate the proposed baseline feature sets, real-life robustness methods, and the quality of the feature extraction algorithms implemented for this thesis systematically on a vast number of speech and music analysis tasks. The goal thereby was to confirm the validity and suitability of the proposed feature sets, highlight the differences of the sets with respect to different tasks and data-sets, and investigate the influence of classifier parameters and variations of the training and evaluation procedure.

1.4 Overview

This thesis is grouped into five main Chaps. (2–6): Chap. 2 describes all the methods applied in this thesis for the automatic classification of speech and music signals. These include in particular the signal processing and audio feature extraction algorithms (Sect. 2.2) and the supra-segmental feature summarisation and large-scale feature brute-forcing (Sect. 2.4), as well as the static and dynamic modelling (machine learning) methods (Sect. 2.5).

The proposed standard baseline acoustic feature sets are introduced in Chap. 3, The description of the incremental processing framework developed for this thesis is given in Chap. 4 and the methods developed to increase real-world robustness of speech and music analysis are described in Chap. 5.

Chapter 6 describes the extensive, large-scale evaluations of the standard acoustic parameter sets conducted on several data-sets for the automatic recognition of speech affect categories, music styles, and fully time and value continuous affect recognition. Moreover, the effectiveness of the real-world robustness methods is evaluated on realistic data.

The findings of this thesis are summarised and concluded in Chap. 7. Further, it is critically evaluated how well the goals described in Sect. 1.3 were achieved and which topics are still open and deserve more attention in follow up work.

References

A. Batliner, D. Seppi, S. Steidl, B. Schuller, Segmenting into adequate units for automatic recognition of emotion-related episodes: a speech-based approach. *Advances in Human Computer Interaction, Special Issue on Emotion-Aware Natural Interaction*, 2010, p. 15 (Article ID 782802). (on-line)

A. Batliner, B. Schuller, D. Seppi, S. Steidl, L. Devillers, L. Vidrascu, T. Vogt, V. Aharonson, N. Amir, The Automatic Recognition of Emotions in Speech, in *Emotion-Oriented Systems: The Humaine Handbook, Cognitive Technologies*, ed. by P. Petta, C. Pelachaud, R. Cowie (Springer, Berlin, 2011), pp. 71–99

J.P. Bello, G. Monti, M. Sandler. Techniques for automatic music transcription. In *Proceedings of the International Symposium on Music Information Retrieval (MUSIC-IR 2000)*. ISMIR, 2000, p. 8

P. Boersma, Praat, a system for doing phonetics by computer. Glot Intern. **5**(9/10), 341–345 (2001)

G. Dahl, D. Yu, L. Deng, A. Acero, Context-dependent pre-trained deep neural networks for large vocabulary speech recognition. IEEE Trans. Audio, Speech Lang. Process. **20**(1), 30–42 (2012)

A. Dhall, R. Goecke, J. Joshi, M. Wagner, T. Gedeon. Emotion recognition in the wild challenge 2013. In *Proceedings of the 15th ACM International conference on multimodal interaction (ICMI) 2013*, ACM. Sydney, Australia, pp. 509–516, December 2013

F. Eyben, A. Batliner, B. Schuller, D. Seppi, S. Steidl, Cross-Corpus Classification of Realistic Emotions—Some Pilot Experiments. In L. Devillers, B. Schuller, R. Cowie, E. Douglas-Cowie, A. Batliner, editors, *Proceedings of the 3rd International Workshop on EMOTION: Corpora for Research on Emotion and Affect, satellite of LREC 2010*, Valletta, Malta, pp. 77–82 May 2010a. European Language Resources Association (ELRA)

F. Eyben, M. Wöllmer, T. Poitschke, B. Schuller, C. Blaschke, B. Färber, N. NguyenspsThien, Emotion on the Road—Necessity, Acceptance, and Feasibility of Affective Computing in the Car. *Advances in Human Computer Interaction (AHCI), Special Issue on Emotion-Aware Natural Interaction*, 2010b. p. 17. doi:10.1155/2010/263593. Article ID 263593

F. Eyben, F. Weninger, E. Marchi, B. Schuller, Likability of human voices: A feature analysis and a neural network regression approach to automatic likability estimation. In *Proceedings of the 14th International Workshop on Image and Audio Analysis for Multimedia Interactive Services (WIAMIS) 2013*, IEEE. Paris, France, pp. 1–4 July, 2013a

F. Eyben, F. Weninger, L. Paletta, B. Schuller, The acoustics of eye contact—Detecting visual attention from conversational audio cues. In *Proceedings of the 6th Workshop on Eye Gaze in Intelligent Human Machine Interaction: Gaze in Multimodal Interaction (GazeIn '13), held in conjunction with the 15th International Conference on Multimodal Interaction (ICMI) 2013*, ACM. Sydney, Australia, pp. 7–12, December 2013b

M. Fingerhut, *Music information retrieval, or how to search for (and maybe find) music and do away with incipits*. In Proceedings of the IAML-IASA Congress, Oslo, Norway, August 2004

S. Furui, *Digital Speech Processing: Synthesis, and Recognition*, 2nd edn., Signal Processing and Communications (Marcel Denker Inc., New York, 1996)

J. Grudin, in *Human-Computer Interaction Handbook*, 2nd edn., A moving target: The evolution of human-computer interaction, ed. by A. Sears, J .A. Jacko (CRC Press, Boca Raton, 2007), pp. 1–24. ISBN 0-8058-5870-9

O. Lartillot, P. Toiviainen, MIR in Matlab (II): a toolbox for musical feature extraction from audio. In *Proceedings of the ISMIR 2007*, ISMIR. Vienna, Austria (2007)

A.V. Oppenheim, R.W. Schafer, *Digital Signal Processing* (Prentice-Hall, Englewood Cliffs, 1975)

A. Pentland, Socially aware computation and communication. IEEE Comput. **38**(3), 33–40 (2005)

L.R. Rabiner, A tutorial on hidden Markov models and selected applications in speech recognition. Proc. IEEE **77**(2), 257–286 (1989)

G. Ruske, *Automatische Spracherkennung. Methoden der Klassifikation und Merkmalsextraktion*, 2nd edn. (Oldenbourg, Germany, 1993)

J. Schenk, G. Rigoll. *Mensch-Maschine-Kommunikation: Grundlagen von sprach- und bildbasierten Benutzerschnittstellen*. Springer, p. 348 (2010). ISBN 978-3-6420-5456-3

E.M. Schmidt , Y.E. Kim, Prediction of time-varying musical mood distributions from audio. In *Proceedings of the ISMIR 2010*, ISMIR. Utrecht, The Netherlands (2010)

M. Schroeder, G. McKeown, Considering social and emotional artificial intelligence. In *Proc.eedings of the AISB 2010 Symposium Towards a Comprehensive Intelligence Test*, SSAISB. Leicester, UK (2010)

B. Schuller, *Automatische Emotionserkennung aus sprachlicher und manueller Interaktion*. Doctoral thesis, Technische Universität München, Munich, Germany, June 2006

B. Schuller, D. Arsić, F. Wallhoff, G. Rigoll, Emotion Recognition in the Noise Applying Large Acoustic Feature Sets. In *Proceedings of the 3rd International Conference on Speech Prosody (SP) 2006*, ISCA. Dresden, Germany, pp. 276–289, May 2006

B. Schuller, F. Eyben, G. Rigoll, Static and dynamic modelling for the recognition of non-verbal vocalisations in conversational speech, in *Proceedings of the 4th IEEE Tutorial and Research Workshop on Perception and Interactive Technologies for Speech-based Systems (PIT 2008)*, vol. 5078, Lecture Notes in Computer Science, ed. by E. André (Springer, Kloster Irsee, Germany, 2008), pp. 99–110

B. Schuller, F. Friedmann, F. Eyben, Automatic Recognition of Physiological Parameters in the Human Voice: Heart Rate and Skin Conductance. In *Proceedings of the ICASSP 2013*, IEEE. Vancouver, Canada, pp. 7219–7223, May 2013

B. Schuller, G. Rigoll, M. Grimm, K. Kroschel, T. Moosmayr, G. Ruske, Effects of In-Car Noise-Conditions on the Recognition of Emotion within Speech. In *Proceedings of the 33. Jahrestagung für Akustik (DAGA) 2007*, DEGA. Stuttgart, Germany, pp. 305–306, March 2007

B. Schuller, S. Steidl, A. Batliner, F. Jurcicek, The INTERSPEECH 2009 Emotion Challenge. In *Proceedings of the INTERSPEECH 2009*, Brighton, UK, pp. 312–315, September 2009

B. Schuller, B. Vlasenko, F. Eyben, M. Wöllmer, A. Stuhlsatz, A. Wendemuth, G. Rigoll, Cross-corpus acoustic emotion recognition: Variances and strategies. IEEE Trans. Affect. Comput. (TAC) **1**(2), 119–131 (2010). doi:10.1109/T-AFFC.2010.8

B. Schuller, Signals and communication technology, *Intelligent Audio Analysis* (Springer, Berlin, 2013)

B. Schuller, A. Batliner, *Computational Paralinguistics: Emotion, Affect and Personality in Speech and Language Processing* (Wiley, Hoboken, 2013), p. 344. ISBN 978-1119971368

G. Sinha, R. Shahi, M. Shankar, Human computer interaction. In *Proceedings of the 3rd International Conference on Emerging Trends in Engineering and Technology (ICETET)*, IEEE. pp. 1–4 (2010)

M. ter Maat, D. Heylen, Using context to disambiguate communicative signals, in *Multimodal Signals: Cognitive and Algorithmic Issues*, Lecture Notes in Computer Science, ed. by A. Esposito, A. Hussain, M. Marinaro, R. Martone (Springer, Berlin, 2009), pp. 67–74. doi:10.1007/978-3-642-00525-1_6. ISBN 978-3-642-00524-4

D. Ververidis, C. Kotropoulos, Emotional speech recognition: resources, features, and methods. Speech Commun. **48**(9), 1162–1181 (2006)

A. Vinciarelli, M. Pantic, H. Bourlard, Social signal processing: Survey of an emerging domain. Image Vis. Comput. **27**(12), 1743–1759 (2009). doi:10.1016/j.imavis.2008.11.007

M. You, C. Chen, J. Bu, J. Liu, J. Tao. Emotion recognition from noisy speech. In *Proceedings of the IEEE International Conference on Multimedia and Expo (ICME 2006)*, IEEE. Toronto, Canada, pp. 1653–1656, July 2006. doi:10.1109/ICME.2006.262865

Z. Zeng, M. Pantic, G.I. Rosiman, T.S. Huang, A survey of affect recognition methods: Audio, visual, and spontaneous expressions. IEEE Trans. Pattern Anal. Mach. Intell. **31**(1), 39–58 (2009)

Chapter 2
Acoustic Features and Modelling

This chapter gives an overview of the methods for speech and music analysis. The methods described, include all the relevant processing steps from an audio signal to a classification result (Fig. 2.1). These steps include pre-processing and segmentation of the input, feature extraction (i.e., computation of acoustic Low-level Descriptors (LLDs) and summarisation of these descriptors in high level segments), and modelling (e.g., classification).

A particular focus is put in this thesis on real-time processing and the capabilities of the methods to work in systems which require incremental on-line processing of the audio input. An in-depth discussion of real-time, incremental processing is given in Chap. 4.

In this chapter, first, the basic concepts of digital audio signal processing and analysis are described in Sect. 2.1 and common terms are defined. Moreover, the most important pre-processing steps, which are commonly used, are presented. The acoustic LLDs that have been implemented for, and investigated in this thesis are covered in Sect. 2.2. Methods for summarising LLDs over a longer time segment are discussed in Sect. 2.4, and static and dynamic modelling methods are described in Sect. 2.5.

2.1 Basics of Signal Processing

The following provides a brief introduction to signal representation theories and defines some important terms of the area of digital signal processing, which will be used throughout the remainder of this thesis. This introduction is kept very brief on purpose, as the focus of this thesis is neither digital signal processing nor signal representation. It shall only serve the purpose of introducing the most important concepts which are required for understanding the following chapters. For an excellent, deeper discussion of signal representation theories the reader is referred to Oppenheim and Schafer (1975) and Oppenheim et al. (1999).

F. Eyben, *Real-time Speech and Music Classification by Large Audio Feature Space Extraction*, Springer Theses, DOI 10.1007/978-3-319-27299-3_2

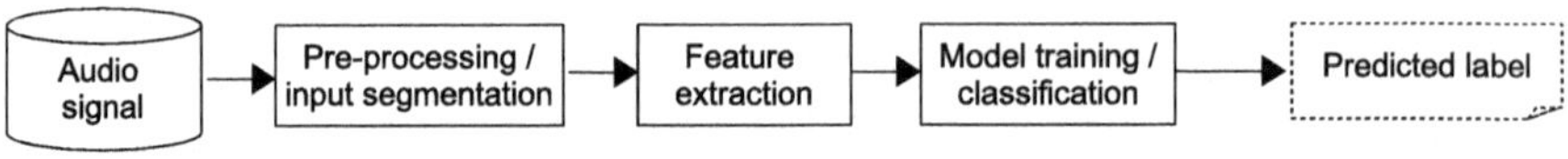

Fig. 2.1 Overview of steps of processing (simplified) for general speech and music analysis methods

2.1.1 Signal Representation

In the following a continuous signal a at time t is defined to have an amplitude of $a(t)$. Such a signal a might be represented by an electrical current in the physical world—e.g., an analogue audio signal, such as the current induced by the vibration of a microphone membrane over a coil. In order to process a signal a in a digital system, the signal must be discretised both in terms of amplitude and time, since processors can represent values only with a finite precision and can only store a finite amount of values.

The discretisation in time is referred to as Nyquist–Shannon sampling (Oppenheim et al. 1996). Thereby the time continuous signal $a(t)$ is represented by a fixed amount of N values $a(n)$ (samples) per unit of time. The sampling rate or sampling frequency f_s is the frequency at which the values $a(n)$ are sampled from the time continuous signal $a(t)$. The relation between the discrete time index $\hat{n}$ and the continuous time t is given by the sampling period T_s

$$T_s = \frac{1}{f_s} \tag{2.1}$$

as:

$$t = n \cdot T_s. \tag{2.2}$$

In order to be able to reconstruct the continuous analogue signal from the finite set of N samples, the following condition for the sampling frequency—called the sampling theorem—has to be met (cf. Oppenheim et al. 1996):

$$f_s \geq 2f_h, \tag{2.3}$$

where f_h is the highest frequency present in the original signal $a(t)$, also referred to as the Nyquist frequency. If this condition is met, the original time continuous signal can be reconstructed from the sampled signal by low-pass filtering, and the signal $a(n)$ contains all the information from $a(t)$. To ensure the sampling theorem for any type of input, in practice an analogue low-pass filter is applied to $a(t)$ before the signal $a(n)$ is sampled from $a(t)$.

Next, each sample $a(n)$ must be representable by a finite set of values, i.e., using a finite precision. This conversion of a continuous amplitude $a(n)$ to a discrete amplitude $x(n)$ is referred to as quantisation (Gregg 1977). A fixed set of discrete values is defined for the expected range of the signal and each continuous amplitude $a(n)$ is mapped to the nearest discrete value of the defined set, yielding $x(n)$. The difference between the actual value $a(n)$ and the resulting discretised value $x(n)$ introduces a small error, referred to as quantisation error. This error is irreversible and—in contrast to the time discretisation—cannot be eliminated when reconstructing the time and value continuous signal $a(t)$. The quantisation error can be seen as additive noise on top of the original signal $a(t)$ and is thus often referred to as quantisation noise.

On typical digital signal processing platforms a precision of $b = 16$ or $b = 24$ bits is used. The number of possible sample values is given by 2^b, and is 65,536 for 16-bit precision and 16,777,216 for 24-bit precision. The samples can be stored as integer values, or as floating point numbers.

For the analysis of speech and music signals it is important for all values to have a common range of values. Thus, it was decided to represent all sample values as 32-bit floating point values in the ongoing and to scale the values to the range $[-1; +1]$. This ensures a common representation regardless of the precision of the input. The scaling of a sample x_b with b-bits precision to the scaled sample x is performed according to:

$$x = \frac{x_b}{2^{b-1}}. \tag{2.4}$$

Note, that this assumes x_b to be represented by a signed integer type, e.g., a range of -32768 – $+32767$ for 16-bit integers.

2.1.2 *Frequency Domain*

The signal $a(t)$ can also be represented in the frequency domain by a superposition of sinusoidal base functions. According to Fourier (1822) (cf. also Lejeune-Dirichlet 1829) any band-limited, finite time signal can be represented by finite superposition of sines and cosines with different frequencies (f), magnitudes (A), and phases (ϕ) (=Fourier Series). I.e., $a(t)$ is represented by $A(f)$ and $\phi(f)$ with a finite set of frequencies f. The process of estimating the magnitudes and phases of the base functions is called Fourier Transformation (FT). As the following deals with signals $x(n)$ which are both discrete in time and value, only the Discrete Fourier Transformation (DFT) will be introduced briefly at this point. For more details and a discussion of the Fourier Transformation the interested reader is referred to Oppenheim et al. (1999) and Lizorkin (2002).

The DFT of a real valued signal $x(n)$ with discrete time index $n = \frac{t}{T_s}$ is defined for the integer discrete frequency (bin index) $m = \frac{f}{f_0}$ and $m \in [0; M]$ $(M = \frac{N}{2} + 1)$ as:

$$X(m) = \sum_{n=0}^{N-1} x(n) e^{\frac{-j2\pi mn}{N}}. \tag{2.5}$$

For more general considerations in the remainder of this thesis, the bin index m shall be converted to a linear frequency f via a function $F(m)$ by Eq. (2.6):

$$f = F(m) \tag{2.6}$$

$$\text{and } m = F^{-1}(f). \tag{2.7}$$

In the linear case of the DFT, the above is:

$$F_{lin,Hz}(m) = mf_0 \tag{2.8}$$

$$F^{-1}_{lin,Hz}(f) = \frac{f}{f_0}. \tag{2.9}$$

The result $X(m)$ is a complex value (j indicates imaginary parts in Eq. (2.5)) which gives both the magnitude and the phase of the corresponding mth base function. The DFT base functions $e^{\frac{-j2\pi mn}{N}}$ are orthogonal if m is chosen as an integer value, i.e., the DFT is computed only for multiples of a base frequency f_0. For the DFT f_0 is given in terms of the sampling frequency T_s and the frame size N as follows:

$$f_0 = \frac{1}{N \cdot T_s}. \tag{2.10}$$

The number of discrete frequency bins M is then given as $M = \frac{N}{2}$. For $m = 0$ the direct current (DC) component of $x(n)$ is returned, while for $m = M$ the magnitude of the Nyquist frequency is returned. The computation of a DFT has an asymptotic complexity of $\mathcal{O}(N^2)$. In practice, an optimised algorithm is implemented: the Fast Fourier Transformation (FFT) (Cooley et al. 1969). The algorithm uses the principle of Divide and Conquer and splits the DFT in two sub-problems of half the size of the original problem. The FFT achieves an asymptotic complexity of $\mathcal{O}(N \log(N))$ but requires the frame size N to be a power of two. If a frame with a size which is not a power of two needs to be transformed, the typical procedure is to apply zero-padding to the frame, i.e., the frame size is increased to the next higher power of two and the additional samples are filled with zeros.[1]

For the human ear only the magnitudes of the components $X(m)$ are relevant, and not the phase. The phase is only necessary in very special analysis applications which require instantaneous frequency estimates, for example, or for a proper reconstruction of $x(n)$ from $X(m)$. For analysis, each component $X(m)$ is therefore converted to the magnitude $X_M(m)$:

$$X_M(m) = |X(m)| = \sqrt{Im(X(m))^2 + Re(X(m))^2}. \tag{2.11}$$

[1] In openSMILE the FFT with complex valued output (and also the inverse FFT) is implemented by the `cTransformFFT` component. Magnitude and Phase can be computed with the `cFFTmagphase` component.

The phase $X_\phi(m)$ in radians is given as:

$$X_\phi(m) = arctan\left(\frac{Im(X(m))}{Re(X(m))}\right). \quad (2.12)$$

The signal $x(n)$ can also be resynthesised from the spectral magnitudes and phases. Even though this is not required for analysis of speech and music directly, it might be required indirectly, e.g., if a filter is to be implemented in the spectral domain and time domain descriptors are to be computed from the resulting signal. Also, in this way, various audio features (cf. Sect. 2.2) can be transformed into an audio signal, i.e., feature trajectories can be made audible.

$X(m)$ is obtained from a magnitude/phase representation as:

$$X_m(m) = X_M(m)e^{jX_\phi(m)} = X_M(m)\left(\cos(X_\phi(m)) + j\sin(X_\phi(m))\right). \quad (2.13)$$

The time domain signal $x(n)$ is obtained with the inverse real-valued DFT:

$$x(n) = \frac{2}{M}\sum_{m=-M}^{M} X(m)e^{\frac{j2\pi mn}{N}}. \quad (2.14)$$

2.1.3 *Short-Time Analysis*

For audio analysis tasks the spectrum contains important information, including very obvious attributes such as information about pitches of musical instruments, or the pitch of a speaker. The time domain signal contains information about amplitude, etc. However, all these attributes change over time and we need to find a way to estimate these attributes periodically, in (quasi-)stationary segments, instead of performing a single global analysis over the whole signal $x(n)$ or the spectrum $X(m)$ computed from the whole signal $x(n)$.

The Concept of Windowing

To solve this problem, commonly the method of short-time analysis (also referred to as framing or windowing) is considered.[2] Thereby a signal $x(n)$ with $n \in [0, N[$ is divided into K short, overlapping frames $x_k(\hat{n})$ ($k \in [0 \ldots K-1]$) of N_f samples or $L_f = N_f \cdot T_s$ seconds length. The discrete time index $\hat{n}$ within each frame is measured relative to the start of the frame, i.e., $\hat{n} \in [0 \ldots N_f - 1]$. The start index $n_{start,k}$ in the signal $x(n)$ which corresponds to the start of the kth frame is given as:

$$n_{start,k} = k \cdot N_f^{(T)}, \quad (2.15)$$

[2] In openSMILE windowing of audio samples (i.e., short-time analysis) can be performed with the `cFramer` component.

where $N_f^{(T)} = \frac{T_f}{T_s}$ is the frame period measured in samples and T_f is the frame period measured in seconds, i.e., the time between the start of two consecutive frames. The names frame step, frame shift, frame period, and frame increment all refer to this same quantity, measured either in seconds or samples.

Typically T_f is smaller than L_f because overlapping frames are used. The percentage of overlap O_f between adjacent frames is defined as:

$$O_f = \frac{L_f - T_f}{L_f}. \tag{2.16}$$

The end index $n_{end,k}$ of the frame is given as:

$$n_{end,k} = k \cdot N_f^{(T)} + N_f. \tag{2.17}$$

For the start and end times of the kth frame relative to the signal $x(t)$ holds:

$$t_{start,k} = n_{start,k} \cdot T_s = k \cdot T_{f,s} \cdot T_s = k \cdot T_f \tag{2.18}$$

$$t_{end,k} = n_{end,k} \cdot T_s = (k \cdot T_{f,s} + N_{f,s}) \cdot T_s = k \cdot T_f + N_f \tag{2.19}$$

In addition to the theoretical framework of short-time analysis laid out above, in practice border conditions have to be addressed. That is in particular, how to deal with N_e excess samples at the end of the signal $x(n)$ for which $N_e < N_f$. Basically three strategies exist:

1. Ignoring these N_e samples,
2. Dealing with a smaller frame at the end, which contains the last N_e samples.
3. Appending $N_f - N_e$ samples to the end of the signal $x(n)$.

In this thesis, strategy (1) is implemented, as the analysed segments are assumed to be much longer than a single frame. It thus can be safely assumed that these discarded samples at the end will not contribute much to the final result. If we choose the frame length L_f to be sufficiently small, we can assume the signal's properties (such as pitch of a voice or an instrument) to vary only minimally or remain constant within a single frame k, i.e., the signal is assumed to be quasi stationary. Typical frame lengths in speech and music analysis range from 20 milliseconds (ms) to 60 ms. The most commonly chosen frame period is 10 ms, which originates from the domain of Automatic Speech Recognition (ASR) (Rabiner 1989; Young et al. 2006). In some applications where a higher temporal accuracy is required, also 5 ms frame period is considered; also 20 ms frame periods are used and are feasible in some analysis tasks.

The concept of framing laid out above is not limited to short-time analysis of audio samples. Any time-value series can be segmented into 'short' time segments. For instance, parameters extracted from short audio frames, such as signal energy (see Sect. 2.2.2), can also be viewed as a time-value series with a sampling period of e.g., 10 ms. In the ongoing, the terms 'frame' and 'window' refer to a frame of audio samples, if not explicitly specified otherwise. The term 'segment' refers to a higher level segment, e.g., one second of signal energy values, or the whole signal. An in depth discussion of such higher level segment features is contained in Sect. 2.4 and a discussion of the choice of the segment length is found in Sect. 4.1.

Note: To simplify the equations in the ongoing and foster generalisation, it will not be discriminated between frames $x^{(k)}(\hat{n})$ and a signal $x(n)$ on the symbolic level, unless explicitly necessary. The symbol $x(n)$ will be used for a general signal of length N in the equations. An optional discrete time index k as superscript $x^{(k)}(n)$ will denote that this signal represents or is derived from a frame at a discrete frame index k. The text will further clarify on whether frames, higher level segments, or the whole signal is considered.

Window Functions

The framing described in the previous section corresponds to a multiplication of the signal $x(n)$ with a rectangular window function $w_r(n)$ for the kth frame x_k as follows:

$$x_k(\hat{n}) = x(k \cdot N_f^{(T)} + \hat{n}) \cdot w_r(\hat{n}). \tag{2.20}$$

The rectangular window function is defined as:

$$w_r(n) = 1 \text{ for } n = 0 \dots N_f \tag{2.21}$$

In the spectral domain (cf. Sect. 2.1.2) such a multiplication corresponds to a convolution. Therefore, the ideal windowing function—i.e., one which does not distort the spectrum—has a dirac impulse shaped spectrum. In the time domain this corresponds to a constant function of infinite duration. Since the window always has a finite duration, a certain amount of spectral distortion due to the framing can never be avoided, regardless of the choice of the windowing function. In fact, a compromise between time and frequency domain properties of the windowing function must be found. In the time domain a finite duration with steep edges is preferred (rectangle like), while in the frequency domain a very narrow main maximum (dirac like) is preferred with near zero side maxima.

In the following a discussion of the most common windowing functions (as they are implemented in openSMILE[3]) and their properties is contained:

Rectangular window The Rectangular Window is defined as a constant function for $n \in [0, N-1]$: $w_{Rec}(n) = 1$. It corresponds to a *sinc*-function in the spectral domain $W(m)$:

$$W_{Rec}(m) = \frac{\sin(m)}{m} = sinc(m) \tag{2.22}$$

[3]http://opensmile.audeering.com.

The Rectangular Window is best suited for analysis in the time domain such as for Zero-Crossing Rate or Amplitude descriptors. It is very efficient to compute, as no multiplications are required. Due to the high side maxima of the sinc function it is not recommended for frequency domain analysis.

Hann(ing) window The Hanning window—also known as Hann-window or the raised cosine window—is named after the Austrian meteorologist Julius von Hann by Blackman and Tukey (1959). It is defined for $n \in [0, N-1]$ as:

$$w_{Han}(n) = 0.5\left(1 - \cos\left(\frac{2\pi n}{N-1}\right)\right) \tag{2.23}$$

The side lobes in the spectrum roll off by approximately 18 dB per octave, which makes this window suitable for spectral analysis. Moreover, the symmetry in the time domain and the fact that the amplitude reaches zero at both sides of the window makes this window perfectly suitable for applications in which $x(n)$ has to be reconstructed from spectra with the overlap-add method when the overlap of the frames is 50 % (Oppenheim and Schafer 1975).

Hamming window A modification of the Hanning window is the Hamming window. It reduces the amplitude of the first sidelobe in the spectrum significantly (by about one fith), at the cost of higher amplitudes for the higher order sidelobes (Enochson and Otnes 1968). In contrast to the Hanning window, it does not reach zero amplitude at the sides. It is the most commonly used window for analysis in the spectral domain, especially in speech recognition and speech analysis. There, according to Young et al. (2006), it is defined as follows for $n \in [0, N-1]$:

$$w_{Ham}(n) = \alpha - \beta\cos\left(\frac{2\pi n}{N-1}\right). \tag{2.24}$$

with $\alpha = 0.54$ and $\beta = 0.46$. A more in depth discussion, as well as a more precisie definition of the coefficients is found in (Harris 1978). According to Harris, the theoretical optimal values for the coefficients are $\alpha = \frac{25}{46}$ and $\beta = \frac{21}{46}$. Later, Nuttal (1981) proposed the following values as optimal with respect to minimizing the sidelobes in the spectrum: $\alpha = 0.53836$ and $\beta = 0.46164$. In the automatic speech recognition community, however, the definition of Young et al. (2006) is widely used. Therefore, only this definition is used in this thesis.

Gaussian window The Gaussian window uses a Gaussian function as windowing function ($n \in [0, N-1]$):

$$w_{Gau}(n) = e^{-\frac{1}{2}\left(\frac{n-(N-1)/2}{\sigma(N-1)/2}\right)^2} \tag{2.25}$$

The Gaussian function has the special property of being an eigenfunction of the Fourier Transform, i.e., when—transformed to the spectral domain—it remains Gaussian shaped (Oppenheim et al. 1999). It therefore has no sidelobes; the bandwidth (defined by the standard deviation of the Gaussian in the spectral domain) is inversely proportional to standard deviation of the Gaussian in the time domain. In the time domain it does not touch zero at the ends of the window due to the infinite length of the Gaussian function. In some applications it might be necessary to have zeroes at the ends of the window. This can be achieved by multiplying the Gaussian window with another window, e.g., a Hanning window.

Sine/Cosine window In contrast to a raised cosine (Hanning) window, the sine window consists of the first sine half-wave ($n \in [0, N-1]$) (Oppenheim et al. 1999):

$$w_{Sin}(n) = \sin\left(\frac{\pi n}{N-1}\right) \tag{2.26}$$

Due to its steepness at the window ends it has large side maxima in the spectrum and thus is less preferred.

Triangular window For some applications where overlapping windows need to be resynthesised in the time domain, triangular windows might be considered due to their symmetry. However, they have less favorable spectral properties than, e.g., a Hann window. A general Triangular window with non zero-valued end points is defined as (for $n \in [0, N-1]$) (Oppenheim et al. 1999):

$$w_{Tri}(n) = \begin{cases} \frac{2(n+1)}{N} & \text{if } n < \frac{N}{2} \\ \frac{2(N-n)}{N} & \text{if } n \geq \frac{N}{2} \end{cases}. \tag{2.27}$$

Bartlett window If the window must have zeroes at the endpoints, a Bartlett window can be used. The Bartlett window is a Triangular window with zero-valued end points defined as (for $n \in [0, N-1]$) (Oppenheim et al. 1999):

$$w_{Bar}(n) = \begin{cases} \frac{2n}{N-1} & \text{if } n < \frac{N}{2} \\ \frac{2(N-n-1)}{N-1} & \text{if } n \geq \frac{N}{2} \end{cases}. \tag{2.28}$$

A triangular window with zero valued endpoints can be expressed by the convolution of two rectangular windows in the time domain. Thus, the resulting spectral shape is that of a squared $sinc(m)$ function: $sinc^2(m) = \left(\frac{sin(m)}{m}\right)^2$ function.

Lanczos window For the completeness of this overview on windowing functions, the following contains a list of further functions without discussion. These functions are used for specific digital signal processing applications, but are only of minor importance for analysis of speech and music signals.

The Lanczos window represents a sinc function in the time domain, trimmed to a finite length $n \in [0, N-1]$, which includes only the main maximum:

$$w_{Lac}(n) = \frac{\sin\left(\pi \frac{2n}{N-1} - 1\right)}{\frac{2n}{N-1} - 1} \tag{2.29}$$

It was introduced for the purpose of Lanczos resampling (Turkowski and Gabriel 1990).

Blackmann window A Blackmann window is defined for $n \in [0, N-1]$ as:

$$w_{Bla}(n) = \frac{1-\alpha}{2} - \frac{1}{2}\cos\left(\frac{2\pi n}{N-1}\right) + \frac{\alpha}{2}\cos\left(\frac{4\pi n}{N-1}\right) \tag{2.30}$$

The suggested default for the α parameter is 0.16. It is based on the Blackman Function (Blackman and Tukey 1959).

2.1.4 Pre-processing

Before audio features (see Sect. 2.2) are extracted from speech or music signals, several pre-processing steps can be applied to the signal. In general, pre-processing refers to everything done to the signal (in time or frequency domain) before LLDs are extracted. The typical pre-processing for speech and music analysis, however, is limited to the time domain and concerns the following steps:

Down-mixing, i.e., conversion of multi-channel signals to a single channel (mono). Instead of extracting near redundant features for both channels individually, most often a stereophonic signal is converted to a monophonic signal prior to feature extraction. The most simple procedure is linear down-mixing, where the samples of all C channels are linearly averaged to a single channel x_0:

$$x_0(n) = \frac{1}{C}\sum_{c=1}^{C} x_c(n). \tag{2.31}$$

Down-mixing is not performed when multi-channel features are to be extracted, such as spatial features, or when reconstructing a single source signal from a multi-channel recording (source separation). Both aspects are, however, not in the scope of this thesis, thus monophonic down-mixing is always applied here.

Re-sampling, i.e., change of the sampling frequency to a common value for inputs with different sampling frequencies, or reduction of the sampling frequency in order to speed up the analysis. In speech, the relevant frequency range is from ≈50 Hz to ≈6 kHz, and in music from ≈50 Hz to ≈8 kHz. A sampling frequency of 16 kHz is thus sufficient for analysis of most speech and music signals.

Pre-emphasis,[4] i.e., filtering of the signal to attenuate frequency bands which carry important information. For speech analysis typically a 1st order high-pass filter is applied to a signal $x(n)$ in order to emphasise information on formants (Young et al. 2006), yielding the pre-emphasised signal $x_p(n)$:

$$x_p(n) = x(n) - kx(n-1), \tag{2.32}$$

where $k \in [0; 1]$ is the pre-emphasis coefficient, which controls the strength of the pre-emphasis (1 is most, 0 is least). Typical values for speech processing range from 0.9 to 0.97 Young et al. (2006). In music processing a band pass filter could be used to emphasise certain octaves, or reduce sub-bass effects, for example. Also, the above 1st order high-pass pre-emphasis filter can be inverted to a low-pass "de-emphasis" filter as follows:

$$x_d(n) = x(n) + k_d x(n-1). \tag{2.33}$$

Other pre-processing steps include noise-reduction and echo cancellation. Some of these methods operate in the spectral domain. As the focus of this thesis is on large scale feature extraction and not on signal enhancement and pre-processing, the reader is referred to (Schuller 2013) for further reading on these topics. All acoustic descriptors described in this thesis can be extracted from enhanced and filtered signals, too.

2.2 Acoustic Low-Level Descriptors

In this section all acoustic LLDs which have been implemented for and evaluated in the course of this thesis are described in detail.

An acoustic LLD is defined as a parameter computed from a short time frame $x_k(n)$ (length N_f) from an audio signal at time $t = k \cdot T_f$. The length of the frame should be chosen in a way to a) ensure quasi stationarity of the signal within the frame with respect to the LLD of interest, and b) ensure that the frame contains enough data to compute the LLD (cf. Sect. 2.1.3 on details of the framing). For some LLDs a window function is applied to the frame prior to computing the LLD (cf. Sect. 2.1.3 for a discussion of windowing functions). Typical frame lengths are 25–32 ms for most LLDs.

[4]In openSMILE pre-emphasis can be implemented with the `cPreemphasis` component on a continuous signal, or with the `cVectorPreemphasis` component on a frame base (Hidden Markov Toolkit (Young et al. 2006) (HTK) compatible behaviour).

2.2.1 *Time Domain Descriptors*

Time domain descriptors are computed directly from the time domain signal $x(n)$. Commonly these include, the number of zero-crossings (Sect. 2.2.1.1), amplitude statistics (Sect. 2.2.1.2), and a DC offset. Strictly speaking, other descriptors such as the signal energy (Sect. 2.2.2), or linear predictive coding coefficients (Sect. 2.2.7.1) are also extracted from the time domain signal. However, they can also be extracted from frequency domain representations of the signal and moreover are more related to spectral characteristics of the signal. Thus, they are not considered in this section.

2.2.1.1 Zero- and Mean-Crossing Rate

The Zero-Crossing Rate (ZCR) describes the number of sign changes c of $x(n)$ per unit of time (usually one second) (Chen 1988):

$$ZCR = \frac{c}{1.0\,\mathrm{s}}. \tag{2.34}$$

A sign change is defined to occur when:

$$x(n-1)x(n) < 0 \tag{2.35}$$

or

$$x(n-1)x(n+1) < 0 \quad \text{and} \quad x(n) = 0. \tag{2.36}$$

In analogy to the ZCR, one can define the Mean-Crossing Rate (MCR) as the rate of changes from below to above the mean μ_x of $x(n)$ (or vice versa). To compute the MCR μ_x is subtracted from $x(n)$ resulting in the mean normalised signal $\hat{x}_n$:

$$\hat{x}(n) = x(n) - \mu_x. \tag{2.37}$$

with

$$\mu_x = \frac{1}{N}\sum_{0}^{N-1} x(n) \tag{2.38}$$

The MCR is then computed from $\hat{x}(n)$ using the same algortihm as for ZCR.

A high ZCR or MCR indicates a signal with much high frequency content. Typically harmonic signals have a low zero crossing rate, which is related to the fundamental frequency of the signal. E.g., a single pure sine will have a zero crossing rate of twice its frequency. White Gaussian noise on the other hand will have a rather high zero crossing rate, due to the broadband high-frequency content. ZCR is used to distinguish voiced speech from unvoiced speech (Bachu et al. 2010) as well as percussive from harmonic parts in music (Gouyon et al. 2000). It is, however, also strongly affected by additive noise, especially at low SNRs. Better metrics have thus been proposed for the voiced/unvoiced decision, as will be shown in the ongoing.

2.2.1.2 Amplitude

Other time domain signal descriptors are the maximum and minimum signal amplitudes, or the maximum absolute value of the amplitudes.

Usually the amplitudes of audio signals are symmetric around 0 amplitude, i.e., the range of the amplitude is from $-a$ to $+a$, and the amplitude is 0 if there is no signal at the input. Sometimes, however, an offset is present, due to various effects, such as a electrical DC offset due to faulty or cheap recording equipment, or—when framing—the influence of extremely low frequencies (which have a period larger than the frame length). In these cases the DC offset of the signal can provide information of intereset. The DC offset of $x(n)$ is equivalent to the mean μ_x of $x(n)$ (cf. Eq. (2.38)).

2.2.2 *Energy*

One of the most basic, yet powerful audio descriptors is the signal energy[5] (Oppenheim et al. 1999). If we assume an audio signal to have no DC offset, i.e., a mean value of zero, the signal energy E for a signal $x(n)$ with $n \in [0; N-1]$, is defined as the sum of squared amplitudes of the signal x:

$$E = \sum_{n=0}^{N-1} x^2(n). \tag{2.39}$$

Often the normalised signal energy is used in order to eliminate the influence of the frame length on the descriptor:

$$E_n = \frac{1}{N} \sum_{n=0}^{N-1} x^2(n). \tag{2.40}$$

In speech and music processing two variations of the signal energy are commonly employed. The first is the Root Mean Square (RMS) energy (Kenney and Keeping 1962):

$$E_{rms} = \sqrt{\frac{1}{N} \sum_{n=0}^{N-1} x^2(n)}. \tag{2.41}$$

[5]RMS and logarithmic energy can be computed in openSMILE with the `cEnergy` component.

The second is the logarithmic (log) energy (Young et al. 2006):

$$E_{log} = E_{bias} + E_0 \cdot \log \sum_{n=0}^{N-1} x^2(n), \tag{2.42}$$

where log represents the natural logarithm. E_0 is a scaling factor which is used to scale the logarithmic energy to different unit scales. If $E_0 = 1$, then E_{log} is measured in 'neper', a unit similar to Decibel (dB), only that the natural logarithm is used as a basis. If $E_0 = \frac{10}{\log 10} \approx 4.343$, then E_{log} is measured in dB. For the logarithmic energy it is common to define a 'floor' value, i.e., a minimum value, to avoid very high negative values with high variation due to noise in low energy and silent frames.[6]

The above definitions of the signal energy do not consider any properties of the human hearing and/or human perception of loudness. According to Zwicker and Fastl (1999) it is important to consider human perception for many tasks. An example for identification of stressed syllables in speech is shown where signal amplitude is compared to a loudness measure obtained from a psychoacoustic model. Computing perceptual loudness from a psychoacoustic model is computationally demanding. An approximation $E_{l,approx}$ for the loudness[7] E_l for a narrow-band signal has been used by (Kießling 1997, pp. 156–157):

$$E_{l,approx} = \left(\frac{I}{I_0}\right)^{0.3}, \tag{2.43}$$

where I is the signal intensity defined as the signal energy E of $x(n)$ where $x(n)$ has been weighted with a Hamming window function (cf. Sect. 2.1.3) and I_0 is the reference intensity. For a maximum absolute signal amplitude $|x(n)| = 1.0$ and a reference signal at 60 dB Sound Pressure Level (SPL), $I_0 = 10^{-6}$ is defined (Kießling 1997, pp. 156–157).

The exact measurement of loudness according to a simplified psychoacoustic model is discussed in Sect. 2.2.9.3.

2.2.3 Spectrum

This section introduces various types of magnitude spectra, which can be used as LLDs directly and/or serve as the basis for other descriptors which are computed from the spectra, such as spectral statistics (Sect. 2.2.4) or Cepstral features (Sect. 2.2.10).

---

[6] openSMILE defines 8.674676×10^{-19} as a floor value for the argument of the log, for samples scaled to the range of −1–+1. In case of sample value range from −32767 to +32767 (HTK compatible mode), the floor value for the argument of the log is 1.

[7] The loudness approximation and the signal intensity as defined here can be extracted in openSMILE with the `cIntensity` component.

2.2.3.1 Linear Magnitude Spectrum

The spectrum magnitudes $X_M(m)$, as introduced in Sect. 2.1.2 and computed from a short-time frame, can be used as LLDs directly. This spectrum is both linear in frequency m and magnitude X.

In order to make this LLD independent of the analysis frame size N_f (e.g., if inputs with varying sampling rates are analysed), the magnitudes $X_M(m)$ can be normalised by N_f:

$$X_{M,norm}(m) = \frac{X_M(m)}{N_f}. \tag{2.44}$$

The resulting measure $X_{M,norm}(m)$ is the spectral magnitude density. *Note:* in practice, a scaling by the relative energy of the windowing function must be performed when calibrated magnitude/power measurements are required. As this scaling is a constant, however, it can be neglected for feature extraction or speech/music analysis purposes.

The bins $X_M(m)$ represent a vector of basic acoustic descriptors, which contains almost all relevant information from the original signal and—if suitable phase information was available—would allow for reconstruction of the original signal in the time domain. However, raw spectra are not ideal as LLDs because they contain high amounts of redundancy, i.e., individual descriptors (bins, bands, etc.) are highly correlated with each other but are seldom highly correlated to analysis tasks' targets. Therefore, descriptors derived from the raw spectra are preferred, such as spectral statistics (Sect. 2.2.4).

2.2.3.2 Non-linear Magnitude Scales

The linear values $X_M(m)$ or $X_{M,norm}(m)$ correspond to the physical unit of voltage, as they are computed from signal amplitude values. However, human auditory perception is highly non-linear (Zwicker and Fastl 1999) and thus a non-linear representation of the magnitudes might be better suited.

The first step towards human perception is to use power spectra $X_P(m)$:

$$X_P(m) = X_M(m)^2 \tag{2.45}$$

which—despite the name *power* spectra—represent the quadratic energy in each bin, or power spectral densities, accordingly:

$$X_{P,norm}(m) = \frac{|X_{M,norm}(m)|^2}{N_f}, \tag{2.46}$$

which in fact represent the power in each bin, because of the normalisation with the window length.

The next step is the application of a non-linearity. Commonly, a logarithm is applied as non-linearity, or an exponent of $\approx$0.3 is used as non-linearity (e.g., as for the approximation of loudness in Sect. 2.2.2 or for the auditory spectra in Sect. 2.2.9.3).

The spectral magnitudes are converted to a logarithmic spectral power density representation $X_{dBpsd}(m)$ in dB by the following equation[8] (cf. Spanias et al. 2007):

$$X_{dBpsd}(m) = X^{(0)}_{dBpsd} + 20 \cdot \log_{10}\left(|X_{M,norm}(m)|\right) \tag{2.47}$$

Thereby $X^{(0)}_{dBpsd}$ is a normalisation factor for the logarithmic scale and suggested as 90.302 dB by the psychoacoustic model layer I, defined in the MPEG-1 standard.

2.2.3.3 Non-linear Frequency Scales

The magnitudes $X_M(m)$ are equidistant (factor f_0) on a linear (Hz) frequency scale. Human hearing is also non-linear in terms of frequency perception and a non-linear frequency scale has to be preferred (Zwicker and Fastl 1999). This fact has been successfully exploited in the fields of Automatic Speech Recognition, Speaker Recognition, and Music Information Retrieval for a long time by use of Mel-Frequency Cepstral Coefficient (MFCC) features (Rabiner and Juang 1993). Thereby the spectrum is transferred to a so called Mel-Frequency scale (cf. below) before further processing. More details on MFCC are given in Sect. 2.2.10.1 and Young et al. (2006) as well as in Rabiner and Juang (1993).

This section introduces various non-linear frequency scales and discusses methods for transformation of a linear frequency scale spectrum to a non-linear frequency scale by interpolation.[9] Each frequency scale is defined by a forward transformation function $\Theta_{scale}(f)$ of the linear frequency $f^{(lin)}$ (in Hz) to the non-linear frequency $f^{(scale)}$ in units of the respective scale (e.g., Mel or Bark (critical band rate, cf. below)):

$$f^{(scale)} = \Theta_{scale}(f^{(lin)}) \tag{2.48}$$

For some scales also a backward transformation function Θ^{-1}_{scale} is given analytically:

$$f^{(lin)} = \Theta^{-1}_{scale}(f^{(scale)}). \tag{2.49}$$

A discrete magnitude spectrum given for *scale* is denoted by the symbol $X^{(scale)}_M(m^{(scale)})$ where $m^{(scale)}$ is the integer bin index on the non-linear frequency scale which is mapped to $f^{(scale)}$ by a general function (cf. Sect. 2.1.2) $m^{(scale)} = F^{-1^{(scale)}}(f^{(scale)})$.

---

[8]In openSMILE the option `dBpsd` must be enabled in the `cFftMagphase` component in order to compute logarithmic power spectral densities.

[9]In openSMILE these spectral scale transformations and spline interpolation can be applied with the `cSpecScale` component.

For the transformation of the M linear frequency bins m to $M^{(scale)}$ non-linear frequency bins $m^{(scale)} \in 0 \ldots M^{(scale)}$, first the $M^{(scale)}$ frequencies $f^{(scale)}$ for all the $m^{(scale)}$ bins need to be defined. It is common and convenient to use an equidistant spacing (on the non-linear target scale) between a minimum and a maximum frequency f_{min} and f_{max} and usually $M^{(scale)} \leq M$. In most cases $f_{min}^{(lin)} = 0$ and $f_{max}^{(lin)} = 1/(2T_s)$ (converted to the target scale) is chosen to match the range of the linear scale spectrum. Some scales, such as the semitone scale or some versions of the Bark scale, however, require $f_{min}^{(lin)}$ to be greater than zero because $\Theta_{scale}(0)$ is not defined. With

$$f_{min}^{(scale)} = \Theta_{scale}(f_{min}^{(lin)}) \tag{2.50}$$

$$\text{and } f_{max}^{(scale)} = \Theta_{scale}(f_{max}^{(lin)}) \tag{2.51}$$

$$\text{and } f_0^{(scale)} = \frac{1}{M^{(scale)}} \left(f_{max}^{(scale)} - f_{min}^{(scale)} \right), \tag{2.52}$$

the frequencies $f^{(scale)} = F^{(scale)}(m^{(scale)})$ of the non-linear bins (equidistant on the target scale) are now given as:

$$F^{(scale)} \left(m^{(scale)} \right) = m^{(scale)} f_0^{(scale)} + f_{min}^{(scale)} \tag{2.53}$$

These frequencies $f^{(scale)}$ can be converted to a linear frequency scale with the inverse transformation function:

$$f^{(lin)} \left(m^{(scale)} \right) = \Theta_{scale}^{-1} \left(F^{(scale)}(m^{(scale)}) \right) \tag{2.54}$$

Now the magnitudes $X_M^{(scale)} \left(m^{(scale)} \right)$ can be interpolated from the M linear scale magnitude bins $X_M(m)$ using various interpolation methods such as linear, cubic, or spline interpolation (Steffensen 2012).

Bark-Frequency scale The Bark frequency scale has been developed to numerically describe the loudness perception of human hearing. It was introduced by Zwicker (1961) and named after Heinrich Barkhausen, for his early achievements on subjective loudness measurements. The fundamental assumption in the definition is that a tone which is *perceived* as having twice the pitch than a reference tone also has twice the critical band rate (measured in Bark) of the reference tone. The original scale is defined from 0.2 Bark to 25 Bark. According to Zwicker (1961), below 500 Hz a tone with twice the frequency is perceived as having twice the pitch, while above 500 Hz a logarithmic rule applies for pitch perception versus frequency. This fact makes it hard to find a single exact analytic expression for the Bark frequency scale.

However, multiple analytical approximations of this scale exist. The most common one is the scale proposed by Traunmueller (1990). The critical band rate $f^{(bark)}$ (often called z in the literature) is computed via this core transformation:

$$z'(f) = \frac{26.81}{1 + \frac{1960}{f}} - 0.53, \tag{2.55}$$

followed by corrections for very low and very high frequencies:

$$f^{(bark)} = \Theta_{bark}(f) = \begin{cases} 0.85z'(f) + 0.3 & \text{if } z'(f) < 2 \\ z'(f) & \text{if } 2 \le z'(f) \le 20.1 \\ 1.22z'(f) - 0.22 \cdot 20.1 & \text{if } z'(f) > 20.1 \end{cases}. \quad (2.56)$$

The inverse transform is derived by inverting the above equations:

$$z' = \begin{cases} \frac{f^{(bark)} - 0.3}{0.85} & \text{if } f^{(bark)} < 2 \\ f^{(bark)} & \text{if } 2 \le f^{(bark)} \le 20.1 \\ \frac{f^{(bark)} + 0.22 \cdot 20.1}{1.22} & \text{if } f^{(bark)} > 20.1 \end{cases}. \quad (2.57)$$

and

$$f = \frac{1960}{\frac{26.81}{z' + 0.53} - 1}. \quad (2.58)$$

The Speex audio codec[10] uses a different version of the Bark scale given as:

$$\Theta_{bark,speex}(f) = 13.1 \arctan\left(0.74 \frac{f}{1000}\right) + 2.24 \arctan\left(\frac{1.85}{10^8} f^2\right) + 10^{-4} f, \quad (2.59)$$

which is supposedly based on the original analytic approximation by Zwicker and Terhardt (1980):

$$\Theta_{bark,zwicker}(f) = 13 \arctan\left(0.76 \frac{f}{1000}\right) + 3.5 \arctan\left(\left(\frac{f}{7500}\right)^2\right). \quad (2.60)$$

It is not trivial to invert the Speex approximation, and it is computationally more demanding to compute. For these reasons it is not considered any further in this thesis.[11]

Another approximation of the Bark frequency scale was suggested by Schroeder (1977):

$$\Theta_{bark,schroed}(f) = 6 \log\left(\frac{f}{600} + \sqrt{\left(\frac{f}{600}\right)^2 + 1}\right) \quad (2.61)$$

$$= 6 \sinh^{-1}\left(\frac{f}{600}\right), \quad (2.62)$$

[10] http://www.speex.org/.

[11] The SPEEX version of the Bark transformation is implemented in openSMILE as forward transformation only. Not all components will work, as most components require a backward scale transformation.

which has an inverse of:

$$f = \Theta^{-1}_{bark,schroed}\left(f^{(bark)}\right) = 600 \sinh\left(\frac{f^{(bark)}}{6}\right). \tag{2.63}$$

Thereby the hyperbolic sine ($\sinh(x)$) and the inverse ($\sinh^{-1}(x)$) are defined as:

$$\sinh(x) = \frac{1}{2}\left(e^{x} - e^{-x}\right), \tag{2.64}$$

$$\sinh^{-1}(x) = \operatorname{arsinh}(x) = \log\left(x + \sqrt{x^2 + 1}\right). \tag{2.65}$$

Mel-Frequency scale Another analytical approximation of the critical band rate is the Mel-Frequency scale (Beranek 1949). According to Beranek (1949) and Young et al. (2006) it is defined as:

$$f^{(mel)} = \Theta_{mel}(f) = 1127 \cdot \log\left(1 + \frac{f}{700}\right). \tag{2.66}$$

In theory, one Bark corresponds to 100 Mel, although the scale approximations are different, and thus the scales cannot be converted directly one to the other.

The inverse transformation of the frequency $f^{(mel)}$ to a linear frequency f in Hz is given as:

$$f = \Theta^{-1}_{mel}(f^{(mel)}) = 700\left(e^{\frac{f^{(mel)}}{1127}} - 1\right). \tag{2.67}$$

Semitone-Frequency (Octave) scale For music analysis a frequency scale aligned with music notes is required for some descriptors. A semitone frequency scale for Western/European music is defined for 12 semitones per octave and the tones in the next octave have double the frequency of the tones in the previous octave. With this, the frequency f of each semitone $f^{(oct)}$ is given by:

$$f = f_{o0} \cdot 2^{\frac{f^{(oct)}}{12}}, \tag{2.68}$$

where f_{o0} is the frequency (in Hz) of the first note in the first octave, i.e., the frequency of $f^{(oct)} = 0$. Typically values of $f_{o0} = 27.5$ Hz or $f_{o0} = 55$ Hz are chosen, corresponding to an A_0 or A_1 note of modern pitch, where A_4 is standardised at 440 Hz (ISO16:1975 1975).

From Eq. (2.68) the forward frequency transformation from the linear frequency f in Hz to a real valued semitone number $f^{(oct)}$ can be derived:

$$f^{(oct)} = \Theta_{oct}(f) = 12 \log_2\left(\frac{f}{f_{o0}}\right). \tag{2.69}$$

2.2.3.4 Band Spectra

Human hearing is redundant in a sense that masking takes place in both time and frequency (Zwicker and Fastl 1999). This means that of two sounds which have similar frequency the loudner one can (depending on the type of sound and levels) mask out the other one, rendering it inaudible. This suggests a reduction of the number of frequency bins M to a reduced number of bands B by combining bins within a defined band.[12] This is motivated by loudness perception.

This band division can be done linearly, i.e., each band having a constant width in Hertz, or non-linearly, where the bandwidth increases with the centre frequency of the band. Typically these non-linear bands are preferred because they are closer to non-linear frequency perception of the human hearing system. The methods for reducing M by combining bins into B bands, which will be described in the following, are general methods, which can work with any input frequency scale (linear or non-linear).

The general method for modelling masking and reducing the number of bands/bins at the same time is to define the power spectra of B band filters and then discretely convolve the M bin spectrum with each of the B band filters (cf. e.g., Hermansky 1990).[13] Assuming a general filter with a power spectrum $\Phi_b(m)$ which approximates the masking effects around the centre frequency of band b, the discrete convolution can be expressed as:

$$X_P(b) = \sum_{m=1}^{M} X_P(m)\Phi_b(m). \tag{2.70}$$

Next, a general function $g(x)$ defines the shape of the filter function on a general scale x. This scale can be any scale such as Bark scale, Mel scale, or linear frequency scale (Hertz (Hz)). In the ongoing, a linear Hz frequency scale is assumed, i.e., $x = f_{lin}$. Conversions to and from other scales can be implemented via Eqs. (2.48) and (2.49) from Sect. 2.2.3.3, respectively:

$$g^{(scale)}(f_{scale}) = g^{(lin)}(\Theta_{scale}^{-1}(f_{scale})), \tag{2.71}$$

and

$$g^{(lin)}(f_{lin}) = g^{(scale)}(\Theta_{scale}(f_{lin})). \tag{2.72}$$

The mapping of a continuous frequency value f_{scale} on scale *scale* to a real-valued bin index m_{scale} is given by Eq. (2.6) and the inverse by Eq. (2.7) (Sect. 2.1.2).

For linearly mapping a general filter shape $\Phi(m)$ defined for real-valued bin indices m': $g(m') = g(F^{-1}(f))$ to a discretised bin spectrum $\Phi(m)$ with integer bin

---

[12]For an implementation, see the `cMelspec` component in openSMILE and scale transformation functions in the smileUtil library.

[13]Band spectra can be computed in openSMILE with the `cMelspec` component, which—despite the name Melspec—can compute general band spectra for all supported frequency scales from a linear magnitude or power spectrum.

indices m, Eq. (2.73) can be applied:

$$\Phi(m) = \begin{cases} 0 & \text{for } m < m_l \\ \int\limits_{m-0.5}^{m+0.5} g(x)dx & \text{for } m \geq m_l \text{ and } m \leq m_u \\ 0 & \text{for } m > m_u \end{cases} . \tag{2.73}$$

Equation (2.73) assumes the bins to have a rectangular shape which is centred at the bin frequency $F(m)$.

In this thesis, rectangular and triangular filter shapes $g(f)$ are investigated, for which discrete versions are derived in the following. A rectangular filter function

$$g^{(rect)}(m') = \begin{cases} 1 & \text{for } m' \geq m'_l \text{ and } m \leq m'_u \\ 0 & \text{otherwise} \end{cases} , \tag{2.74}$$

with a lower cut-off frequency f_l and a corresponding real-valued bin number $m'_l = F^{-1}(f_l)$ as well as a upper cut-off frequency f_u with $m'_u = F^{-1}(f_l)$ is considered. Discrete lower and upper bound bin indices are obtained by rounding off the real-valued bin numbers: $m_l = \lfloor m'_l + 0.5 \rfloor$ and $m_u = \lfloor m_u l' + 0.5 \rfloor$. Now, Eq. (2.73) is applied and the rectangular filter can be expressed as:

$$\Phi^{(rect)}_{f_l,f_u}(m) = \begin{cases} 0 & \text{for } m < m_l \\ m_l - m'_l + 0.5 & \text{for } m = m_l \\ 1 & \text{for } m > m_l \text{ and } m < m_u \\ m'_u - m_u + 0.5 & \text{for } m = m_u \\ 0 & \text{for } m > m_u \end{cases} . \tag{2.75}$$

For the triangular filter, a real-valued centre (peak of the triangle) bin number $m'_c = F^{-1}(f_c)$ is required in addition to the real-valued bin number bounds m'_l and m'_u. A general triangle $g^{(tri)}(m')$ as function of the real valued bin number m' is then given as:

$$g^{(tri)}(m') = \begin{cases} 0 & \text{for } m' < m'_l \\ \dfrac{m' - m'_l}{m'_c - m'_l} & \text{for } m > m'_l \text{ and } m' \leq m'_c \\ \dfrac{m'_u - m'}{m'_u - m'_c} & \text{for } m > m'_c \text{ and } m' < m'_u \\ 0 & \text{for } m' > m'_u \end{cases} . \tag{2.76}$$

A triangular filter function $\Phi^{(tri)}_{f_l,f_c,f_u}(m)$ for integer bin indices is then given by integrating the function $g^{(tri)}(m')$ over each bin according to Eq. (2.73). Using $m_c = \lfloor m'_c + 0.5 \rfloor$, the discrete version of the triangular filter's power spectrum can be expressed as:

$$\Phi^{(tri)}_{f_l,f_c,f_h}(m) = \frac{1}{2}\begin{cases} 0 & \text{for } m < m_l \\ g^{(tri)}(m+0.5) & \text{for } m = m_l \\ g^{(tri)}(m+0.5) + g^{(tri)}(m-0.5) & \text{for } m > m_l \text{ and } m < m_c \\ \dfrac{g^{(tri)}(m_c-0.5)+g^{(tri)}(m'_c)}{1/\left(m'_c - m_c + 0.5\right)} + \dfrac{g^{(tri)}(m_c+0.5)+g^{(tri)}(m'_c)}{1/\left(m_c + 0.5 - m'_c\right)} & \text{for } m = m_c \\ g^{(tri)}(m+0.5) + g^{(tri)}(m-0.5) & \text{for } m > m_c \text{ and } m < m_u \\ g^{(tri)}(m_u-0.5) & \text{for } m = m_u \\ 0 & \text{for } m > m_u \end{cases} \tag{2.77}$$

To account for the non-linear frequency resolution of the hearing system, filters with asymmetric slopes and a frequency dependent bandwidth proportional to the critical band rate are suggested by Zwicker and Fastl (1999). According to Zwicker (1970) such filters can be approximated by linear, triangular shaped filters on a non-linear frequency scale (e.g., Bark scale). The filters should be spaced equidistantly on the non-linear scale, which results in a non-linear spacing on the linear scale where the spacing of the filters increases with frequency. The bandwidth of the filters on the non-linear scale is constant. To apply such filters designed on a non-linear scale to a linear scale magnitude spectrum, the spectrum must first be scaled to a non-linear scale before the linear shape filters can be applied. This step has to be performed for every spectrum, i.e., for every audio frame. To save computation time, alternatively, the linear filters can be transformed from the non-linear (Bark or Mel, for example) scale to a linear Hertz scale:

The general filter shape $g^{(scale)}\left(m^{(scale)'}\right)$ on the non-linear frequency scale with $m^{(scale)'} = F^{-1}\left(f^{(scale)}\right) = F^{-1}\left(\Theta_{scale}\left(f^{(lin)}\right)\right)$ can be converted to the filter shape on a linear scale:

$$g^{(lin)}(m^{(lin)'}) = g^{(scale)}\left(F^{-1^{(scale)}}\left(\Theta_{scale}\left(F^{(lin)}\left(m^{(lin)'}\right)\right)\right)\right). \tag{2.78}$$

Equation (2.73) and derived versions thereof can now be used to estimate the shape of the filter on the linear frequency scale if $g^{(lin)}(m^{(lin)'})$ is substituted for $g(m')$ and the boundaries m_l, m_c, and m_u are replaced by the respective bin indices on a linear frequency scale:

$$m^{(lin)'} = F^{(lin)}\left(\Theta^{-1}_{scale}\left(F^{-1^{(scale)}}\left(m'_{scale}\right)\right)\right) \tag{2.79}$$

$$m^{(lin)} = \lfloor m^{(lin)'} + 0.5 \rfloor \tag{2.80}$$

It is important to note that the integer bin indices $m^{(lin)}$ for the bounds m_l, m_c, and m_u must be generated by converting the real-valued bin numbers $m^{(scale)'}_{l,c,u}$ (not the rounded bin indices $g^{(lin)}(m^{(lin)'})$) and rounding them according to Eq. (2.80).

The generalised filter equations derived here differ from the ones commonly implemented for Mel-frequency filter banks (Young et al. 2006) or Bark filter banks (Hermansky 1990), which are approximations optimised for improved computation time decades ago and have not been changed for compatibility reasons.

For compatibility with Young et al. (2006), the following triangular filterbank is used in this thesis for all non-linear frequency scales.[14] A filterbank of B filters with centres evenly spaced between f_{min} and f_{max} is created as follows: A constant bandwidth β for the filters on the target scale *scale* is assumed:

$$\beta = \frac{1}{B+1}\left(f_{max}^{(scale)} - f_{min}^{(scale)}\right). \tag{2.81}$$

The centre frequencies $f_c^{(scale)}(b)$ on the target scale for B filters $b = 1 \ldots B$ and the lower and upper bounds ($b = 0$ and $b = B + 1$) are computed via:

$$f_c^{(scale)}(b) = f_{min}^{(scale)} + b\beta. \tag{2.82}$$

The power spectrum shape of each filter is then given by:

$$\Phi_b(m) = \begin{cases} 0 & \text{for } f^{(scale)}(m) \leq f_c^{(scale)}(b-1) \\ \dfrac{f^{(scale)}(m) - f_c^{(scale)}(b-1)}{f_c^{(scale)}(b) - f_c^{(scale)}(b-1)} & \text{for } f_c^{(scale)}(b-1) < f^{(scale)}(m) \leq f_c^{(scale)}(b) \\ \dfrac{f_c^{(scale)}(b+1) - f^{(scale)}(m)}{f_c^{(scale)}(b+1) - f_c^{(scale)}(b)} & \text{for } f_c^{(scale)}(b) < f^{(scale)}(m) \leq f_c^{(scale)}(b) \\ 0 & \text{for } f^{(scale)}(m) > f_c^{(scale)}(b+1) \end{cases}, \tag{2.83}$$

with

$$f^{(scale)}(m) = \Theta_{scale}\left(F^{(lin)}(m)\right). \tag{2.84}$$

Figure 2.2 shows plots of the power spectrum $\Phi(m)$ of two triangular filters with different centre frequencies which were designed as linear triangles on the Mel scale and converted to a discretised linear frequency scale (Eqs. (2.83) and (2.77)). A very low resolution for the DFT is assumed to highlight differences between the two implementations. 24 linear scale bins from $m = 1 \ldots 24$ are shown in the plot, each bin 32.5 Hz wide. The x-axis labelling indicates the corresponding Mel frequency. It can be seen that both implementations produce nearly similar results, except for a small deviation at the centre frequency and at the left and right borders of the filter. It will be left to future work to empirically find out which implementation of the filters is better. Due to the minor differences, which are even less for a higher DFT resolution, it is fully justified that for the experiments in this thesis the simplified

[14]In openSMILE the `cMelspec` component implements these filterbanks for various frequency scales (not only Mel).

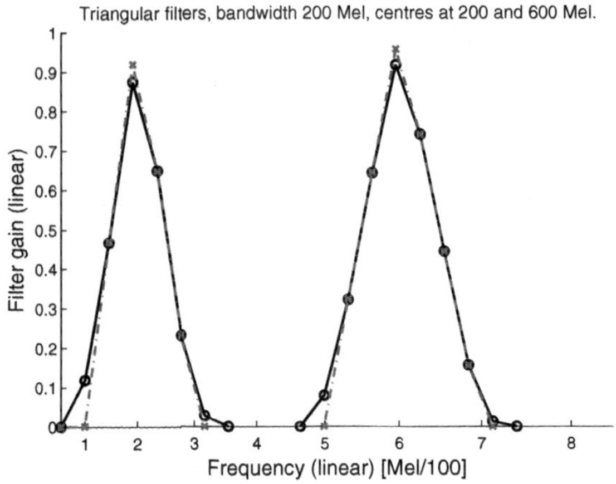

Fig. 2.2 Spectral shapes of two triangular filters designed with a centre frequency of 200 and 600 Mel and a symmetric (on the Mel scale) bandwidth of 200 Mel; *Dashed (blue) line* with (x) showing Eq. (2.83) and *black solid line* shows the version from Eq. (2.77) as derived in this thesis based on integration (Eq. (2.73))

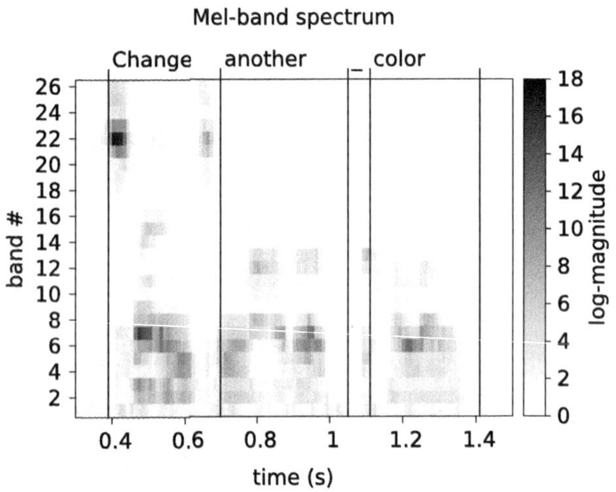

Fig. 2.3 Mel-band power spectrogram plot of a sample sentence from the AVIC database (Sect. 6.1.3); female speaker, words: "change another color"

implementation of Eq. (2.83) is used, which has also been used in other related work (e.g., Young et al. 2006). An example plot of a Mel-spectrogram obtained with this triangular filterbank is shown in Fig. 2.3.

2.2.3.5 Filterbank Spectra

An alternative to DFT based band spectra—as described in the previous sections—are time domain filterbank spectra. These are obtained by passing the unwindowed signal $x(n)$ through a bandpass filterbank with M filters and then computing the envelope of each of the filterbank outputs. To compute the envelope, windowing with size $N_f^{(m)}$ is performed on each of the filterbank outputs with a rate $T_f^{(m)}$ and the energy (Sect. 2.2.2) is computed for each window. The result is the spectral magnitude $X_M(m)$ for each band m at a rate of $T_f^{(m)}$. Thereby, for each band m a different envelope sampling rate $T_f^{(m)}$ as well as a different window size $N_f^{(m)}$ can be chosen. In practice, however, the sampling rate $T_f^{(m)}$ is a constant T_f for each band, with $T_f = 10$ ms or $T_f = 20$ ms in most cases.

Depending on the type, order, and number of bandpass filters, computation of such spectra can be slow when compared to FFT based spectra. Advantages, on the other side, are the improved frequency selectivity of time domain band-pass filters as well as the possibility of using a frequency dependent time resolution $T_f^{(m)}$.

For this thesis specifically Gabor filterbanks and gammatone filterbanks were implemented to approximate a critical band filterbank.[15] For practical implementation reasons, all filters were implemented as Finite Impulse Response (FIR) filters with discrete convolution. Thereby an impulse response $h(n)$ of finite length N_h is convolved with the input signal $x(n)$ to obtain the time domain output signal $y(n)$ (cf. Damelin and Miller 2011, p. 232):

$$y(n) = x(n) * h(n) = \sum_{m=-\frac{N_h}{2}}^{\frac{N_h}{2}} x(n-m) \cdot h(m). \tag{2.85}$$

To optimise the computational complexity, especially with long impulse response lengths N_h, the convolution can be carried out as a multiplication in the frequency domain. This approach is more efficient than the time domain implementation of the convolution because applying twice a FFT for long frames is more efficient than computing the full length convolution for every sample.

The one dimensional Gabor-filter impulse response $h_{gabor}(n)$ (for $N = -\frac{N}{2} \dots \frac{N}{2} - 1$) is given by (cf. Feichtinger and Strohmer 1998):

$$h_{gabor}(n) = e^{\beta\sqrt{2\pi}T_s n^2} \cos(2\pi f_c T_s n), \tag{2.86}$$

with bandwidth β (in Hz), centre frequency f_c (in Hz), and sampling period T_s.

The gammatone filter impulse response $h_{gammatone}(n)$ (for $N = -\frac{N}{2} \dots \frac{N}{2} - 1$) is given by (cf. Slaney 1993):

[15] In openSMILE the FIR filterbanks with Gabor, gammatone, high- and low-pass filters can be applied with the `cFirFilterbank` component.

$$h_{gammatone}(n) = aT_s n^{(o-1)} e^{-2\pi\beta n T_s} \cos(2\pi f_c n T_s), \tag{2.87}$$

where $n = 0 \dots N_h - 1$, f_c is the centre frequency of the filter in Hz, β is the bandwidth of the filter in Hz, a is a linear gain factor which most often defaults to 1, and T_s is the sampling period.

The impulse responses for the gammatone and Gabor filters are infinite in time. In order to implement them with a FIR filter, a windowing must be performed, which introduces artefacts. This windowing is also referred to spectral shaping (Oppenheim et al. 1999). A windowing in the time domain corresponds to a convolution of the filter's frequency response with the spectrum of the windowing function. A good frequency response without side maxima can be achieved by multiplying the filter with a zero-endpoint Gaussian window $w_{gauss0}(n)$ (i.e., a Gaussian window scaled to have zero valued endpoints):

$$w_{gauss0}(n) = \frac{1}{1-\Gamma}\left(e^{-0.5\left(\frac{n}{\sigma\frac{N}{2}}\right)^2} - \Gamma\right) \tag{2.88}$$

with Γ being the minimum value of the unscaled Gaussian at the endpoints:

$$\Gamma = e^{-0.5\left(\frac{-\frac{N}{2}+1}{\sigma\frac{N}{2}}\right)^2} \tag{2.89}$$

An approximation of an ideal high or low-pass filter can be implemented via a windowed *sinc* function. The frequency response of the unwindowed, infinite duration *sinc* function is a perfect rectangle in the frequency domain, which corresponds to a low-pass filter. The steepness of the filter depends on the length of the window, i.e., the number of taps of the impulse response. The cut-off frequency f_c is determined by the *sinc* function, which leads to the following equation for the *sinc* impulse response h_{lowp} (for $N = -\frac{N}{2} \dots \frac{N}{2} - 1$) of a low-pass filter:

$$h_{lowp}(n) = 2f_c \frac{sin\,(2\pi f_c T_s n)}{2\pi f_c T_s n}. \tag{2.90}$$

The high-pass filter can be constructed from the low-pass filter in the spectral domain by subtracting the low-pass filter spectrum from a flat spectrum which is the constant 1. In the time domain (for $N = -\frac{N}{2} \dots \frac{N}{2} - 1$) this corresponds to:

$$h_{highp}(n) = \begin{cases} -h_{lowp}(n) & \text{for } n \neq 0 \\ -h_{lowp}(n) + 1 & \text{for } n = 0 \end{cases} \tag{2.91}$$

2.2.4 Spectral Descriptors

Now, a general spectrum $X(m)$ is defined, which can be a magnitude spectrum $X_M(m)$, a power spectrum $X_P(m)$, power spectral densities in dB, or a band or filterbank spectrum. With such a general spectrum $X(m)$ and a relation $f = F(m)$ (and $m = F^{-1}(f)$ between the linear frequency f in Hz and the bin index m (cf. Sect. 2.1.2), spectral statistics LLDs are defined in this section.[16] Most spectral descriptors can be computed from an arbirtrary sub-band range defined by the lower and upper bin indices m_l and m_u. The full range of the spectrum is covered when $m_l = 1$ and $m_u = M$. For the case of a constrained sub-band frequency range with respective lower and upper border frequencies of f_l and f_u, the respective integer valued bin indices are $m_l = \lfloor F^{-1}(f_l) + 0.5 \rfloor$ and $m_u = \lfloor F^{-1}(f_u) + 0.5 \rfloor$.

2.2.4.1 Band Energies

When computed from a high resolution FFT spectrum $X(m)$, this LLD allows to consider the energy in arbitrary bands. Similar to the concept of band spectra and triangular filterbanks, described in Sect. 2.2.3.4, the band energy is computed by a rectangular filter here, i.e., by summation of all the magnitudes within the interval $[f_l; f_u]$, where f_l and f_u are the lower and upper frequency bounds of the band, respectively. *Note:* This descriptor can be computed from any type of spectrum $X(m)$ in theory. However, the equations given below rely on a power spectrum $X_P(m)$ as input. Other spectral representations must be converted to a power spectrum, in order to be able to sum up the energies in each band with Eq. (2.92).

At the band borders interpolation is performed to consider partial bins appropriately, as derived in Sect. 2.2.3.4. With general lower and upper bound frequencies of f_l and f_u, respectively, according to Eq. (2.75) respective weighting factors $\alpha_l = m_l - m'_l + 0.5$ and $\alpha_u = m'_u - m_u + 0.5$ for the lower and upper integer bin indices m_l and m_u are found.

The energy for a band bounded by f_l and f_u is then computed as the following sum over the power spectrum X_P:

$$E_{f_l}^{f_u} = \alpha_l X_P(m_l) + \left(\sum_{m=m_l+1}^{m_u-1} X_P(m) \right) + \alpha_u X_P(m_u). \tag{2.92}$$

2.2.4.2 Spectral Slope

The overall shape of a spectrum $X(m)$ can be expressed by its linear slope. To compute the slope, a minimum quadratic error approximation of the spectrum $X(m)$ by a line

[16] In openSMILE these spectral descriptors can be extracted with the `cSpectral` component.

$$\hat{y} = ax + b \tag{2.93}$$

is attempted, where a is the slope, and b is the vertical displacement (Tamarit et al. 2008). In order to keep the following equations generic, the spectrum $X(m)$ is represented as a general function $y = f(x)$ with $y = X$ and $x = m$. The function $f(x)$ is defined for a finite set of N points x_i with $i = 0 \ldots N-1$.

The minimisation of the quadratic error e^2 between the linear approximation of the function and the function itself is expressed as:

$$\begin{aligned} e^2 &= \sum_{i=0}^{N-1} \left(y(x_i) - \hat{y}(x_i)\right)^2 = \sum_{i=0}^{N-1} (y_i - ax_i - b)^2 \\ &= \sum_{i=0}^{N-1} \left(y_i^2 - 2ax_iy_i - 2by_i + 2abx_i + a^2x_i^2 + b^2\right)^2 \stackrel{!}{=} min. \end{aligned} \tag{2.94}$$

From this, the following differential equations for a and b for the points x_i with $i = 0 \ldots N$ are obtained:

$$\frac{\delta}{\delta a} e^2 = \sum_{i=0}^{N-1} \left(-2x_iy_i + 2bx_i + 2ax_i^2\right) \stackrel{!}{=} 0, \tag{2.95}$$

$$\frac{\delta}{\delta b} e^2 = \sum_{i=0}^{N-1} (-2y_i + 2ax_i + 2b) \stackrel{!}{=} 0. \tag{2.96}$$

Rewritten as:

$$-\sum_{i=0}^{N-1} x_iy_i + b\sum_{i=0}^{N-1} x_i + a\sum_{i=0}^{N-1} x_i^2 \stackrel{!}{=} 0, \tag{2.97}$$

$$-\sum_{i=0}^{N-1} y_i + a\sum_{i=0}^{N-1} x_i + Nb \stackrel{!}{=} 0, \tag{2.98}$$

yields the solution for a:

$$a = \frac{N\sum_{i=0}^{N-1} x_iy_i - \sum_{i=0}^{N-1} x_i \sum_{i=0}^{N-1} y_i}{N\sum_{i=0}^{N} x_i^2 - \left(\sum_{i=0}^{N-1} x_i\right)^2} \tag{2.99}$$

For simplification, the following substitutions can be made in Eq. (2.99):

$$\begin{aligned} &\sum_{i=0}^{N-1} x_i = \Sigma_x, \quad \sum_{i=0}^{N-1} y_i = \Sigma_y, \\ &\sum_{i=0}^{N-1} x_i^2 = \Sigma_{x^2}, \quad \sum_{i=0}^{N-1} x_iy_i = \Sigma_{xy}, \end{aligned} \tag{2.100}$$

and Eq. (2.99) can be rewritten as:

$$a = \frac{N\Sigma_{xy} - \Sigma_x \Sigma_y}{N\Sigma_{x^2} - \Sigma_x^2} \tag{2.101}$$

For computational efficiency, when using a linear frequency scale, the following substitutions can be further made (by applying exponential sum properties from (Rade et al. 2000, p.189):

$$\Sigma_x = \frac{1}{2}N(N-1) \tag{2.102}$$

$$\Sigma_{x^2} = \frac{1}{6}N(N-1)(2N-1). \tag{2.103}$$

The spectral slope can be computed over the full range of the spectrum, or over a sub-band. When computing the spectral slope in a sub-band, the border points must be linearly interpolated from the neighbouring bins, if they do not exactly match a bin. Let us assume respective lower and upper band border frequencies of f_l and f_u, which correspond to real valued bin values $m'_l = F^{-1}(f_l)$ and $m'_u = F^{-1}(f_u)$. Then, the substitutions in Eq. (2.100) can be rephrased as (replacing $x_i = x_m = F(m)$ and $y_i = y_m = X(m)$):

$$\Sigma_x = f_l + \left(\sum_{m=\lceil m'_l \rceil}^{\lfloor m'_u \rfloor} F(m) \right) + f_u, \tag{2.104}$$

$$\begin{aligned} \Sigma_y = {} & X\left(\lfloor m'_l \rfloor\right) + \left(m'_l - \lfloor m'_l \rfloor\right)\left(X\left(\lceil m'_l \rceil\right) - X\left(\lfloor m'_l \rfloor\right)\right) \\ & + \left(\sum_{m=\lceil m'_l \rceil}^{\lfloor m'_u \rfloor} X(m) \right) \\ & + X\left(\lfloor m'_u \rfloor\right) + \left(m'_u - \lfloor m'_u \rfloor\right)\left(X\left(\lceil m'_u \rceil\right) - X\left(\lfloor m'_u \rfloor\right)\right), \end{aligned} \tag{2.105}$$

$$\Sigma_{x^2} = f_l^2 + \left(\sum_{m=\lceil m'_l \rceil}^{\lfloor m'_u \rfloor} F(m)^2 \right) + f_u^2, \tag{2.106}$$

$$\begin{aligned} \Sigma_{xy} = {} & f_l\Big(X\left(\lfloor m'_l \rfloor\right) + \left(m'_l - \lfloor m'_l \rfloor\right)\left(X\left(\lceil m'_l \rceil\right) - X\left(\lfloor m'_l \rfloor\right)\right)\Big) \\ & + \left(\sum_{m=\lceil m'_l \rceil}^{\lfloor m'_u \rfloor} F(m)X(m) \right) \\ & + f_u\Big(X\left(\lfloor m'_u \rfloor\right) + \left(m'_u - \lfloor m'_u \rfloor\right)\left(X\left(\lceil m'_u \rceil\right) - X\left(\lfloor m'_u \rfloor\right)\right)\Big) \end{aligned} \tag{2.107}$$

The spectral slope a computed with Eqs. (2.104)–(2.107) and Eq. (2.101) has the unit magnitude/Hz, if computed from a magnitude spectrum $X_M(m)$, energy/Hz if computed from a power spectrum $X_P(m)$, and dB/Hz if computed from log-power spectral densities $X_{P,norm}(m)$, for example.

Often, the spectral slope is computed from a linear frequency power spectrum over the full spectrum range (typically 0–8,000 Hz), as in, e.g., (Eyben et al. 2010a; Schuller et al. 2010, 2011, 2012a, 2013b). Alternatively, as suggested by Scherer et al. (2015), the spectral slope should be computed in three different bands, which are 0–1, 1–5, and 0–5 kHz. Thereby a logarithmic band spectrum is used, with constant linear bandwidth of 400 Hz.

2.2.4.3 Hammarberg Index

Besides computing the exact spectral slope directly, as was outlined above, features closely related to the spectral slope can be used. Tamarit et al. (2008) mention the *Hammarberg index* in this context. The measure was defined by Hammarberg et al. (1980) as the ratio of the strongest energy peak in the 0–2 kHz region to that of the strongest peak in the 2–5 kHz region. Hammarberg defined a fixed static pivot point of 2 kHz where the low and high frequency regions are separated. Symbolically the Hammarberg index η is defined here as:

$$\eta = \frac{\max_{m=1}^{m_{2k}} X(m)}{\max_{m=m_{2k}+1}^{M} X(m)}, \tag{2.108}$$

where m_{2k} is the highest spectral bin index where $f \leq 2$ kHz is still true.

According to more recent findings (Tamarit et al. 2008) it could be beneficial to pick the pivot point dynamically based upon the speaker's fundamental frequency. This is, however, on purpose not considered in this thesis because it would break the strictly static nature of all the feature extraction algorithms described. It will be left an open issue for future work where a multi-level feature extraction could be investigated, i.e., features are organized in different hirarchies and extraction algorithms for higher level features change their parameters according to decisions based on values of lower level features.

2.2.4.4 Alpha Ratio

Similar to the Hammarberg index, the Alpha Ratio (Patel et al. 2010) is defined as the ratio between the energy in the low frequency region and the high frequency region. More specifically, it is the ratio between the summed energy from 50 to 1000 Hz and 1 to 5 kHz, expressed as ρ_α:

$$\rho_\alpha = \frac{\sum_{m=1}^{m_{1k}} X(m)}{\sum_{m=m_{1k}+1}^{M} X(m)}, \tag{2.109}$$

where m_{1k} is the highest spectral bin index where $f \leq 1$ kHz is still true.

In applications of emotion recognition from speech, this parameter most often—like other spectral slope related paramters—is computed from a logarithmic representation of a band Long-Term Average Spectrum (LTAS) (cf. Scherer et al. 2015; Patel et al. 2010). However, the definition—in my opinion, holds for any sort of spectrum—and thus also for linear short time magnitude spectra, for example, as used in this thesis. Which spectral representation is best has to be determined empirically for every use-case.

2.2.4.5 Spectral Flatness

Spectral flatness ($S_{flatness}$) is computed as the ratio of the geometric mean of the spectral bins to the arithmetic mean of the spectral bins (Johnston 1988):

$$S_{flatness} = \frac{\sqrt[(m_u-m_l+1)]{\prod_{m=m_l}^{m_u} X(m)}}{\frac{1}{(m_u-m_l+1)} \sum_{m=m_l}^{m_u} X(m)}, \tag{2.110}$$

with lower and upper bound bin indices m_l and m_u, respectively. As suggested by the name this descriptor describes the flatness of a spectrum, i.e., it will have a higher value for a spectrum with strong peaks (e.g., a harmonic spectrum, but also a spectrum with non equidistant peaks such as multiple tones or modulated noise). This descriptor is part of the MPEG-7 audio content description standard (Manjunath et al. 2002).

2.2.4.6 Spectral Centroid

Spectral centroid ($S_{centroid}$) is computed as the centre of gravity of a spectrum $X(m)$ (cf. Peeters 2004):

$$S_{centroid} = \frac{\sum_{m=m_l}^{m_u} F(m)X(m)}{\sum_{m=m_l}^{m_u} X(m)}, \tag{2.111}$$

with lower and upper bound bin indices m_l and m_u, respectively. $F(m)$ is the frequency (in Hz) corresponding to bin m (cf. Sect. 2.1.2). It is correlated to the brightness or sharpness of an audio signal, according to Zwicker and Fastl (1999). Other studies (e.g., Kendall and Carterette 1996) suggest that the ratio between F_0 and the spectral centroid is better correlated to brightness than the spectral centroid alone. This descriptor is part of the MPEG-7 audio content description standard (Manjunath et al. 2002), also known as spectrum centroid.

The sums in Eq. (2.111) re-appear in equations for the spectral slope (Sect. 2.2.4.2), thus—in efficient code—spectral slope and spectral centroid should be computed in the same component in order to share the results and avoid computing these sums twice.[17]

2.2.4.7 Spectral Moments

Spectral moments include the second, third, and fourth order moments, which are named spectral variance (also referred to as spectral spread in the MPEG-7 standard (cf. Peeters 2004 and Manjunath et al. 2002)), the spectral skewness, and the spectral kurtosis, respectively. In order to compute these statistical moments, the spectrum $X(m)$ must be converted to a Mass Function (PMF) $p_X(m)$:

$$p_X(m) = \frac{X(m)}{\sum_{i=m_l}^{m_u} X(i)}. \tag{2.112}$$

Thereby usually the full spectral range is considered, i.e., $m_l = 1$ and $m_u = M$ and $X(m)$ is chosen to be the power spectrum, i.e., $X(m) = X_P(m)$.

Based on the above PMF, spectral variance $S_{variance}$ is defined as:

$$S_{variance} = S_\sigma^2 = \sum_{m=m_l}^{m_u} \left(F(m) - S_{centroid}\right)^2 p_X(m), \tag{2.113}$$

with the spectral centroid $S_{centroid}$ as defined in Sect. 2.2.4.6. The spectral standard deviation S_σ is given as:

$$S_\sigma = \sqrt{S_{variance}}. \tag{2.114}$$

Accordingly, spectral skewness $S_{skewness}$ is defined as:

$$S_{skewness} = \frac{1}{S_\sigma^3} \sum_{m=m_l}^{m_u} \left(F(m) - S_{centroid}\right)^3 p_X(m), \tag{2.115}$$

and spectral kurtosis $S_{kurtosis}$ as:

$$S_{kurtosis} = \frac{1}{S_\sigma^4} \sum_{m=m_l}^{m_u} \left(F(m) - S_{centroid}\right)^4 p_X(m). \tag{2.116}$$

[17] In openSMILE, this is implemented in the `cSpectral` component.

2.2.4.8 Spectral Entropy

The spectral entropy is similar to the spectral flatness, as it relates to the peakedness of the spectrum. Again, the spectrum must be converted to a PMF $p_X(m)$ using Eq. (2.112). Then, according to Misra et al. (2004), the spectral entropy $S_{entropy}$ is defined as:

$$S_{entropy} = -\sum_{m=m_l}^{m_u} p_X(m) \cdot \log_2 p_X(m). \tag{2.117}$$

The definition is based on the original definition of the Shannon Entropy by Shannon (1948).

2.2.4.9 Spectral Roll-Off Point

The $n\,\%$ spectral Roll-off Point (RoP) $S_{rop(n)}$ is defined as the frequency below which $n\,\%$ of the total energy is concentrated (generalised version of Peeters 2004). Thus, the spectral RoPs must be computed from the power spectrum $X_P(m)$. The total energy E (in the band from bin m_l to m_u) is computed as:

$$E = \sum_{m=m_l}^{m_u} X_P(m). \tag{2.118}$$

The $n\,\%$ RoPs is then estimated by iteratively increasing r in the following equation until the equation becomes valid:

$$\sum_{m=1}^{r} X_P(m) \geq \frac{n}{100} E. \tag{2.119}$$

The RoPs is then $S_{rop(n)} = F(r)$. Typical values for n are 95, 90, 75, and 50 %.

2.2.4.10 Psychoacoustic Sharpness

According to Zwicker and Fastl (1999), the spectral centroid is better correlated to the perceived sharpness of a sound, if it is computed on a Bark frequency scale and a frequency dependent auditory weighting for frequencies greater than 16 Bark (function $g(f^{(bark)})$) is performed on the magnitudes $X(m)$. This leads to the following definition of perceptual sharpness:

$$S_{sharpness} = 0.11 \frac{\sum_{m=1}^{M} \Theta_{bark}\big(F(m)\big) X(m) g\Big(\Theta_{bark}\big(F(m)\big)\Big)}{\sum_{m=1}^{M} X(m)}, \tag{2.120}$$

with

$$g\left(f^{(bark)}\right)=\begin{cases}1 & \text{if} f^{(bark)}\leq 16\\ \left(\frac{f^{(bark)}-16}{4}\right)^{\log 3/\log 2}+1 & \text{otherwise}\end{cases}. \tag{2.121}$$

2.2.4.11 Spectral Differences

The spectral descriptors described so far are static descriptors in a sense that they can be applied to single spectra, e.g., short time spectra of a single frame. In order to consider changes in the spectrum over time, Spectral Difference (SD) features can be used. The general idea is to treat each short-time spectrum $X^{(k)}(m)$ at discrete time (frame index) k as a point X in an M-dimensional space and to compute the distance between two successive spectra. For distance computation the L_2 norm is considered in this thesis and the spectral difference $SD^{(k)}$ at frame index k is defined as:

$$SD^{(k)}=\sqrt{\sum_{m=m_l}^{m_u}\left(X^{(k)}(m)-X^{(k-1)}(m)\right)^2}. \tag{2.122}$$

This descriptor is useful for detecting sudden changes in the overall energy as well as sudden changes in the spectral shape of the signal without large changes in signal energy. It is widely used in Music Information Retrieval (MIR) for detection of instrument onsets (Masri 1996; Duxbury et al. 2002; Eyben et al. 2010b). A variation of the SD function is common in this field: the positive spectral difference $SD_+^{(k)}$, where only positive differences are considered in the sum:

$$SD_+^{(k)}=\sqrt{\sum_{m=m_l}^{m_u}\left(\frac{X^{(k)}(m)-X^{(k-1)}(m)+|X^{(k)}(m)-X^{(k-1)}(m)|}{2}\right)^2}. \tag{2.123}$$

This function emphasises areas of fast rising spectral energy, which in music corresponds to onsets of instruments and vocals. Decreasing energy is neglected, which is also of no importance of an onset detection algorithm, for example.

2.2.4.12 Spectral Flux

The spectral flux S_{flux} represents a quadratic, normalised version of the simple spectral difference (Sect. 2.2.4.11). With general normalisation coefficients μ_k the definition of spectral flux is as follows:

$$S_{flux}^{(k)}=\sum_{m=m_l}^{m_u}\left(\frac{X^{(k)}(m)}{\mu_k}-\frac{X^{(k-1)}(m)}{\mu_{k-1}}\right)^2. \tag{2.124}$$

The normalisation coefficients μ_k and μ_{k-1} can either be one[18] which corresponds to no normalisation, or chosen in a way that normalises the L_1 or L_2 norm of the spectrum vector to one:

$$\mu_k^{(L_1)} = \sum_{m=m_l}^{m_u} |X^{(k)}(m)|, \tag{2.125}$$

$$\mu_k^{(L_2)} = \sqrt{\sum_{m=m_l}^{m_u} \left(X^{(k)}(m)\right)^2}. \tag{2.126}$$

In the latter case (L_1 or L_2 vector normalisation) we talk about *normalised* spectral flux, otherwise, about *unnormalised* spectral flux.

2.2.4.13 Harmonicity

The *Harmonicity* descriptor implemented in openSMILE[19] describes the amount and quality of the harmonics in a signal, i.e., it can be used to discriminate harmonic and non-harmonic signals. It is thus related to the more common Harmonics-to-Noise ratio (see Sect. 2.2.13.3), however, not the same. Harmonicity is computed directly from a magnitude spectrum by applying a simple peak picking algorithm based on identification of local minima and maxima by looking at the 2 left and right neighbours of the current frame. Then, the ratio between the minima and the maxima in relation to the amplitude of the maxima is computed.

In detail, the maxima and minima of the magnitude spectrum $X_M(m)$ are found according to the following rule:

$$Max(m) = 1, \quad \text{if} \begin{cases} \quad\;\; X_M(m-2) < X_M(m) \\ \text{and } X_M(m-1) < X_M(m) \\ \text{and } X_M(m+1) < X_M(m) \\ \text{and } X_M(m+2) < X_M(m) \end{cases} \tag{2.127}$$

$$Min(m) = 1, \quad \text{if} \begin{cases} \quad\;\; X_M(m-2) > X_M(m) \\ \text{and } X_M(m-1) > X_M(m) \\ \text{and } X_M(m+1) > X_M(m) \\ \text{and } X_M(m+2) > X_M(m) \end{cases}. \tag{2.128}$$

[18] This is the current default in all openSMILE feature sets up to version 2.0. An option for normalisation might appear in later versions.

[19] In the `cSpectral` component.

Now, with an array $L(p)$ of the amplitudes of P alternating maxima and minima $p = 1 \dots P$ a sum Σ_P of maxima to minima distances is given as:

$$\Sigma_P = \sum_{p=2}^{P} |L(p) - L(p-1)|. \tag{2.129}$$

The harmonicity descriptor HM is computed by normalising with the number of bins in the spectrum:

$$HM = \frac{\Sigma_P}{m_u - m_l + 1}, \tag{2.130}$$

or by the total energy[20] of the frame in the sub-band:

$$\overline{HM} = \frac{\Sigma_P}{\sum_{m=m_l}^{m_u} X_M(m)}. \tag{2.131}$$

2.2.5 *Autocorrelation*

Complementary to a short-time spectrum representation, the short-time Autocorrelation function (ACF) has a high resolution for low frequency periodicities within a frame or segment of interest. The autocorrelation describes the signal's self similarity at given discrete time lags $\tau \in [-\tau_{max} \dots \tau_{max}]$.

For a discrete signal $x(n)$ of infinite duration, the ACF can be computed in the time domain as:

$$ACF_e(\tau) = \sum_{n=-\infty}^{\infty} x(n)x(n+\tau). \tag{2.132}$$

This energy based definition yields infinite values for signals of infinite duration. Thus, in practice the signal length is limited to N (frame length or period of an infinite duration signal):

$$ACF_e(\tau) = \sum_{n=0}^{N-1} x(n)x(n+\tau). \tag{2.133}$$

Further, the energy ACF can be normalised by the length N in order to obtain a power ACF:

$$ACF_p(\tau) = \frac{1}{N}\sum_{n=0}^{N-1} x(n)x(n+\tau). \tag{2.134}$$

[20]Enabled by the option *normBandEnergies* of the cSpectral component of openSMILE.

The definitions in Eqs. (2.133) and (2.134) assume that the signal is either periodic with N or has at least a duration of $N + 2\tau_{max}$. In the latter case (non-periodic), a signal of length N can also be zero-padded (i.e., filled with zeros) with $\tau_{max}/2$ zeroes at the left and $\tau_{max}/2$ zeroes at the right of the actual N samples.

Instead of computing the ACF via the sums in Eq. (2.133), the discrete FFT can be used to speed up computations (Wiener–Khintchine Theorem, cf. Wiener 1964 which bases on original publications in Wiener 1930, Khintchine 1934, and Levinson 1947a, b for discrete signals). Due to the fact that the autocorrelation can be seen as a convolution of $x(n)$ with a reversed version of $x(n)$, the autocorrelation of $x(n)$ can be expressed as multiplication in the frequency domain[21]:

$$ACF_e(x) = FFT^{-1}(FFT(x) \cdot FFT^*(x)), \tag{2.135}$$

with $FFT^*(x)$ denoting the complex conjugate of the discrete FFT of the signal $x(n)$ and $FFT^{-1}(.)$ representing the inverse discrete FFT. The above produces equivalent output to Eq. (2.133). This definition implicitly assumes the signal $x(n)$ to be continued periodically to the left and right. To avoid artefacts when the signal is not periodic in N, zero-padding to a length of $2N$ must be performed by placing $N/2$ zeroes to the left and $N/2$ zeroes to the right of $x(n)$. The valid range for τ is from $-N$ to $+N$ after zero-padding.

Designing the ACF in this way will avoid artefacts, however the magnitude of the ACF function will linearly decay towards the border because more zeroes will be part of the sum (Eq. (2.134)) for higher τ. To avoid this, two signals $x_1(n)$ and $x_2(n)$ can be defined, where $x_1(n) = x(n)$ for all $n \in [0; N[$ and $x_2(n) = x(n)$ for $n \in [\frac{N}{4}; \frac{3N}{4}[$. x_1 is zero-padded with $N/4$ zeroes to the left and right, and x_2 is zero-padded with $N/2$ zeroes to the left and right. The autocorrelation is then expressed as the cross-correlation between x_1 and x_2:

$$ACF_e(\tau) = \sum_{n=0}^{N-1} x_1(n)x_2(n+\tau). \tag{2.136}$$

This ensures a constant amplitude of the ACF for τ from $-N/4$ to $+N/4$ and a valid range of τ (i.e., non-zero output) from $-3N/4$ to $3N/4$.

In practice, most signals are not periodic with the frame length. Thus, symmetric zero-padding with $-N/2$ samples at both ends before applying the ACF is the most common method. The modification presented in Eq. (2.136) is not commonly used because it is not an autocorrelation in a strict sense any more and has slightly different properties. Instead, a τ-adaptive energy normalisation can be applied to obtain a corrected power ACF:

$$ACF_p(\tau) = \frac{1}{N-\tau} \sum_{n=0}^{N-1} x(n)x(n+\tau). \tag{2.137}$$

[21] ACF according to this equation is implemented in openSMILE in the cAcf component.

According to the above definitions, the ACF is defined for negative and positive values of τ. However, by definition the ACF is always symmetric and only the positive half, i.e., $\tau \in [0; \tau_{max}[$ is relevant.

The ACF constitutes a basis from which further descriptors can be computed. These include the Harmonics to Noise Ratio (HNR) (Sect. 2.2.13.3), fundamental frequency (Sect. 2.2.11), and ACF peak ratios (not considered here), for example.

2.2.6 Cepstrum

The Cepstrum (CEP) was introduced by Bogert et al. (1963). It is closely related to the ACF with respect to the way it is computed. Equation (2.135) is extended by a natural logarithm before the inverse FFT:

$$CEP_e = FFT^{-1}\{\ln\left(FFT(x) \cdot FFT^*(x)\right)\}, \tag{2.138}$$

$$CEP_e = FFT^{-1}\{\ln\left(|FFT(x)|^2\right)\}. \tag{2.139}$$

The effect of the logarithm is the temporal separation of the source and filter parts of the signal x in the Cepstrum for which the following will give a brief derivation and detailed explanation.

A source signal s is assumed which passes through a Linear Time Invariant (LTI) system which has an impulse response h. The signal y at the output of the LTI system is then defined by:

$$y(n) = \{s * h\}(n), \tag{2.140}$$

where $*$ denotes a convolution of the two discrete signals s and h, which is analytically defined as:

$$y(n) = \{s * h\}(n) = \sum_{n_h=0}^{N_h-1} s(n - n_h)h(n_h), \tag{2.141}$$

where N_h is the length of the impulse response h.

By exploiting properties of the FFT, the convolution in Eq. (2.140) can be expressed in the frequency domain as:

$$Y(m) = S(m)H(m), \tag{2.142}$$

where Y, S, and H are the results of the discrete FFT of the signals y, s, and h, respectively. If we now apply the power operator we obtain:

$$\begin{aligned} |Y(m)|^2 &= Y(m)Y^*(m) = S(m)S^*(m)H(m)H^*(m), \\ &= |S(m)H(m)|^2 = |S(m)|^2|H(m)|^2. \end{aligned} \tag{2.143}$$

Applying the natural logarithm to both sides of Eq. (2.143) yields:

$$\ln\left(|Y(m)|^2\right) = \ln\left(|S(m)|^2|H(m)|^2\right). \tag{2.144}$$

Now the algebraic property of logarithms that a multiplication of the arguments of the logarithm function is equivalent to a summation of the logarithms of the individual arguments can be exploited to transform the multiplicative (in the frequency domain) or convolutive (in the frequency domain) relation between the signals s and h into an additive relation:

$$\ln\left(|Y(m)|^2\right) = \ln\left(|S(m)|^2\right) + \ln\left(|H(m)|^2\right). \tag{2.145}$$

After inverse FFT the above becomes:

$$FFT^{-1}\{\ln\left(|Y(m)|^2\right)\} = FFT^{-1}\{\ln\left(|S(m)|^2\right)\} + FFT^{-1}\{\ln\left(|H(m)|^2\right)\}, \tag{2.146}$$

where

$$FFT^{-1}\{\ln\left(|Y(m)|^2\right)\} = FFT^{-1}\{\ln\left(|FFT(y)|^2\right)\} = CEP_e(y). \tag{2.147}$$

Thus, the Cepstrum of the output of a LTI system is the sum of the Cepstrum of the source signal and the Cepstrum of the system's impulse response.

In the case of a human speech signal the vocal tract can be seen—in a very simplified way—as a LTI system with the glottal excitation signal as the input $s(n)$ and the vocal tract together with the sound wave propagation from the mouth to adjacent air and the room as filter with the impulse response $h(n)$ (linear source-filter model) (Parsons 1987; Rosen and Howell 1991; Fant 1973). An example cepstrogram (plot) of a real speech signal is shown in Fig. 2.4. The impulse response h in this case (if room reverberation with high delay constants is not considered) is of rather short duration, thus, the Cepstrum is also short. In contrast, the excitation signal s is of longer duration, as it is either a periodic signal for voiced sounds (created by opening and closing of the vocal cords—in the ideal model it is a sequence of Dirac pulses) or a noise like signal for unvoiced sound (created by airflow through the glottis—white Gaussian noise in the ideal model). In the case of a periodic excitation this fact allows to separate the impulse response Cepstrum from the excitation function by taking the first K samples of the Cepstrum as impulse response and then filling them with zeroes leaving only the excitation signal. Thereby K must be smaller than the period length of the excitation signal. The fist two pulses of the excitation signal are clearly visible in Fig. 2.4.

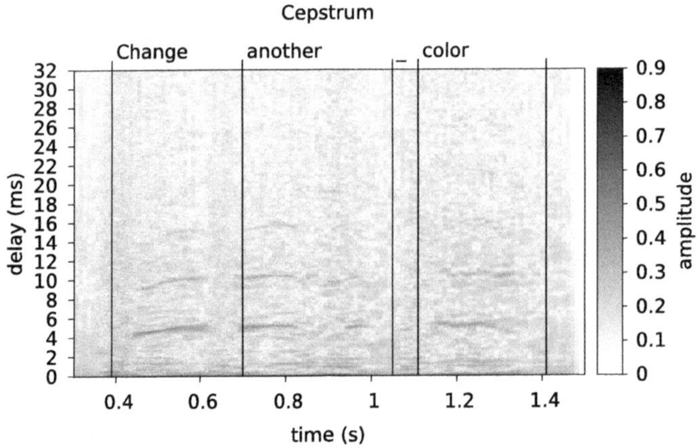

Fig. 2.4 Cepstrogram plot of a sample sentence from the AVIC database (Sect. 6.1.3); female speaker, words: "change another color"

2.2.7 Linear Prediction

Similar to Sect. 2.2.6, Linear Prediction (LP) analysis exploits the properties of the simple speech production model based on a LTI system. In this case the assumption is that if voiced speech is produced by a LTI system excited with periodic Dirac like signal, a sample at discrete time n will have a high correlation with the previous sample at $n-1$, i.e., one sample can be used to predict the following sample. Details on LP can be found in many other excellent sources, e.g., in a review by Makhoul (1975), in Ruske (1993), Furui (1996), and Schuller (2013). Therefore, in the following only the basic equations are summarised, which are required for implementation of LP for acoustic feature extraction.[22] In signal processing the term Linear Predictive Coding (LPC) is commonly used when talking about LP analysis.

In linear prediction an approximation $\hat{x}(n)$ of the original signal $x(n)$ is constructed by a weighted sum of p previous samples (Schuller 2013):

$$\hat{x}(n) = -\sum_{i=1}^{p} a_i x(n-i). \tag{2.148}$$

Thereby, p—the number of previous samples considered—is also referred to as the order of the linear predictor and the weighting coefficients a_i are called the linear predictor coefficients. The inversion in front of the sum is arbitrary, but simplifies the equations for the computation of the linear predictor coefficients.

Because the order p is finite and in practice much smaller than the length of the signal, an ideal estimation of an arbitrary signal $x(n)$ with Eq. (2.148) is not possible. An error term $e(n)$ is therefore included in Eq. (2.148) to capture the part of the signal

[22] In openSMILE linear predictive coding is supported via the cLpc component.

that cannot be modelled by the linear predictor (Schuller 2013):

$$x(n) = \hat{x}(n) + e(n) = -\sum_{i=1}^{p} a_i x(n-i) + e(n). \tag{2.149}$$

The signal $x(n)$ itself can now be included in the summation (starting at $i = 0$) which yields the following equation for the prediction error:

$$e(n) = \sum_{i=0}^{p} a_i x(n-i), \quad with \;\; a_0 = 1. \tag{2.150}$$

For acoustic parameter extraction based on LP analysis, primarily the predictor coefficients a_i are of interest. From Eq. (2.150) it can be seen that the predictor with the coefficients a_i can be seen as a linear transversal filter with the finite length (p) impulse response $h_{inv} = a_0, a_1, \ldots, a_p (a_0 = 1)$, which filters the input signal $x(n)$ and yields the output signal $e(n)$. Because this filter resembles the inverse of the vocal tract filter (which produces the speech signal $x(n)$ from the excitation signal $e(n)$), it is referred to as the *inverse* filter, hence the subscript *inv*. *Important:* It is to note here that the filter h_{inv} does not model only the vocal tract, it also includes the effects of the sound wave radiating from the mouth into the air, and the response of the room. For simplification, however, h_{inv} will be referred to as the vocal tract impulse response, and the reader is advised to keep in mind this approximation.

In the source-filter model of human speech production (cf. Sect. 2.2.6), the a_i coefficients correspond to the impulse response of the vocal tract and the prediction error signal $e(n)$ corresponds to the source excitation signal (Schuller 2013). The vocal tract impulse response depends on the configuration of the vocal tract, which in turn depends on the current phone that is being spoken. In particular, the impulse response contains vocal tract resonance frequencies (referred to as *formants*, cf. Deller et al. 1993 and Sect. 2.2.8).

In order to compute the predictor coefficients a_i, the constraint of quadratic minimisation of the error signal $e(n)$ is applied to Eq. (2.150). A sum squared error α is defined as:

$$\alpha = \sum_{n=0}^{N-1} e(n), \tag{2.151}$$

for the analysis interval $n = 0 \ldots N-1$, which is typically a short-time window (frame) of 10–20 ms. For a predictor of order p, α resembles the energy of the residual signal, which is related to the mean amplitude of the signal. It is thus also referred to as the LP gain.

From the quadratic minimisation condition:

$$\frac{\delta}{\delta x}\alpha \stackrel{!}{=} 0, \tag{2.152}$$

a system of linear equations is obtained, which can be solved for the p coefficients a_i with various methods (Ruske 1993).

2.2.7.1 LP Coefficients—Autocorrelation Method

The LP coefficients a_i are also referred to as the autoregressive (AR) coefficients because they are coefficients of an AR model. This section deals with the autocorrelation method for computing these AR coefficients (autoregressive modelling) as described and derived in more detail by Ruske (1993).

As a result of the derivation of the autocorrelation method, a recursive algorithm[23] for computing the predictor coefficients is found (according to Ruske 1993) as follows: The algorithm is initialised with a predictor of order $j = 0$, and with a sum quadratic prediction error of α_0:

$$\alpha_0 = r(0), \quad a_{00} = 1. \tag{2.153}$$

Then, a predictor of the next higher order j is iteratively estimated via the following equations, known as Levinson-Durbin recursion (cf. Levinson 1947b and Durbin 1960):

$$k_j = -\frac{1}{\alpha_{j-1}} \sum_{i=0}^{j-1} a_{i,j-1} r(j-i) \tag{2.154}$$

$$a_{jj} = k_j \tag{2.155}$$

$$a_{0j} = 1 \tag{2.156}$$

$$a_{ij} = a_{i,j-1} + k_j a_{j-i,j-1} \quad \text{with } i = 1 \ldots j-1 \tag{2.157}$$

$$\alpha_j = \alpha_{j-1}(1 - k_j^2), \tag{2.158}$$

where $j = 1 \ldots p$ indicates the iteration number, a_{ij} is the ith predictor coefficient after iteration j, $r(d)$ is the autocorrelation coefficient with associated lag d. The iteration terminates if a predictor of the requested order p is found, i.e., when $j = p$. The temporary variables k_j with $k \in [1; p]$ are—without further explanation at this point—named the reflection coefficients or the *partial correlation coefficients (PARCOR)* (cf. Ruske 1993).

The autocorrelation coefficients $r(d)$ for lags d are estimated via the autocorrelation function (cf. Sect. 2.2.5), assuming a window (frame) of N samples length:

$$r(d) = \sum_{n=d}^{N-1} x(n)x(n-d), \tag{2.159}$$

with $d \in [0; p]$.

[23] As implemented in openSMILE in the `cLpc` component.

2.2.7.2 LP Residual

The error signal $e(n)$ is also referred to as the LP residual. In general, it is computed from the speech signal by applying the inverse filter h_{inv} which is composed of the predictor coefficients (see Sect. 2.2.7). A preferred implementation of this filter is via a lattice filter structure. This is briefly outlined in this section.

The reflection coefficients k_j introduced in the previous section constitute an alternative representation of the LP filter and are restricted to the following condition (Ruske 1993):

$$|k_j| \geq 1. \tag{2.160}$$

Thus, they can be used to implement stable lattice filters for a forward vocal tract model filter (producing the speech signal from the excitation/error signal), and an inverse vocal tract model filter (producing the excitation/error signal from the speech signal).

For computing the residual $e(n)$ of $x(n)$, the inverse LP filter must be applied, of which the efficient lattice filter implementation is defined by the following equations (Ruske 1993):

$$\text{Initialisation:} \quad y_t^{(0)}(n) = y_b^{(0)}(n) = x(n) \tag{2.161}$$

$$\text{for } j = 1 \dots p: \quad y_t^{(j)}(n) = y_t^{(j-1)}(n) + k_j \cdot y_b^{(j-1)}(n-1) \tag{2.162}$$

$$y_b^{(j)}(n) = y_b^{(j-1)}(n-1) + k_j \cdot y_t^{(j-1)}(n) \tag{2.163}$$

$$\textit{Output}: \quad e(n) = y_t^{(p)}(n), \tag{2.164}$$

where p is the order of the linear predictor and k_j are the reflection coefficients (see Sect. 2.2.7.1).

2.2.7.3 LP Spectrum

The LP coefficients are related to the impulse response of the vocal tract (more exactly, the whole speech production system which includes, vocal tract, radiation into the air, and the room—however, for simplification it is referred to as vocal tract here), as mentioned before. The coefficients a_i constitute the impulse response of the inverse vocal tract filter (h_{inv}), which produces the excitation signal from the speech signal. In the source-filter model of speech production, however, the glottal excitation signal is convolved with the vocal tract system's impulse response h. Therefore, h must be obtained from h_{inv}. The spectrum H of h represents the frequency response of the vocal tract. In the following it is shown that H can be computed from the LP coefficients and is thus referred to as LP spectrum. Applying the DFT to h_{inv} yields (Schuller 2013):

$$H_{inv}(m) = DFT(h_{inv}). \tag{2.165}$$

The forward vocal tract filter H is defined by a recursive all-pole filter, which is obtained by inversion of H_{inv}:

$$H(m) = \frac{1}{H_{inv}(m)}. \tag{2.166}$$

Thus, the LP spectrum is defined as $H(m)$, which is, in words, the inverse of the DFT of the LP coefficients.

Given a predictor length p, a p-point DFT would be applied which yields $p/2$ frequency bins. For typical LP analysis p is rather small ($\approx$10–15), which leads to a bad frequency resolution in the LP spectrum. The resolution can be improved by non-linear interpolation (spline or *sinc*) or by zero-padding the LP coefficients to a length $N > p$, which is equivalent to the *sinc* interpolation (Oppenheim and Schafer 1975). The spectral resolution in the latter case is determined by N and the sampling rate of the original speech signal. For M bins $(2M - p)$ zeros need to be appended to the predictor coefficients.

2.2.7.4 Relation of LP Coefficients and the Cepstrum

A strong link between Cepstrum and LP AR coefficients exists due to their common relation to the impulse response of the vocal tract system in the linear speech production model. The link between cepstral coefficients $C(i)$ and the linear predictor coefficients a_i is defined by the following recursion (Young et al. 2006) (starting with $i = 0$):

$$C(i) = -a_i - \frac{1}{i}\sum_{j=1}^{i-1}(i-j)a_j C(i-j). \tag{2.167}$$

2.2.7.5 Line Spectral Pairs

Besides the direct AR coefficient and the reflection coefficient representation of the LP coefficients, an alternate parametrisation as Line Spectral Pairs (LSPs) or Line Spectral Frequencies (LSFs) exists (Kabal and Ramachandran 1986). This representation is favoured for low bandwidth channel transmissions, as it was shown to be less sensitive to quantisation noise (Soong and Juang 1984; Kang and Fransen 1985).

According to Kabal and Ramachandran (1986), the LSPs are obtained by a decomposition of the LP coefficient polynomial in the z-domain ($H(z)$) with the help of the following two constructed polynomials[24]:

$$P(z) = H(z) + z^{-(p+1)}H(z^{-1}) \tag{2.168}$$

[24]In openSMILE the cLsp component implements LSP computation based on code from the Speex codec library (www.speex.org).

$$Q(z) = H(z) - z^{-(p+1)}H(z^{-1}), \tag{2.169}$$

$$\text{with } z = e^{j\omega} = e^{j2\pi f}, \tag{2.170}$$

where $P(z)$ and $Q(z)$ represent the vocal tract system with the glottis closed and opened, respectively (Schuller 2013). More details on the computation of LSPs are given by Kabal and Ramachandran (1986) and Schuller (2013) and are thus not repeated here. In short, the roots of the polynomials $P(z)$ and $Q(z)$ (of order $p + 1$) are determined empirically. The name Line Spectral Pairs stems from the fact that these roots are all complex symmetrical pairs with respect to positive and negative frequencies f (Schuller 2013; Furui 1996). In total, p roots can be found—the same number as LP coefficients.

The LSPs are related to the formants of the speech signal, which are estimated from the roots of the polynomial $H(z)$. Precisely, two LSFs enclose a formant as left and right boundaries.

2.2.8 Formants

Formants are resonance frequencies of the vocal tract system (Ruske 1993), which characterise phonemes, esp. vowels. They have also been discussed in the context of music, in particular the singing voice (Sundberg 1987). Hence, they are visible in speech spectra as maxima of the envelope. From this, one method for identification of formants is the application of a peak picking algorithm directly to speech power spectra. However, this approach suffers from distortions by the fundamental frequency and other peaks caused by other components of the speech production system (room transfer function, for example) as well as additive noise (Ruske 1993). The LP spectrum (Sect. 2.2.7.3) can be used to eliminate the influence of the fundamental frequency and to obtain a smoothed spectral envelope for formant estimation. An even more robust way of formant estimation bases on finding the roots of the LP coefficient polynomial (see Sect. 2.2.7), which will be outlined in the following section.[25] For all methods, again, one has to keep in mind that in the spectral envelope and also in the LP coefficients other resonance frequencies (room) might be contained and one must determine which peaks (or roots of the LP polynomial) belong to formants and which can be attributed to external influences (McCandless 1974).

Most recent methods build on the fundamental extraction principles outlined above, but consider temporal context for smoothing the formant trajectories and correcting local errors. For instance, the Viterbi algorithm is used to smooth trajectories (Yan et al. 2007) and/or Kalman filters are applied (Yan et al. 2007; Mehta et al. 2012). Few methods, such as (Glaser et al. 2010) implement alternative algorithms

[25] In openSMILE formant extraction is implemented via this method in the cFormant component, which processes the AR LP coefficients from the cLpc component.

for identifying formant trajectories directly from spectrograms, i.e., not basing on linear prediction. Temporal smoothing for formants is not considered in this thesis.

As for the nomenclature, the symbol F_i with $x \geq 1$ will be used for formants. This takes into account the fact that the fundamental frequency is denoted by the symbol F_0 and formants are, loosely speaking, higher orders of resonance frequencies excited by the fundamental frequency.

2.2.8.1 Estimation from Linear Prediction

When computing formant frequencies from LP coefficients, it is assumed that the formants are the resonance frequencies of the forward LP transfer function $H(m)$ (Eq. (2.165)). Thus, they represent zeros in the inverse transfer function, and hence can be determined from a factorisation of the inverse transfer function polynomial $H_{inv}(z)$ in the z-domain (McCandless 1974).

Given a set of predictor coefficients $h_{inv} = a_0, a_1, \ldots, a_p$ with $a_0 = 1$, the z-domain inverse transfer function of the LP filter can be expressed as (Schuller 2013; Ruske 1993):

$$H_{inv}(z) = 1 + a_1 z^{-1} + a_2 z^{-2} + \cdots + a_p z^{-p}. \tag{2.171}$$

The factorisation of $H_{inv}(z)$ is given by:

$$H_{inv}(z) = \left(1 - \frac{p_1}{z}\right)\left(1 - \frac{p_1}{z}\right) \cdot \ldots \cdot \left(1 - \frac{p_M}{z}\right) \tag{2.172}$$

The factorisation is estimated with numerical methods, such as the Newton-Rhapson iteration (cf. Deuflhard 2011; Schuller 2013): the algorithm is initiated with an estimate of the first zero; then, the value of the polynomial and the numerical derivative at this location are computed. Iterative improvements of the estimate are computed by adjusting the location of the zero in the direction of the derivative, until a convergence criterion is reached (the delta from one iteration to the next is smaller than a threshold). A polynomial division is performed to remove this zero from the polynomial, and the algorithm is re-initialised for the next zero. This procedure is repeated until all zeros have been found, i.e., the degree of the remaining polynomial is 1. Because an error made for the first zero will propagate as an increased error to the subsequent zeros, it is recommended to refine the estimates of all the zeros in a small number of iterations on the whole polynomial.

The following algorithm to compute the formant frequencies from the transfer function poles is described loosely after the implementation in (Boersma 2001). Before computing the formant frequencies, the poles p_i are mapped into the unit circle by the following equation:

$$p_i = \frac{1}{p_i^*} \text{ if } |p_i| > 1. \tag{2.173}$$

Since all poles which are found for a polynomial $H_{inv}(z)$ with real-valued coefficients are symmetric on the imaginary axis, i.e., for each pole a complex conjugate partner pole exists, the poles with negative imaginary parts are discarded. Next, the phase of each pole p_i (with positive imaginary part) is converted to the frequency of the formant F_i:

$$F_i = \frac{1}{T_s 2\pi} \left| \arctan \left(\frac{Im(p_i)}{Re(p_i)} \right) \right| . \tag{2.174}$$

The corresponding bandwidth ("width of the spectral envelope peak") is estimated as:

$$F_i^{(bw)} = -\log \left(\frac{|p_i|}{T_s 2\pi} \right) . \tag{2.175}$$

2.2.8.2 Estimation from the Spectral Envelope

Formants can theoretically be estimated directly from the spectral envelope by peak picking. It is crucial though to choose an accurate method for estimating the spectral envelope. Basically two approaches exist:

1. Direct method: smoothing of the magnitude spectrum with a low-pass filter, or
2. computing the LP spectrum from LP AR coefficients.

The direct method has numerous disadvantages, hence, the LP spectrum method is the preferred way. The greatest disadvantage of the direct method is the influence of the fundamental frequency F_0 of voiced sounds. Because the range of possible F_0 values overlaps with the range of the first formant F_1, in some cases (especially for female voices) these two peaks might not be separable, or F_0 might be mistaken for F_1.

The LP spectrum (Sect. 2.2.7.3) contains a spectral envelope where the influence of F_0 has been completely removed by the linear predictor, if the length of the linear predictor $p \ll N_{F_0}^{(T)}$ ($N_{F_0}^{(T)}$ is the length of the fundamental frequency period in samples). Given that the LP spectrum has a sufficiently high resolution, the resonance frequencies of the vocal tract can be estimated by identifying peaks in the LP spectrum (McCandless 1974).

2.2.9 *Perceptual Linear Prediction*

The concept of Perceptual Linear Prediction (PLP) was introduced by Hermansky (1990). It extends the standard LP analysis by a psychoacoustic model of sound perception in the human auditory system. Hermansky (1990) defines the five steps of the PLP procedure:

1. Spectral analysis: 20 ms Hamming window with zero-padding to nearest larger power of 2 for FFT,
2. Critical-band power spectrum,
3. Equal-loudness pre-emphasis,
4. Intensity loudness power law,
5. Autoregressive modelling.

The result of the first four steps is a discrete auditory band spectrum $X_{p,aud}(b)$. With this, Hermansky (1990) suggests autoregressive modelling (step 5) with the autocorrelation method (as described in Sect. 2.2.7.1) to obtain AR coefficients. These coefficients can be then converted to a cepstral representation (Perceptual Linear Prediction Cepstral Coefficients (PLP-CC)) with Eq. (2.167). Similarly, PLP-CC could be computed directly from $X_A(m)$ by taking the natural logarithm of $X_A(m)$ and applying the Discrete Cosine Transformation (DCT). In this thesis, however, the original method via autoregressive modelling is adopted.[26]

The following sections describe the steps of PLP as they have been implemented in openSMILE for this thesis. Further, an extension to PLP, which accounts for temporal properties of speech is summarised in Sect. 2.2.9.5.

2.2.9.1 Critical-Band Spectrum

A critical-band power spectrum $X_P(b)$ is obtained for B Bark or Mel bands, as outlined in Sect. 2.2.3.4. Hermansky (1990) suggests the use of a Bark scale for frequency warping according to Eq. (2.62) (cf. Schroeder 1977).[27] Yet, any other non-linear frequency scale can also be used theoretically to obtain non-standard PLP-like coefficients.[28]

In this thesis triangular filters for the band spectrum are used, as implemented by Young et al. (2006). Although, Hermansky (1990) gives a piecewise linear shape for the band filters the triangular shape approximation has in practice been used more frequently due to the implementation provided by Young et al. (2006).

2.2.9.2 Equal-Loudness Pre-emphasis

The band spectrum $X_P(b)$ obtained in the previous step is now weighted by an equal loudness curve function $E(b)$ to attribute for the frequency dependent loudness sensitivity of the human ear:

---

[26]PLP via this method is implemented in openSMILE via the `cPlp` component.

[27]In openSMILE this Bark scale can be selected in the cMelspec component by setting the specScale option to 'bark_schroed'.

[28]openSMILE allows for this flexibility because the PLP procedure builds on a chain of components: cTransformFFT, cFFTmagphase, cMelspec (for the non-linear band spectrum), and cPlp (for equal loudness and intensity power law and autoregressive modelling and cepstral coefficients).

$$X_{P,eql}(b) = E(b) \cdot X_P(b) \tag{2.176}$$

The equal loudness function adopted by Hermansky (1990) is originally from Makhoul and Cosell (1976). It is given for a linear frequency f by:

$$E(f) = \frac{10^{31} \cdot \left((2\pi f)^2 + 56.8 \times 10^6\right)(2\pi f)^4}{\left((2\pi f)^2 + 6.3 \times 10^6\right)^2 \cdot \left((2\pi f)^2 + 0.38 \times 10^9\right)\left((2\pi f)^7 + 1.7 \times 10^{31}\right)}. \tag{2.177}$$

To evaluate the equal loudness function for frequencies $f^{(bark)}$ on the Bark scale, one substitutes f by:

$$f = \Theta_{bark}^{-1}\left(f^{(bark)}\right). \tag{2.178}$$

For the band power spectrum $X_P(b)$ the Bark scale band centre frequencies $f_c^{(bark)}(b)$ are used to compute the weights $E(b)$ for each band $b = 0 \ldots B-1$. To speed up computations, the equal loudness weights can be applied directly to the triangular band filters (cf. previous section).

For compatibility with the Hidden Markov Toolkit (Young et al. 2006), the following approximation for the equal loudness curve is also investigated[29]:

$$E_{htk}(f) = \left(\frac{f^2}{f^2 + 1.6 \times 10^5}\right)^2 \frac{f^2 + 1.44 \times 10^6}{f^2 + 9.61 \times 10^6}. \tag{2.179}$$

The same conversions from linear to bark frequencies and to band indices as above can be applied to obtain the weighting factor for each band b.

2.2.9.3 Intensity Loudness Power Law (Compression)

To model humans' non-linear perception of sound power (Zwicker and Fastl 1999), a cubic root amplitude compression[30] is performed (Hermansky 1990), and a spectrum, referred to as *auditory spectrum* herein, is obtained:

$$X_{P,aud}(b) = X_{P,eql}(b)^{0.33}. \tag{2.180}$$

See Fig. 2.5 for an example auditory spectrogram. $X_{P,aud}(b)$ can be seen as a perceptually corrected band spectrum. The sum of the auditory spectrum over all bands is a perceptual *loudness* measure, which approximates the Zwicker loudness (Zwicker and Fastl 1999) and should be used for polyphonic wide-band signals as a substitute for the narrow-band loudness approximation (Sect. 2.2.2) in all cases.

---

[29] In openSMILE it is enabled by setting `htkcompatible` to 1 in the `cPlp` component.

[30] Configurable via the option `compression` in the openSMILE component `cPlp`.

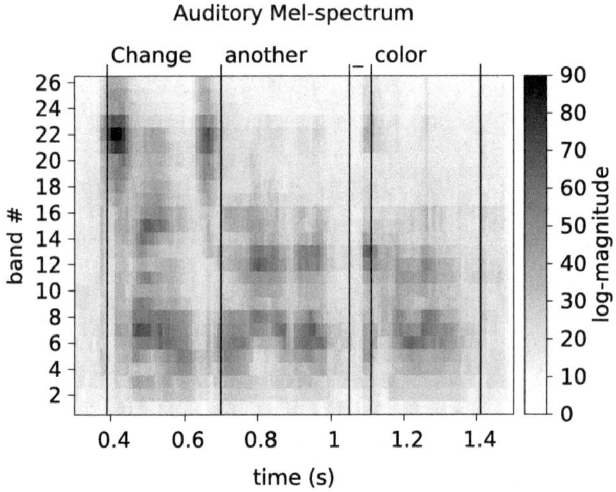

Fig. 2.5 Auditory spectrogram (based on 26-band Mel-band power spectrum) of a sample sentence from the AVIC database (Sect. 6.1.3); female speaker, words: "change another colour"

2.2.9.4 Autoregressive Modelling

The auditory spectrum $X_{P,aud}(b)$ is now used to estimate linear prediction coefficients, which—due to the perceptual/auditory nature of this spectrum—are called *perceptual* linear prediction (PLP) coefficients. These coefficients are estimated via the autocorrelation method, as described in Sect. 2.2.7.1. Instead of computing the autocorrelation coefficients $r(d)$ from the time domain signal $x(n)$, these coefficients are computed by applying the inverse DFT to the auditory spectrum. The auditory spectrum is a power spectrum, and thus, the inverse DFT of $X_{P,aud}(b)$ resembles an autocorrelation function. As only very few autocorrelation coefficients are required (order of the linear predictor 10), computing only the first few coefficients via a DFT is faster than using a FFT to compute the full autocorrelation.

2.2.9.5 Temporal Filtering: RASTA-PLP

RelAtive Spectral TrAnsform (RASTA) PLP is an extension to PLP which was first presented by Hermansky et al. (1992). It considers temporal properties of the human hearing and speech production systems. In particular, it exploits the fact that speech is primarily composed of modulations around 4 Hz (Zwicker and Fastl 1999; Hermansky et al. 1992). In RelAtive Spectral TrAnsform Perceptual Linear Prediction (RASTAPLP) a bandpass filter adapted to a range around 4 Hz is applied to the bands $X_P(b)$ before the auditory processing steps of PLP are applied.

An Infinite Impulse Response (IIR) filter for the RASTA bandpass is given by Hermansky et al. (1992) in the z-domain:

$$H(z) = 0.1z^4 \cdot \frac{2 + z^{-1} - z^{-3} - 2z^{-4}}{1 - 0.98z^{-1}}. \tag{2.181}$$

Adapting this filter for arbitrary lower (f_l) and upper (f_u) cut-off frequencies, and splitting it into an array of 5 FIR coefficients $\underline{a}$ and one IIR coefficient b gives the following set of equations for these filter coefficients:

$$\underline{a} = \left[\frac{2}{d}, \frac{-4o}{d}, 0, -\frac{-4o}{d}, -\frac{2}{d}\right], \tag{2.182}$$

$$b = 1 - \sin\left(2\pi f_l T_f\right), \tag{2.183}$$

where the helper variables o and d are defined as:

$$o = \cos\left(2\pi f_u T_f\right), \tag{2.184}$$

$$d = \sqrt{10\left(32o^2 + 8\right)}. \tag{2.185}$$

T_f is the period (in seconds) of the frames $X_P(b)$ (typically 10 ms).

Figure 2.6 shows the effect of the RASTA filter applied to the auditory spectrum shown in Fig. 2.5.

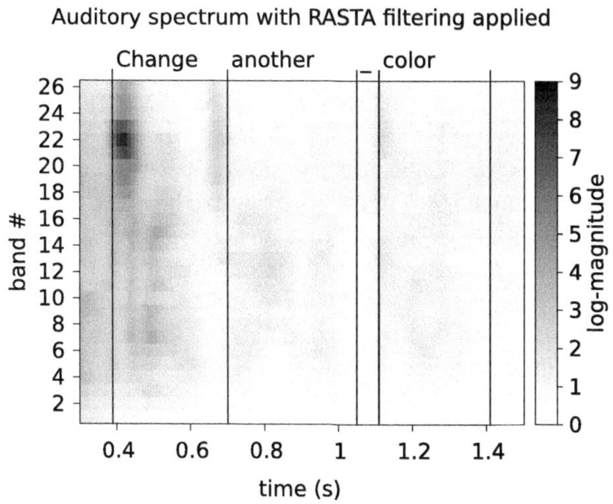

Fig. 2.6 RASTA filtered auditory spectrogram (based on 26-band Mel-band power spectrum) of a sample sentence from the AVIC database (Sect. 6.1.3); female speaker, words: "change another colour"

2.2.10 Cepstral Features

The idea of separating source and filter parts of a speech signal by Cepstral analysis, i.e., by applying a logarithm to the spectral magnitudes (cf. Sect. 2.2.6) is exploited by the group of cepstral features. These features are among the most successful and probably the most often used acoustic features for Automatic Speech Recognition (ASR) (Young et al. 2006; Rabiner and Juang 1993). They have also been applied successfully to various other audio tasks, such as speaker identification (Martinez et al. 2012), acoustic gunshot detection (Suman et al. 2014), music mood recognition (Nalini and Palanivel 2013), acoustic geo-sensing (Schuller et al. 2013a), and Computational Paralinguistics (Schuller 2013).

The general procedure for computing cepstral features is basically the same procedure as for computing the Cepstrum (Sect. 2.2.6):

1. Power spectrum representation (typically non-linear band spectrum)
2. Application of logarithm to power spectrum
3. Inverse spectral transformation (typically DCT)
4. Filtering of cepstral coefficients, called *liftering*.

However, in contrast to Sect. 2.2.6 for the cepstral descriptors discussed below, a non-linear frequency scale band spectrum serves as basis for Cepstrum computation. Further, the inverse FFT is replaced by a DCT.

In the following two sections the most prominent cepstral descriptors, namely MFCC and PLP-CC are described.

2.2.10.1 Mel-Frequency Cepstral Coefficients

With the building blocks described in the previous sections, MFCC as used by Young et al. (2006) are quickly described[31]: A 26 band spectrum $X_P^{(mel)}(b)$ (Sect. 2.2.3.4) is computed from a linear scale magnitude or power spectrum using triangular filters which are equidistant on the Mel-frequency scale and have 50 % overlap, i.e., the left and right endpoints of each triangular filter match the centres of the left and right bands. The filterbank is typically designed for the range from 20 Hz (in order to avoid the influence of DC components) to 8,000 Hz, in which case the bands have centre frequencies and half bandwidths as listed in Table A.1 in Appendix.

Next, the logarithm is applied to $X_P^{(mel)}(b)$:

$$X_P^{(log,mel)}(b) = \begin{cases} \log\left(X_P^{(mel)}(b)\right) & \text{if } X_P^{(mel)}(b) < X_{floor} \\ X_{floor} & \text{otherwise} \end{cases}. \tag{2.186}$$

[31]In openSMILE MFCC are computed via `cMelspec` (taking FFT magnitude spectrum from cFFTmagphase as input) and `cMfcc`.

The value X_{floor} is used to floor very small values (resulting from quantisation noise) in order to avoid large negative log values. If the input samples have been scaled to a range $[-1; +1]$, then $X_{floor} = 10^{-8}$ is assumed as default.[32]

On the log Mel-spectrum $X_P^{(log,mel)}(b)$, a Discrete Cosine Transformation Type-II (DCT-II) is performed (Young et al. 2006),[33] resulting in K Mel-cepstral coefficients $C^{(mel)}(k)$:

$$C^{(mel)}(k) = \sqrt{\frac{2}{B}} \sum_{b=0}^{B-1} X_P^{(log,mel)}(b) \cos\left(\frac{\pi k}{B}\left(b + \frac{1}{2}\right)\right). \tag{2.187}$$

Typically $K = 12 \ldots 16$ is chosen for most speech and music tasks. The advantage of these Cepstral coefficients is that they are decorrelated due to the orthogonal base of the DCT transformation. This has made them very popular for the use in Hidden Markov Model (HMM) systems with diagonal covariance matrices (Young et al. 2006).

In the final step, the Cepstral coefficients $C^{(mel)}(k)$ are filtered in a way to emphasize lower order coefficients. Because this happens in the Cepstral domain, it is called—in analogy to the wordplay of Cepstrum versus spectrum—*liftering*. Given a liftering coefficient L, the liftering is expressed as a weighting with a biased sine function:

$$C^{(mel)'}(k) = C^{(mel)}(k)\left(1 + \frac{L}{2}\sin\frac{\pi k}{L}\right). \tag{2.188}$$

The final coefficients $C^{(mel)'}(k)$ are those that are widely known and used as MFCC.

2.2.10.2 PLP Cepstral Coefficients

Similar to the MFCC, Cepstral coefficients can be computed based on the auditory spectrum of PLP analysis (Sect. 2.2.9) or RASTA-PLP (Sect. 2.2.9.5). These can either be computed in analogy to the MFCC by using the auditory spectrum from the PLP procedure instead of the log Mel-spectrum, or by applying the cepstral recursion from Eq. (2.167) to the PLP AR coefficients from Sect. 2.2.9.4. The latter method is the one implemented by Young et al. (2006) and in openSMILE.[34]

---

[32]In openSMILE the floor value is also 10^{-8} by default, and 1 when `htkcompatible=1` in `cMfcc`.

[33]Please note, that the DCT equation given in Young et al. (2006) and here differ because Young et al. (2006) start the summation at $b = 1$ for the first Mel-spectrum band, while here the first band is set at $b = 0$.

[34]PLP-CC can be computed in openSMILE by creating a chain of `cFFTmagphase`, `cMelspec`, and `cPlp` and setting the appropriate options for cepstral coefficients in the `cPlp` component.

2.2.11 Pitch

The previous descriptors have focussed on the vocal tract transfer function and have ignored the excitation signal. However, for applications besides ASR—such as Computational Paralinguistics and Music Information Retrieval—prosody (i.e., tonality, melody, dynamics of speech) plays a major role (Batliner et al. 2007; Schuller 2013). Prosody is mainly expressed through pitch and loudness.

This section deals with the estimation of pitch, or for most methods, an approximation of pitch via estimation of the fundamental frequency F_0. Pitch is a perceptual term which refers to the perceived tonality, i.e., frequency of a tone (Zwicker and Fastl 1999). It is related to the F_0, i.e., the lowest frequency in the harmonic series of the tone, however, not identical. Various effects contribute to the perception of pitch, leading to phenomena where the pitch of a tone is potentially perceived as higher or lower than the actual fundamental frequency. The most prominent case is the case of the missing fundamental (Zwicker and Fastl 1999). If a tone of 100 Hz (including higher order harmonics) is transmitted over a channel with a high-pass or band-pass characteristic (e.g., a telephone line), then the first harmonic might be the dominant fundamental frequency on the receivers side, while the actual fundamental of 100 Hz has been almost fully removed from the signal by the channel. Our ear reconstructs the original pitch based on the structure of the harmonics, which are nonetheless 100 Hz apart and not 200 Hz as they would have to be for a tone which actually has a pitch of 200 Hz.

Many automatic extraction methods aim at detecting periodicities in the signal, and thus constitute estimators of the fundamental frequency and not necessarily of pitch. Despite this fact, the topic is often commonly referred to as pitch detection in the literature (Hess 1983). Thus, the methods are also referred to as Pitch Detection Algorithms (PDAs) in the literature and in this thesis. In general two groups of PDAs exist: methods operating in the time domain, and methods operating in the short-time domain (frames) (cf. Schuller 2013). Short-time domain methods provide an F_0 estimate for each frame (typically 20–60 ms), which represents an average over $\approx$2–10 individual F_0 periods. Time domain methods detect individual fundamental periods directly from the waveform, and thus have a better temporal resolution than short-time domain methods, which is required, e.g., for estimation of micro-perturbations of F_0 (Jitter, cf. Sect. 2.2.13.1 and Schuller 2013). However, the frame-wise estimate from a short-time domain method can be used to initialise a refined search in the time domain, as will be shown in Sect. 2.2.13. Most PDAs also provide a measure of voicedness, the so called voicing probability p_v, which indicates how close the signal is to an ideal harmonic signal (high probability) or to a noise like signal (low probability). In the case of the time domain methods (Sect. 2.2.11.1), an additional method for estimating the voicing probability must be used, e.g., via autocorrelation (Sect. 2.2.11.2). Since the value of F_0 is not well defined for unvoiced parts of the signal, a value of 0 will be returned by all PDAs implemented in the scope of this thesis if p_v of a frame is smaller than a pre-defined threshold (specific for the respective PDA used).[35]

In the ongoing, four methods for pitch detection are described: in the time domain via peak picking (Sect. 2.2.11.1), via short-time autocorrelation (Sect. 2.2.11.2), and short-time Cepstral analysis (Sect. 2.2.11.2), as well as a spectral method based on subharmonic summation (Sect. 2.2.11.4). All four methods provide estimates for F_0 on a frame or sub-frame level, without considering the context of neighbouring frames. The context of a frame can be used, however, to eliminate errors a PDA makes in the presence of noise-bursts or irregular phonation. Such a context-based smoothing using the Viterbi algorithm is presented in Sect. 2.2.11.5. More advanced pitch detection algorithms such as Talkin (1995) and Cheveigne and Kawahara (2002) are not considered in this thesis, as they are not well suited for simple, fast, and efficient real-time on-line use due to their more complex smoothing methods and context requirements.

2.2.11.1 Time-Domain Based Estimation

The earliest methods for pitch detection were based on algorithms which attempted to directly estimate the fundamental period T_0 of the speech signal from the time domain waveform by smoothing and peak picking as reviewed by e.g., Ruske (1993) and Schuller (2013).

Since the fundamental period can only be determined for voiced signals, for each frame a decision is made, whether the frame contains a voiced (harmonic) or unvoiced (noise-like) signal, based on, e.g., the ZCR (Sect. 2.2.1.1) or any other suitable method.

In voiced speech segments, maxima which have a certain minimum and maximum distance are found: for a range of expected F_0 values from $F_{0,\,min}$ to $F_{0,\,max}$ a search window of length

$$T_{0,\,max} = \frac{1}{F_{0,\,min}} \tag{2.189}$$

and

$$T_{0,\,min} = \frac{1}{F_{0,\,max}}, \tag{2.190}$$

is assumed and the position of the maximum positive value within this window is found. If multiple values which are of approximately equal height are found, the first one is used. The position of this value is stored, and the start of the search window is shifted to $T_{0,\,min}$ samples after this position and the search is carried out for the next peak until the end of the voiced segment is reached.

---

[35]In openSMILE this behaviour is implemented in the pitch smoother components and in the `cPitchACF` component; the output F_0 final contains F_0 with values forced to 0 for unvoiced regions. See the documentation for more details.

The accuracy of such a method is very sensitive to all sorts of signal distortions, especially additive noise, digital clipping, and high-pass transfer characteristic channels. The noise robustness can be improved by low-pass filtering the signal with a cut-off frequency of $F_{0,\,max}$ or by applying the method to the (low-pass filtered) LP residual signal. The method fails, however, to detect the correct F_0 when the actual F_0 is missing (virtual pitch), or other harmonics are more dominant than F_0 (influence of strong first and second formants). The advantage, on the other hand, is that locations of individual pitch periods are obtainable (Schuller 2013).

2.2.11.2 Autocorrelation Based Estimation

The ACF of a frame (cf. Sect. 2.2.5) can be used to robustly estimate the fundamental frequency (Rabiner 1977; Boersma 1993). Peaks are found in the ACF at the period lengths of the harmonics of a harmonic signal. The ACF method assumes that T_0 is given by the location highest peak in the ACF in the search window from $T_{0,\,min}$ to $T_{0,\,max}$ (from the previous section). The amplitude of this peak, normalised by the amplitude of the 0th ACF coefficient gives a measure for the voicedness of the signal, i.e., the similarity of the pitch periods in the given frame. Thus, the probability of voicing p_v is estimated from the ACF as follows:

$$p_v = \frac{ACF_{max}}{ACF_0}. \tag{2.191}$$

where ACF_{max} is the maximum value in the range $T_{0,\,min} \ldots T_{0,\,max}$ and ACF_0 is the frame energy (0th coefficient of the ACF).

2.2.11.3 Cepstrum Based Estimation

In early studies related to this thesis (e.g., Schuller et al. 2009a, 2010), a PDA based on both autocorrelation and Cepstrum was used. Thereby the frequency F_0 is determined by the Cepstrum based method described in the following, and the probability of voicing is determined by the ACF method described in the previous section.

In a Cepstrum, as described in Sect. 2.2.6, the excitation signal (source) and the impulse response related to the filter transfer function of the linear model of speech production are overlaid in an additive way. Because the impulse response rapidly decays (due to the size of the human vocal tract), the higher order cepstral coefficients contain approximatively only the source signal. In the case of voiced speech signals, this is a Dirac impulse series. Due to the limited size of the analysis frame, often only one (for male voices) or two (for female voices) Dirac impulses are contained in the Cepstrum. F_0 is thus again determined by finding the location of the highest peak in the range $T_{0,\,min}$ to $T_{0,\,max}$ (the Quefrency axis of the Cepstrum has the units of seconds, thus the location of the peak corresponds directly to the fundamental period T_0 in seconds) (Noll 1967).

For this thesis, a hybrid autocorrelation/Cepstrum pitch detector has been implemented in openSMILE (Eyben et al. 2010a) and openEAR (Eyben 2009a).[36] First the probability of voicing is estimated from the ACF as described by Eq. (2.191). Next, F_0 is computed from the Cepstrum by searching for prominent peaks (extreme values) of the Cepstrum in the range $T_{0,min} \dots N$ and favouring peaks with a lower T. To this extent, the mean value of the Cepstrum over the full range is computed (C_{mean}) and the absolute maximum value C_{max} in the range $T_{0,min} \dots N$ is found. Then, a peak search starting at $T = T_{0,min}$ up to $T = N$ is performed, and the first peak found which meets the following criterion is chosen as T_0 peak:

$$C(T_0) > 0.6\,(C_{mean} + C_{max})\,. \tag{2.192}$$

The voicing probability p_v is estimated from the ACF via Eq. (2.191). In Fig. 2.8 (left), the pitch computed with this algorithm from the Cepstrum in Fig. 2.4 is shown. Octave jumps (double the actual frequency), discontinuities, and unvoiced segments which are wrongly classified as voiced (before 'change' and at the end and after 'color') are visible in the plot.

2.2.11.4 Subharmonic Summation (SHS)

A pitch detection method based on human perception principles (cf. Terhardt's virtual pitch theory (Terhardt 1974, 1979; Zwicker and Fastl 1999)) is shown in (Hermes 1988).[37] The method makes use of the harmonic structure of a signal to identify the correct pitch, even if the fundamental frequency is missing. It is thus called Subharmonic Summation (SHS) by the author. This method presented by Hermes (1988) is numerically more efficient compared to other earlier methods, such as the spectral-compression method (Schroeder 1968; Noll 1970) and the spectral-comb method (Martin 1982), which both also exploit the subharmonic structure. Therefore, the method by Hermes (1988) is adapted in this thesis and implemented in openSMILE. The following text briefly summarises the method.

First, the audio signal $x(n)$ is windowed by applying a 60 ms Gaussian window with $\sigma = 0.4$. The Gaussian window is chosen because of its Gaussian spectral shape with no side maxima, i.e., it does not distort the subharmonic structure. The larger size (60 vs. 40 ms as by Hermes 1988) is chosen to compensate for the flatter slope of the Gaussian window (as compared to the Hamming window), and in order to not distort low male pitch frequencies down to 50 Hz by the windowing.

[36] In the `cPitchACF` component, which requires combined ACF and Cepstrum input from two instances of the `cAcf` component.

[37] The method is implemented in openSMILE in two components: `cSpecScale` which performs spectral peak enhancement, smoothing, octave scale interpolation, and auditory weighting; `cPitchShs` which expects the spectrum produced by `cSpecScale` and performs the shifting, compression, and summation as well as pitch candidate estimation by peak picking.

A magnitude spectrum $X_M(m)$ is computed for the window. From there, the steps of the SHS algorithm (Hermes 1988) are as follows:

(1) Spectral peak enhancement (cf. Martin 1981)
(2) Spectral smoothing (optional)
(3) Interpolation from linear- to log-scale spectrum
(4) Auditory weighting
(5) Summation of shifted versions of the spectrum
(6) Peak picking

For (1), all local maxima are detected and their positions are stored. Then, all spectral bins which are more than two frames away from a local maximum are set to 0, to preserve only the peaks and their immediate surroundings. The spectrum $X_M^{(peak)}(m)$ is then smoothed (2) by convolving with a symmetric 3-tap filter in the spectral domain:

$$X_M^{(smooth)}(m) = \frac{1}{4}X_M^{(peak)}(m-1) + \frac{1}{2}X_M^{(peak)}(m) + \frac{1}{4}X_M^{(peak)}(m+1). \tag{2.193}$$

Next, the spectrum is transformed to a logarithmic octave (log_2) frequency scale (cf. Sect. 2.2.3.3). $B = M$ points on the target scale are assumed here, in contrast to (Hermes 1988), where a fixed number of $B_{oct} = 48$ points per octave is used. Spline interpolation (cf. Dalquist et al. 1974) is used to compute the magnitudes at the frequencies belonging to the log-scale bins $X_M^{(oct)}(m^{(oct)})$ (3). The minimum and maximum frequency (f_{min} and f_{max}) of the target scale are set to 20 Hz and the maximum frequency of the source scale, respectively. The number of octaves N_{oct} is then given as:

$$N_{oct} = \log_2\left(\frac{f_{max}}{f_{min}}\right) \tag{2.194}$$

and the number of points per octave computed as:

$$B_{oct} = \frac{B}{N_{oct}}. \tag{2.195}$$

In order to model the sensitivity of the human auditory system for low frequencies, a raised arctangent function is applied to the log-scale spectra $X_M^{(oct)}\left(m^{(oct)}\right)$ as weighting function $W\left(m^{(oct)}\right)$ (with $m^{(oct)} = 0 \ldots B-1$) (4):

$$W(m^{(oct)}) = 0.5 + \frac{1}{\pi}\arctan\left(3\frac{m^{(oct)} + 1 - \alpha_s}{B_{oct}}\right), \tag{2.196}$$

with

$$\alpha_s = B_{oct}\log_2\left(\frac{65}{50}\right) - 1. \tag{2.197}$$

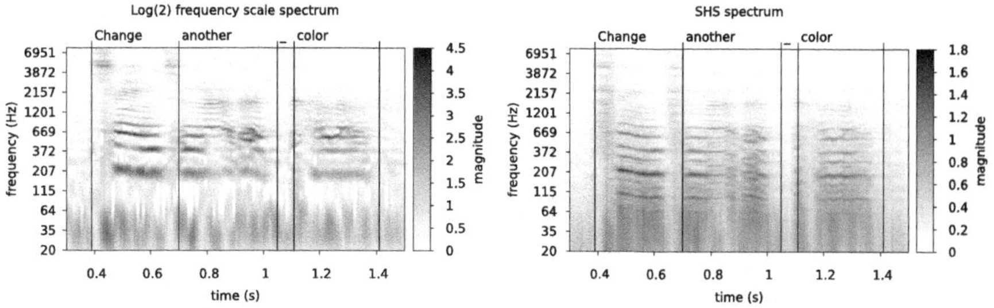

Fig. 2.7 *Left* octave (log(2)) scaled spectrum with peak enhancement, smoothing, and auditory weighting applied ($X_M^{(oct,w)}$); *right* subharmonic sum spectrum X_H. Sample sentence from the AVIC database (Sect. 6.1.3); female speaker, words: "change another colour"

The resulting spectrum $X_M^{(oct,w)}(m^{(oct)}) = W(m^{(oct)})X_M^{(oct)}(m^{(oct)})$ (see Fig. 2.7, left) is then shifted by a constant factor along the octave frequency axis (5), scaled by a factor h_i and all I scaled versions of the spectrum are summed:

$$X_H(m^{(oct)}) = \sum_{i=1}^{I} h_i X_M^{(oct,w)} \left(m^{(oct)} + \log_2 i\right) \tag{2.198}$$

Thereby, I is the number of harmonics which are considered, i.e., the number of harmonics to add up. The factor h_i is given by Hermes (1988) as $h_i = \gamma^{i-1}$ with the compression factor[38] $\gamma = 0.84$. The result $X_H(m)$ is called the subharmonic sum spectrum. In Fig. 2.7 (right) it can be seen that, in the subharmonic sum spectrum the peak corresponding to the pitch (around 200 Hz) is stronger than in the octave scale spectrum (left, in Fig. 2.7) and it is more narrow. A more narrow peak gives a better spectral resolution for pitch estimation. In SHS the spectral resolution in the low-frequency regions of the spectrum is increased by the influence of the higher harmonics.

Hermes (1988) now defines the pitch estimate as the location of the maximum of $X_H(m)$. Further, a method based on sub-frame correlation coefficients is given to perform a voiced/unvoiced decision for each frame. In this thesis a more complex method for picking multiple pitch candidate peaks is applied, and an alternative method for performing voiced/unvoiced decisions directly from the subharmonic sum spectrum is introduced below (Eq. (2.211)). First, all possible pitch candidates are identified:

N_c (typically 5–6) highest peaks are found by iterating through all the local maxima of $X_H(m^{(oct)})$ and retaining the N_c ones with highest magnitude. For each of these peaks the magnitude X_i and the discrete octave scale frequency $m_i^{(oct)}$ are stored. In the next step the exact frequency and magnitude of each peak candidate i is refined by quadratic interpolation from three points. The peak candidates are stored sorted by

---

[38] γ can be changed in openSMILE via the *compressionFactor* option of the `cPitchShs` component.

their magnitude, not by their frequency. This algorithm is the "greedy" peak picking algorithm, because it finds all N_c candidates with highest magnitude. The non-greedy version (in old versions of openSMILE, and in feature sets up to the INTERSPEECH 2012 Speaker Trait Challenge (IS12) set—see Sect. 3.4) only detected multiple candidates if the candidates with higher frequencies than the first had a greater magnitude than the first.[39] This is acceptable when the best candidate (the first one) is selected directly without Viterbi smoothing (as in the INTERSPEECH 2011 Speaker State Challenge (IS11) set, for example—see Sect. 3.3), otherwise the greedy algorithm should be used. This is implemented in the ComParE feature set (Sect. 3.5).

Quadratic interpolation tries to fit a parabola to three points (x_1, y_1), (x_2, y_2), (x_3, y_3) where the x-values meet the condition $x_1 < x_2 < x_3$ and the y-values satisfy $y_1 < y_2$ and $y_3 < y_2$, i.e., the point (x_2, y_2) represents a local maximum. The interpolation algorithm computes the values of a, b, and c of the quadratic equation:

$$y = ax^2 + bx + c. \tag{2.199}$$

With the three given points a set of three linear equations is constructed, which is solved for the variables a, b, and c. The closed form solution yields the following equations for the parabola parameters:

$$a = \frac{y_1x_2 + y_2x_3 + y_3x_1 - y_3x_2 - y_2x_1 - y_1x_3}{d}, \tag{2.200}$$

$$b = \frac{x_1^2y_2 + x_2^2y_3 + x_3^2y_1 - x_3^2y_2 - x_2^2y_1 - x_1^2y_3}{d}, \tag{2.201}$$

$$c = \frac{x_1^2x_2y_3 + x_2^2x_3y_1 + x_3^2x_1y_2 - x_3^2x_2y_1 - x_2^2x_1y_3 - x_1^2x_3y_2}{d}, \tag{2.202}$$

with the denominator:

$$d = x_1^2x_2 + x_2^2x_3 + x_3^2x_1 - x_3^2x_2 - x_2^2x_1 - x_1^2x_3. \tag{2.203}$$

The refined estimates of the original local maximum in (x_2, y_2) are now defined by the vertex of the parabola given by a, b, and c. By applying elementary calculus (solving for extreme values of functions), the coordinates x_v and y_v of the vertex are found as:

$$x_v = \frac{-b}{2a}, \tag{2.204}$$

$$y_c = c - ax_v^2. \tag{2.205}$$

[39]The greedy peak picking algorithm behaviour is achieved in openSMILE when the `greedyPeakAlgo` option is set to 1. The old (non-greedy) version of the algorithm searched through the peaks from lowest to highest frequency and considered the first peak found as the first candidate. Another candidate was only added if the magnitude was higher than that of the previous first candidate. This behaviour was sub-optimal for Viterbi based smoothing, which requires multiple candidates to evaluate the best path among them.

Here,

$$x_{1,i} = F^{(oct)}\left(m_i^{(oct)} - 1\right), \tag{2.206}$$

$$x_{2,i} = F^{(oct)}\left(m_i^{(oct)}\right), \tag{2.207}$$

$$x_{3,i} = F^{(oct)}\left(m_i^{(oct)} + 1\right), \tag{2.208}$$

where $F^{(oct)}\left(m_i^{(oct)}\right)$ is the octave scale frequency of the bin index of the ith peak candidate. The parabolic refined estimates of the pitch candidate frequency and amplitude are computed from the exact parabola vertex $(x_{v,i}, y_{v,i})$ with the following equations (assuming a $\log_2$ scale):

$$f_i^{(lin)} = 2^{x_{v,i}}, \tag{2.209}$$

$$X_i' = y_{v,i}. \tag{2.210}$$

Next, the arithmetic mean μ_H of $X_H(m^{(oct)})$ is computed. The voicing probability p_v associated with each pitch candidate i is then given as a function of the pitch candidate's refined amplitude X_i':

$$p_{v,i} = 1 - \frac{\mu_H}{X_i'}. \tag{2.211}$$

The above equation is based on the assumption that a subharmonic sum spectrum of a voiced signal will have a higher peak with respect to the mean than that of a noise like signal. Practical experiments have shown that this assumption holds true for many types of speech signals, except for signals with very low energy. Thus, a RMS energy (Eq. (2.41)) threshold is defined as 0.001 (assuming that the input signal $x(n)$ is in the range of $[-1; +1]$). Every frame with a RMS energy below this threshold is assumed to be unvoiced (i.e., $p_{v,i} = 0$ is assigned).[40]

From the N_c pitch candidates, now, a single—most probable—candidate must be selected. A simple solution[41] is to choose the candidate with the highest refined magnitude X_i', i.e., the highest voicing probability (see Fig. 2.8, right). It is better, however, to perform this choice in context with past and future pitch candidates in order to smooth the pitch trajectory and eliminate errors which are evident as rapid jumps. Such a smoothing method is presented in the next section.

---

[40] In openSMILE this behaviour is not implemented in the `cPitchShs` component, but is rather implemented via the configuration, e.g., for the `smileF0_base.conf` and `IS13_ComParE.conf` configurations. Thereby, the `cValbasedSelector` component is used to force F0 values to 0 (indicating unvoiced parts) if the energy falls below the threshold.

[41] Available in openSMILE via the `cPitchSmoother` component.

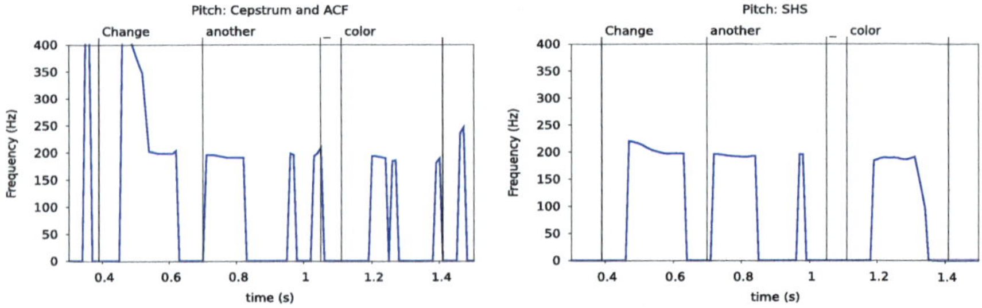

Fig. 2.8 *Left* pitch computed with the Cepstrum/ACF algorithm (Sect. 2.2.11.2); *right* pitch computed with the SHS algorithm (Sect. 2.2.11.4). Sample sentence from the AVIC database (Sect. 6.1.3); female speaker, words: "change another colour"

2.2.11.5 Post Smoothing with Dynamic Programming

To N_c pitch candidates $i = 1 \ldots N_c$ and associated scores (magnitudes, voicing probabilities, etc.) estimated by an arbitrary PDA, dynamic programming can be applied to find the best trajectory of pitch values and associated voicing probabilities. In order for the algorithm to also perform the voiced/unvoiced decision dynamically, an additional pitch candidate for the unvoiced case is added ($i = N_c + 1$). The algorithm described by Luengo (2007) is adopted in this thesis.[42] It implements a Viterbi algorithm based least cost search. Thereby each pitch candidate i at time (frame) n has an associated cost value $c_i(n)$. This cost has two components: the *local cost* $c_i^{(l)}(n)$, i.e., a cost value based on the likelihood of the current pitch candidate being a valid candidate, and the *transition cost* $c_i^{(t)}(n)$, i.e., a cost value which considers the current candidate in the context of possible past and future candidates. The total cost is given as: $c_i(n) = c_i^{(l)}(n) + c_i^{(t)}(n)$.

The local cost $c_i^{(l)}(n)$ can be split into two parts (Luengo 2007):

$$c_i^{(l)}(n) = c_i^{(V)}(n) + w_{thr} c_i^{(thr)}(n). \tag{2.212}$$

The first part ($c^{(V)}$) is related to the costs of voiced parts. It is composed of a cost derived from the voicing probability $p_{v,i}(n)$ and a cost derived from the frequency $f_i^{(lin)}(n)$ of the candidate with respect to the expected range of pitch frequencies:

$$c_i^{(V)}(n) = \begin{cases} w_{local}\left(-\log(p_{v,i}(n)) + p_{thr,i}(n)\right) + w_{range} c_i^{(frange)}(n) & i \le N_c \\ 0 & i = N_c \end{cases}, \tag{2.213}$$

[42] In openSMILE the Viterbi based pitch smoothing is implemented in the `cPitchSmoother Viterbi` component.

with the statically defined frequency weighting cost:

$$c_i^{(frange)}(n) = \begin{cases} 2 & \text{if } f_i^{(lin)}(n) = 0 \\ 1 - \frac{f}{100} & \text{if } 0 < f_i^{(lin)}(n) < 100 \\ 0 & \text{if } 100 \le f_i^{(lin)}(n) < 350 \\ \frac{f-350}{250} & \text{if } 350 \le f_i^{(lin)}(n) < 600 \\ 1.2 & \text{if } f_i^{(lin)}(n) \ge 600 \end{cases}, \qquad (2.214)$$

$$p_{thr,i}(n) = \begin{cases} 0 & p_{v,i}(n) < \theta_v \\ w_{thr} & \text{otherwise} \end{cases}, \qquad (2.215)$$

where θ_v is the voicing probability threshold for voiced/unvoiced frames. The threshold depends on the PDA used: for ACF based voicing probabilities, it is typically in the range of 0.4–0.6 (default 0.5), for the SHS algorithm it is typically 0.6–0.8 (default 0.7). As an extension to (Luengo 2007), in this thesis the frequency range weighting cost $c_i^{(frange)}(n)$ was added to the term for voiced candidates ($i \le N_c$) in order to account for different strengths of the candidates. As improvement, a speaker dependent frequency range cost function $c_i^{(frange)}(n)$ could be constructed basing on the incrementally updated mean and standard deviation of a given speaker. A static case of speaker dependent frequency range post-processing was suggested by Luengo (2007), but is not considered in this thesis. Such an adaptation requires speaker ID knowledge, which was not always present or easily accessible in the studies conducted.

The second part of Eq. (2.212) is related to the cost associated with the voicing probability crossing a pre-defined threshold for voiced/unvoiced:

$$c_i^{(thr)}(n) = \begin{cases} 0 & \forall i \le N_c : p_{v,i}(n) < \theta_v \\ 1 & \text{otherwise} \end{cases}. \qquad (2.216)$$

The transition costs for a transition from frame $n-1$ and pitch candidate j to frame n and pitch candidate i are given for all possible transition categories, i.e., voiced–voiced (V–V), voiced–unvoiced (V–U), unvoiced–voiced (U–V), and unvoiced–unvoiced (U–U) by the following equation (Luengo 2007):

$$c_{i,j}^{(t)}(n, n-1) = \begin{cases} c_{i,j}^{(t,VV)}(n, n-1) & \text{for } V{-}V \\ w_{Tvuv} & \text{for } V{-}U \text{ or } U{-}V \\ w_{Tuu} & \text{for } U{-}U \end{cases}, \qquad (2.217)$$

with

$$c_{i,j}^{(t,VV)}(n, n-1) = w_{Tvv} \left| \log \frac{f_i^{(lin)}(n)}{f_j^{(lin)}(n-1)} \right| + w_{Tvvd} \left| \frac{f_i^{(lin)}(n)}{f_j^{(lin)}(n-1)} - \frac{f_j^{(lin)}(n-1)}{f_{j*}^{(lin)}(n-2)} \right|, \qquad (2.218)$$

where j^* is the pitch candidate at time $n - 2$ of which the frequency is closest to the frequency of the pitch candidate j at time $n - 1$. This second *delta continuity constraint* (weight w_{Tvvd}) has also been introduced here as an extension to (Luengo 2007).

Sensible defaults for the cost weights have been tuned empirically by inspecting the final pitch detection results on various speech signals. These are—as used in the ComParE 2013 feature set: $w_{local} = 2$, $w_{thr} = 4$, $w_{range} = 1$, $w_{Tvv} = 10$, $w_{Tvvd} = 5$, $w_{Tvuv} = 10$, and $w_{Tuu} = 0$. A higher value thereby means a higher weight for the corresponding parameter in the cost function, i.e., penalising transitions of the respective kind, or emphasising local cost parameters such as the correct frequency range and correct voicing probability.

In order to find the least cost path, the costs for all possible paths are evaluated recursively. Thereby, if paths meet in the same node, only the one with lowest cost is kept. For details see (Viterbi 1967). In theory, for some worse cases, the whole sequence of inputs has to be processed before the best path can be chosen. In other, more ideal cases, the paths converge after a few frames and can be reduced to one as described above. In this case the best path up to the meeting point is known. In order to make the algorithm work with a fixed maximum delay in on-line systems, a maximum buffer length l_b is introduced. If paths have not merged to a single path at time $n - l_b$, the path with the lowest cost at time $n - l_b$ is selected as the winning path (potentially making a small error) and all other paths are discarded.

2.2.12 F_0 Harmonics

A voiced speech signal is composed of a sinusoid at the fundamental frequency and a series of sinusoids at its harmonics, as well as some noise signal. In the linear speech production model, the fundamental frequency is represented by a Dirac pulse series, which corresponds to a comb spectrum with peaks at integer multiples of F_0. In theory, in the source spectrum all peaks have the same amplitude. The envelope of the comb spectrum is then shaped by the dynamic transfer function of the vocal tract to the spectrum of specific vowels. Thus, the amplitudes of the F_0 harmonics do contain valuable information about the state of the vocal tract. This information is also reflected to some degree in other features such as spectral LLDs (Sect. 2.2.4) or Cepstral LLDs (Sect. 2.2.10). A more direct approach to encoding harmonics information, on the other hand, might reveal additional information and—most important—produces parameters which can be assigned a direct physical meaning, thus making the interpretation of results easier. An example is the ratio of the first harmonic (H_1) to the highest harmonic in the third formant range (A_3) as reported on by Hanson and Chuang (1999) and Hanson (1997).

First, all harmonics need to be identified. The algorithm implemented in openSMILE for this thesis performs the following steps to find the first I harmonics[43]:

1. Starting at the estimated location of the ith harmonic ($i = 0 \dots I$), which is the closest magnitude spectrum bin to the frequency $i \cdot F_0$, the closest local maximum is found in the range $[(i - 0.5) \cdot F_0; (i + 0.5) \cdot F_0]$. The amplitude of this maximum is used as the amplitude H_i of the ith harmonic. If no local maximum is found within the search range, the ith harmonic is marked as a non-existant/invalid harmonic,
2. Duplicates, i.e., where the frequency of the ith harmonic is equal to the frequency of harmonic $i + 1$ are removed by marking the harmonic $i + 1$ as an invalid harmonic,
3. The amplitudes H_i are converted to logarithmic relative amplitudes, by normalising with the amplitude of the 0th harmonic (i.e., amplitude of the F_0 component):

$$H_i^{(rel,\, dB)} = 20 \log_{10} \left(\frac{H_i}{H_0} \right). \tag{2.219}$$

Next, harmonic amplitude differences (ratios in the linear scale) are defined as follows:

$$H_{i-j} = H_i^{(dB)} - H_j^{(dB)}. \tag{2.220}$$

Formant amplitudes A_i are estimated by finding the highest spectral peak (local maximum) in the range $[0.8 \cdot F_i; 1.2 \cdot F_i]$ around the formant centre frequency F_i (estimated via LPC analysis, for example).

2.2.13 *Voice Quality*

In contrast to prosodic features, such as energy, loudness, and pitch, which describe the "global" melody and intensity of a speech utterance, voice quality features describe micro-prosodic variations within short-time frames (Schuller 2013).

2.2.13.1 Jitter

Jitter (cf. Schuller 2013) describes the variation of the length of the fundamental period from one single period to the next. The length of the first period $n' - 1$ is $T_0(n' - 1)$ and the length of the second period n' is $T_0(n')$. Then, the absolute period to period jitter, also referred to as absolute local jitter, is given as (Schuller 2013):

$$J_{pp}(n') = \left| T_0(n') - T_0(n' - 1) \right| \text{ for } n' > 1 \tag{2.221}$$

---

[43] In openSMILE version 2.0 and above, these parameters are implemented by the `cHarmonics` component.

This definition yields one value for J_{pp} for every pitch period, except the first one. Often, however, short-time analysis is used, and descriptor values are expected at the rate of the short-time frames (typically 10 ms). Thus, the average Jitter per frame is used in this thesis. For N' pitch periods $n' = 1 \ldots N'$ within one analysis frame the average local Jitter is given as:

$$\overline{J_{pp}} = \frac{1}{N'-1} \sum_{n'=2}^{N'} \left| T_0(n') - T_0(n'-1) \right|. \tag{2.222}$$

In order to make the Jitter value independent of the underlying pitch period length, it is normalised by the average pitch period length. This yields the average relative Jitter[44]:

$$\overline{J_{pp,rel}} = \frac{\frac{1}{N'-1} \sum_{n'=2}^{N'} \left| T_0(n') - T_0(n'-1) \right|}{\frac{1}{N'} \sum_{n'=1}^{N'} T_0(n')}. \tag{2.223}$$

The variance of Jitter across frames can be described via the "Jitter of the Jitter", i.e., the delta period to period Jitter J_{ddp}. It is defined as the difference between two consecutive J_{pp} values:

$$J_{ddp}(n') = \left| J_{pp}(n') - J_{pp}(n'-1) \right| \text{ for } n' > 2. \tag{2.224}$$

The delta Jitter is also expressed as average over a short-time frame, and is normalised by the average period length[45]:

$$\overline{J_{ddp,rel}} = \frac{\frac{1}{N'-2} \sum_{n'=3}^{N'} \left| \left| T_0(n') - T_0(n'-1) \right| - \left| T_0(n'-1) - T_0(n'-2) \right| \right|}{\frac{1}{N'} \sum_{n'=1}^{N'} T_0(n')}. \tag{2.225}$$

For local Jitter at least two, and for delta Jitter at least three pitch periods per analysis frame are required. Since for low pitch (e.g., male voices) this might not always be the case for typical analysis frame lengths (20–50 ms), in the implementation developed for this thesis, the last frame period length and the last local Jitter from the previous frame are stored for use on the next frame. Further, the last audio samples x of the previous frame, which belong to an incomplete period, are also stored and appended to the beginning of the next frame. The stored values are all set to 0 or deleted (in case of the previous audio samples) when an unvoiced frame is encountered.

---

[44]This definition of Jitter is implemented in openSMILE in the `cPitchJitter` component. It can be enabled via the `jitterLocal` option.

[45]This definition of delta Jitter is implemented in openSMILE in the `cPitchJitter` component. It can be enabled via the `jitterDDP` option.

In order to determine the exact length of pitch periods, a correlation based waveform matching algorithm is implemented. An extended and modified version of the basic matching algorithm by Boersma (2001) is implemented. The algorithm is guided by a frame based estimate of $T_0 = 1/F_0$ determined by a spectral domain PDA as described in Sect. 2.2.11. The estimate of T_0 is used to limit the range of the period cross-correlation to improve both the robustness against noise and computation time. The waveform matching algorithm operates directly on unwindowed frames of the speech signal $x(n)$.

The algorithm performs the following three steps for voiced frames:

(a) A range is defined for the allowed pitch period length in samples, based on the estimated frame T_0 and the sampling rate T_s:

$$N^{(T)}_{0,\,min} = (1-\alpha)\,\frac{T_0}{T_s},$$
$$N^{(T)}_{0,\,max} = (1+\alpha)\,\frac{T_0}{T_s}, \tag{2.226}$$

where α is the relative search range[46] for the correlation from 0.01–0.99, typically 0.1–0.25.

(b) For each integer (sample) period length $N^{(T)} = N^{(T)}_{0,\,min} \dots N^{(T)}_{0,\,max}$, the normalised cross-correlations (Pearson Correlation Coefficient—cf. Eq. (5.8)) of two periods ranging from a start index n to $(n + N^{(T)} - 1)$ and from $(n + N^{(T)})$ to $(n + 2N^{(T)} - 1)$ are computed and stored in a vector $\underline{c}$. The highest local maximum of $\underline{c}$ is now found, parabolic interpolation is applied (Eq. (2.199) in Sect. 2.2.11.4) to enhance the peak, and the resulting period length for the first period is computed from the (real-valued) period length $N^{(T)'}$ corresponding to the interpolated peak location. The corresponding period length in seconds is given as $T_0 = N^{(T)'}T_s$. The amplitude $CC = \underline{c}(N^{(T)'})$ of the correlation peak is used to verify the voicedness of the signal[47]: only if the amplitude is higher than 0.5, the current period is considered for the average Jitter computation (Eqs. (2.223) and (2.225)).

The average, minimum, and maximum CC for all periods within a frame constitute further acoustic descriptors[48]: they describe the variation from period to period, i.e., the regularity of speech signal periods. The average CC will be higher for a regular voice signal. It will be low for a voice signal with laryngalisation (irregular phonation; creaky voice—cf. Schuller and Batliner 2013).

(c) algorithm step (b) is repeated for the next two periods, starting at the new start index $n_1 = n + N^{(T)}$ (end of the previous period), until the end of the analysis frame is reached. The correlation search range defined by $N^{(T)}_{0,\,min} \dots N^{(T)}_{0,\,max}$ is kept constant for the whole frame, i.e., *not* updated on a period to period base.

---

[46] `searchRangeRel` option of the `cPitchJitter` component in openSMILE.

[47] `minCC` option in openSMILE.

[48] `sourceQualityMean` and `sourceQualityRange` options in `cPitchJitter` of openSMILE.

It is to note at this point, that the waveform matching algorithm does not estimate the phase of the periods. The beginning of the first period in the first voiced frame is chosen randomly as the first sample in that frame. If a correct tracking of all periods in the voiced segment is assumed, the phase should be constant throughout the voiced segment, though.

2.2.13.2 Shimmer

Shimmer describes amplitude variations of consecutive voice signal periods (Boersma 2001).

The waveform matching algorithm described in Sect. 2.2.13.1 is used to identify the exact period boundaries. As the phase of the segments determined with this algorithm is random, the maximum and minimum amplitude ($x_{max,n'}$ and $x_{min,n'}$) within each period are found. In analogy to Jitter, the period to period amplitude Shimmer is then expressed as (Schuller 2013):

$$S_{pp}(n') = \left|A(n') - A(n' - 1)\right|, \tag{2.227}$$

with the peak to peak amplitude $A(n') = x_{max,n'} - x_{min,n'}$.

This type of Jitter is very sensitive to additive noise, as this introduces large, random peak amplitude variations which are not related to the original speech signal. To mitigate this effect, the signal $x(n)$ can be low-pass filtered with a cut-off frequency between 600 and 800 Hz. This roughly preserves the first and—partially—the second formant, and high-pass filtered with a cut-off of 50 Hz to remove low frequency noise and slowly varying static offsets which might be introduced by low quality recording equipment. Alternatively, as implemented in this thesis, the period to period RMS amplitude variation can be used. Thereby the RMS amplitude (=unnormalised RMS energy, cf. Eq. (2.41)—without normalising by the frame length) $E_{rmsU,n'}$ is computed for each period n':

$$E_{rmsU,n'} = \sqrt{\sum_{n=0}^{N-1} x^2(n)}, \tag{2.228}$$

where N is the period length $N^{(T)}$. The period to period RMS Shimmer is then expressed as:

$$S_{pp,rms}(n') = \left|E_{rmsU,n'} - E_{rmsU,n'-1}\right|. \tag{2.229}$$

Delta Shimmer, like delta Jitter, can also be computed by replacing Jitter with Shimmer in Eq. (2.224).

As for Jitter, the period to period Shimmer values are averaged over the scope of a short-time frame in order to synchronise the rate of this descriptor with the constant rate of all other short-time descriptors. The averaged, relative Shimmer values are referred to as $\overline{S_{pp(,rms)}}$ for the local Shimmer and $\overline{S_{ddp(,rms)}}$ for the delta Shimmer. They are expressed as amplitude ratios, i.e., the per period amplitude values are normalised to the per frame average amplitude (peak or rms, respectively):

$$\overline{S_{pp(,rms),rel}} = \frac{\frac{1}{N'-1}\sum_{n'=2}^{N'} S_{pp(,rms)}(n')}{\frac{1}{N}\sum_{n'=1}^{N'} A(n')}, \tag{2.230}$$

$$\overline{S_{ddp(,rms),rel}} = \frac{\frac{1}{N'-1}\sum_{n'=3}^{N'} S_{ddp(,rms)}(n')}{\frac{1}{N}\sum_{n'=1}^{N'} A(n')}, \tag{2.231}$$

where the peak to peak amplitude $A(n')$ has to be substituted by the RMS amplitude in case of the $\overline{S_{*,rms,rel}}$ versions.

Amplitude ratios in audio signal processing are commonly expressed on a logarithmic scale, i.e., in dB. For converting the Shimmer ratios to dB the following equation is implemented:

$$S_{dB} = \begin{cases} 20\log_{10}(S_{rel})\ \text{dB} & S_{rel} > 10^{-50} \\ -1000 & \text{otherwise} \end{cases}. \tag{2.232}$$

2.2.13.3 Harmonics-to-Noise Ratio (HNR)

The HNR is defined as the ratio of the energy of harmonic signal components to the energy of noise like signal components. In an early study by Yumoto and Gould (1981), it is referred to as H/N ratio, and an algorithm for computing the ratio in the time domain is given: a sequence of N' periods with the same period length is considered and the average period waveform is computed. The average period waveform is assumed to strongly reflect the harmonic components because these are assumed to be constant over the N' periods, while the noise is uncorrelated and is theoretically cancelled out by the averaging. The noise energy is then computed by subtracting the average waveform from each individual waveform and computing the RMS energy E_{noise} of the remaining signal over all N' periods. The harmonic energy E_{harm} is computed as RMS energy of the average waveform. The HNR is then expressed as HNR_{wf}:

$$HNR_{wf} = \frac{E_{harm}}{E_{noise}}, \tag{2.233}$$

where the subscript *wf* denotes the direct waveform method. In this thesis, the above algorithm is implemented for voice quality analysis based on the waveform matching algorithm in Sect. 2.2.13.1. The average waveform is thereby computed over the N' fundamental periods in a short-time frame. The advantage of the method is that under ideal conditions (no jitter and shimmer, white noise) the result will be very accurate. However, ideal conditions do not exist for real world speech signals, and, moreover, the algorithm is vulnerable to errors from the pitch period detection process.

Another method for computing the HNR, which does not require individual pitch periods in the time domain, is based on the ACF. The method is similar to the estimation of the voicing probability as used in ACF based PDAs (cf. Eq. (2.191)). The HNR is given as HNR_{acf} (cf. Schuller 2006):

$$HNR_{acf} = \frac{ACF_{T_0}}{ACF_0 - ACF_{T_0}}, \tag{2.234}$$

where ACF_{T_0} is the amplitude of the autocorrelation peak at the fundamental period (see Sect. 2.2.11.2) and ACF_0 is the 0th ACF coefficient (equivalent to the quadratic frame energy).

Since the HNR is an energy ratio, it is best expressed on a logarithmic scale in dB (Schuller 2013) as $HNR_{wf,log}$ and $HNR_{acf,log}$:

$$HNR_{wf,log} = 20 \log_{10} \left(HNR_{wf}\right) \text{dB} \tag{2.235}$$

$$HNR_{acf,log} = 10 \log_{10} \left(HNR_{acf}\right) \text{dB} \tag{2.236}$$

Both logarithmic HNR values are floored to -100 dB.

An alternative method for computing the HNR is given by Krom (1993), for example. It exploits the Cepstral domain for separating the harmonic and noise energy components by applying comb-liftering.

2.2.14 Tonal Features

Tonal features are spectral features which are related to music theory. The tonal features which are described in this section are based on the 12-tone scale of western popular music. The features were originally introduced by Fujishima (1999) as Pitch Class Profiles (PCP) for the purpose of automatic chord recognition. They were later also referred to as CHROMA features because they reflect the "colour", i.e., the tonal shape, of the spectrum (Müller 2007). Moreover, PCP were improved by several tweaks and extensions, such as the post-processing method CHROMA Energy-distribution Normalised Statistics (CENS) suggested by Müller et al. (2005b) (Sect. 2.2.14.3).

2.2.14.1 Semitone Spectrum

The first step in the computation of PCP is to obtain a magnitude spectrum $X_M(m)$ (Schuller 2013) or power spectrum $X_P(m)$ (Fujishima 1999)—usually via a FFT. Then, this spectrum is mapped to an octave scale with 12 semitones per octave (cf. Sect. 2.2.3.3) and bins are combined to one bin per semitone, resulting in a semitone band spectrum (Sect. 2.2.3.4). Alternatively, a semitone band spectrum

$X_M^{(oct)}(b)$ can be computed directly from the linear scale magnitude spectrum—i.e., without mapping to a semitone frequency scale—as described in Sect. 2.2.3.4.

For the semitone band spectrum a first note with a positive non-zero frequency has to be defined on which the scale is built. Typically the note **A** with 110 or 55 Hz is used in this context. The range of the spectrum typically covers 6–8 octaves (72–96 semitones), which—when the first note is an **A** with 55 Hz—corresponds to a frequency of 3.52, 7.04, and 14.08 kHz for the last note on the scale, respectively for 6, 7, or 8 octaves.

The semitone band spectra used in this thesis are implemented via a spectral domain filterbank with rectangular filters, as described in Sect. 2.2.3.4. Variations exist where only peaks in the magnitude spectrum are considered and all other bins are set to 0 (cf. Schuller 2013, p. 64). Such variations are not considered here.

2.2.14.2 Pitch Class Profiles

The goal of PCP is to describe the tonality of a piece of music independent of the actual pitch. These descriptors are also known as CHROMA features (Müller 2007).[49] In the most basic case, for example, the semitone spectrum is warped to a single octave resulting in a 12-dimensional feature vector regardless of how many octaves the semitone spectrum spans.

The mapping of a semitone band spectrum $X_M^{(oct)}(b)$ to a PCP vector $PCP(s)$ with $s = 1..S$, and S being the number of semitones to warp to, is given by the following equation:

$$PCP(s) = \frac{1}{O}\sum_{i=0}^{O-1} X_M^{(oct)}(s + i \cdot S), \tag{2.237}$$

where O is the number of octaves to warp.

In general, PCP are not constrained to a 12 dimensional vector. Gómez (2006) describes Harmonic Pitch Class Profiles (HPCP), an extended version of the original PCP, where sampling at sub-semitone level (half or third semitone) is possible, resulting in 24 or 36 dimensional vectors. Also, warping not to a single octave but to two or more consecutive octaves is possible (Schuller 2013), yielding 24 or 36 dimensional feature vectors with semi-tone resolution.

In order to make the PCP features independent of the sound level, each PCP feature vector can be normalised to a length of one:

$$PCP'(s) = \frac{PCP(s)}{\sqrt{\sum_{s=0}^{S-1} PCP(s)^2}}. \tag{2.238}$$

[49]In openSMILE CHROMA features are supported by the `cChroma` component, which requires a semi-tone band spectrum as input, which can be generated by the `cTonespec` component (preferred) or by the (more general) `cMelspec` component.

This normalisation, however, emphasises noise, when the frame has very low energy (Schuller 2013). To avoid problems in this respect, the PCP values should be set to 0 when the frame energy falls below a (low) pre-defined threshold.

2.2.14.3 CENS

PCP features are frame-wise descriptors and thus do not consider context of left and right frames. However, in music rate of change of the tonal structure (240 Beats per Minute (BPM), for example, equals 4 Hz) is well below the frame rate (100 Hz). Thus, it is beneficial to consider the context of neighbouring frames. Above that, noise from non-tonal signal parts (percussion, unvoiced vocals, etc.) might cause differences between subsequent PCP vectors which are not related to tonal changes. CENS features (Müller et al. 2005a) attempt to compensate for these two deficiencies of PCP features by implementing a smoothing over time and a quantisation of the PCP values in order to improve robustness against non-tonal influences.[50]

The quantisation of the PCP amplitude $a = PCP(s)$ is defined as (Schuller 2013):

$$Q(a) = \begin{cases} 4 & 0.4 \leq a \leq 1 \\ 3 & 0.2 \leq a \leq 0.4 \\ 2 & 0.1 \leq a \leq 0.2 \\ 1 & 0.05 \leq a \leq 0.1 \\ 0 & 0 \leq a \leq 0.05 \end{cases} \tag{2.239}$$

Next, a smoothing is applied by convolving the quantised PCP frames $PCP_q(s)$ with a Hanning window of length 11 (Schuller 2013). To reduce redundant data, according to Schuller (2013) this is followed by a downsampling of factor 4. However, in order to be able to process all features at the same rate in the on-line system presented in this thesis, the downsampling is *not* applied here.

2.2.15 *Non-linear Vocal Tract Model Features*

In contrast to the linear model of speech production, recently, the modelling of non-linearities in the human speech production system has gained interest. Especially studies on detection of stress from the voice have investigated methods to model non-linearities in the glottal airflow (Zhou et al. 2001; Zuo and Fung 2011).

---

[50] In openSMILE CENS features can be computed from CHROMA (PCP) features with the `cCens` component.

2.2.15.1 Critical Band Filterbanks

Instead of approximating the frequency sensitivity and masking of the human auditory system in the frequency domain (as in Sect. 2.2.9.1, for example), a critical band filterbank is constructed in the time domain. In this way, the filterbank of the human auditory system can be approximated very closely with all its non-linear aspects.

In analogy to the frequency domain triangular filter banks for Bark- or Mel-scales (Sect. 2.2.3.4), the centre frequencies of the filters are approximately equidistant on a Bark scale, and the bandwidths are approximately constant there. In this thesis the filterbank from (Zhou et al. 2001) has been adopted. The centre frequencies and bandwidths of the filters are given in Table 2.1. Zhou et al. (2001) show the centre frequencies for bands 1–16. Here, the table was extended for the full auditory

Table 2.1 Critical band filter bank according to Zhou et al. (2001) (up to band 16, bands 17–24 calculated for a filter spacing of 1 Bark, see text for details); the bandwidths (f_{bw}) and linear scale centre frequencies ($f_c^{(lin)}$) are rounded to the closest multiple of 10

Band #	$f_c^{(lin)}$ (Hz)	$f_c^{(bark)}$ (Bark)	f_{bw} (Hz)
1	150	1.5	100
2	250	2.5	100
3	350	3.5	100
4	450	4.5	110
5	570	5.5	120
6	700	6.5	140
7	840	7.5	150
8	1000	8.5	160
9	1170	9.5	190
10	1370	10.5	210
11	1600	11.5	240
12	1850	12.5	280
13	2150	13.5	320
14	2500	14.5	380
15	2900	15.5	450
16	3400	16.5	550
17	4050	17.5	680
18	4800	18.5	870
19	5800	19.5	1150
20	7000	20.5	1290
21	8500	21.5	1710
22	10,500	22.5	2450
23	13,500	23.5	3800
24	18,500	24.5	6670

frequency range up to band 24, assuming a spacing of the centre frequencies $f_c^{(bark)}$ of 1 Bark and a bandwidth f_{bw} as is given by the following equation:

$$f_{bw} = \Theta_{bark}^{-1}(f_c^{(bark)} + 0.5) - \Theta_{bark}^{-1}(f_c^{(bark)} - 0.5). \tag{2.240}$$

Two types of filters are found in the literature (cf. e.g., Zhou et al. 2001) for the implementation of critical band filterbanks: gammatone filters and Gabor filters, both as introduced in Sect. 2.2.3.5.

All filters in this thesis are implemented as FIR filters for stability and implementation reasons. In order to implement filters which are described by an infinite impulse response with FIR filters, the impulse response needs to be truncated. This introduces a small error compared to the ideal IIR filter. The amount of error can be controlled by setting the length of the truncated impulse response. A longer impulse response will lead to a smaller error.

2.2.15.2 Teager Energy Operator Envelopes

The Teager energy operator (TEO) is an energy operator where a correction term is added to the quadratic energy operator (Sect. 2.2.2). It is also known as the Teager-Kaiser energy operator because it was introduced in papers by Kaiser (1993). This term accounts for non-linearities which are supposed to occur in the human vocal tract (Teager and Teager 1990).

The TEO $\Phi\{x(n)\}$ is defined for a discrete time signal $x(n)$ as (Kaiser 1993):

$$\Phi\{x(n)\} = x^2(n) - x(n+1)x(n-1). \tag{2.241}$$

This TEO is now applied to each of the filterbank output signals $x_b(n)$ (b is the band number; see previous section) and the envelopes of the resulting signals $x_b^{(\Phi)}(n)$ are computed, i.e., by computing the RMS energy for short-time frames or by applying a maximum operator to each short-time frame.

2.3 Derived Features and Post-processing of Low-Level Descriptors

Almost all of the LLDs shown in the previous sections are computed from isolated frames of audio. No context from previous or future frames is considered, i.e., the descriptors do not capture any signal dynamics beyond the frame boundaries.

To mitigate this issue, post-processing of the LLD signals is performed and derived features are computed. The post-processing implemented for this thesis consists of moving average smoothing as suggested by Schuller (2006) (see Sect. 2.3.4). The derived features consist of jth order simple differences and jth order delta regression coefficients (Sect. 2.3.1 and Sect. 2.3.2, respectively).

Other, more specific, variants of derived features have already been described along with the respective LLDs, such as the spectral differences (Sect. 2.2.4.11) or the CENS features (Sect. 2.2.14.3), for example. In a strict sense, these should not be named LLDs, but derived descriptors. However, as they are so specific to a certain descriptor (esp. CENS), they will be herein treated as LLDs nonetheless even though they have been computed from multiple subsequent frames.

2.3.1 Differences

The most simple derived feature which captures signal dynamics beyond a single frame, is the first order simple difference function.[51] It corresponds to the first order differential $\frac{\delta x}{\delta t}$ of a continuous signal $x(t)$. For a discrete signal $x(n)$, which describes the value of an arbitrary LLD over the frame index n, the simple difference function $d_1(n)$ is expressed as:

$$d_1(n) = x(n) - x(n-1) \tag{2.242}$$

The above function is in principle only defined for $n \geq 1$, if $x(n)$ is defined for $n \geq 0$. In practical implementations it is desirable, however, to obtain one vector of derived features for every vector of LLDs, which especially concerns the first frame. Thus, the signal $x(n)$ is padded for negative $x(n)$ (here: up to $n = -1$) either with zeros or with the value of $x(0)$. Padding with zeros is unsuitable for difference features because the difference $d_1(0)$ will be very large compared to the subsequent differences, if the magnitude of $x(n)$ is rather large. In this thesis, therefore, the padding with the value of $x(0)$ is implemented, i.e., $d_1(0) = 0$ always.

2.3.2 Delta Regression Coefficients

A different approach to the difference function is suggested by Young et al. (2006).[52] Thereby, a context window of length W is defined, over which the difference function is computed and smoothed according to the following regression formula:

$$\delta_1^W(n) = \frac{\sum_{i=1}^{W} i \cdot (x(n+i) - x(n-i))}{2\sum_{i=1}^{W} i^2} \tag{2.243}$$

---

[51] In openSMILE the simple difference function can be applied with the `cDeltaRegression` component with the delta window size set to 0 (option `deltaWin` = 0).

[52] In openSMILE these delta regression coefficients can be computed with the `cDeltaRegression` component.

Hence, these coefficients $\delta_1^W(n)$ are called *delta regression coefficients*. The size of the context window[53] determines the amount of context—and with that the amount of smoothing—that is considered in the computation of the difference around the pivot point at n. When using a larger window, the coefficients will reflect more the mid-term and long-term dynamics of the signal, while with a shorter window, the coefficients will capture the short-term dynamics. Thus, in theory, it makes sense to use delta regression coefficients with different window sizes in the same feature set, although this has not yet been considered in this thesis. The default window size suggested by Young et al. (2006) of $W = 2$ is adopted here.

2.3.3 Higher Order Delta Regression Coefficients and Differences

The simple difference and the delta regression coefficients shown in the previous two sections can be extended to higher order differences, which capture higher order dynamics of the signal. These correspond to higher order (j) differentials $\frac{\delta^j x}{\delta t}$ of the continuous signal $x(t)$. The second order difference, for example, is commonly known as acceleration coefficient (from physics, where the first order differential of the distance as a function of time represents speed and the second order differential represents acceleration).

The jth order discrete difference (d) or delta regression (δ) function is expressed recursively as the difference/delta of the ($j - 1$)th order difference/delta function (Young et al. 2006):

$$d_j(n) = d_{j-1}(n) - d_{j-1}(n-1), \tag{2.244}$$

$$\delta_j^W(n) = \frac{\sum_{i=1}^{W} i \cdot \left(\delta_{j-1}^W(n+i) - \delta_{j-1}^W(n-i)\right)}{2\sum_{i=1}^{W} i^2}. \tag{2.245}$$

2.3.4 Temporal Smoothing

The short-time analysis (Sect. 2.1.3) creates artefacts, which can be reduced by averaging descriptors over a small number of neighbouring frames.[54] Schuller (2006) suggests the use of a moving average filter with a window length of $W = 3$. The filter can be expressed (for odd W only) as:

---

[53]Option `deltaWin` in openSMILE component `cDeltaRegression`.

[54]In openSMILE the smoothing via a moving average window is implemented in the `cContourSmoother` component. Feature names often carry the suffix `_sma`, which stands for 'smoothed (with) moving average (filtering)'.

$$x_{sma}(n) = \frac{1}{W} \sum_{i=-(W-1)/2}^{(W-1)/2} x(n+i), \tag{2.246}$$

where $x_{sma}(n)$ is the smoothed output (*sma* stands for 'smoothed (with) moving average (filtering)'). A reasonable default for removing windowing artefacts is $W = 3$.

2.4 Supra-Segmental Features

In contrast to ASR which deals with short-term phenomena such as phonemes and words, other speech analysis tasks such as affect recognition, speaker state and trait analysis, and music analysis tasks such as mood recognition or chord and key recognition deal with rather long-term phenomena. E.g., the affective state of a person does not change every second, rather even a few seconds of speech material are required in most cases (even by humans) to asses the emotional state of a person reliably.

To enable machines to robustly model such long-term phenomena, either the classification framework must be able to handle long-range temporal dependencies between inputs (Sect. 2.5.2.2), or the features must summarise information over a meaningful unit of time (where the length depends on the analysed phenomenon). The latter approach has been mainly followed in this thesis and it was shown to be highly effective (cf. e.g., Schuller 2013; Schuller and Batliner 2013 and Chap. 6).

The following sections will show various mechanisms to summarise features over a segment of given length. Two basic categories are distinguished hereby:

1. methods which map a segment of a fixed length to a feature vector of a fixed length, where the length of the resulting vector is proportional to the length of the segment, and
2. methods which map a segment of variable length to a feature vector of a fixed length.

2.4.1 Stacking of Low-Level Descriptors

Probably the most straightforward approach to generate a single feature vector $\underline{X}$ from a series of N LLD feature vectors $\underline{x}(n)$ with $n = 0 \ldots N-1$ is to stack all the LLD vectors to a single, large vector:

$$\underline{X} = \begin{bmatrix} \underline{x}(0) \\ \underline{x}(1) \\ \ldots \\ \underline{x}(N-1) \end{bmatrix} \tag{2.247}$$

In $\underline{X}$ all the information of the original features is contained. However, the size of $\underline{X}$ is proportional to the length N of the series of LLD vectors. If the dimensionality of $\underline{x}(n)$ and N are large, the dimensionality of $\underline{X}$ grows rapidly, which makes this approach unsuitable for high dimensional LLD vectors and long segments.

This approach can be used for the sub-second segments and a dimensionality of the LLD vector of approximately 50 or smaller. It cannot be used to summarise LLDs in segments of variable length.

2.4.2 Statistical Functionals

In order to handle segments with variable length and get rid of the dependency of the feature vector dimensionality on the segment length, statistical functionals can be applied to the time series of LLDs. A functional $\mathscr{F}$ maps a series of values $x(n)$ to a single value $X_{\mathscr{F}}$ (Schuller 2013):

$$x(n) \xrightarrow{\mathscr{F}} X_{\mathscr{F}} \tag{2.248}$$

Thus, the result is independent of the length of the input. Examples of commonly used functionals are the arithmetic mean, standard deviation, maximum value, and the minimum value. Typically these functionals are applied to each LLD individually, i.e., they are referred to as *univariate functionals*[55] by the author of this thesis. Functionals can also be applied to multiple descriptors at the same time, such as the covariance or correlation between two descriptors. Such functionals are named *multivariate functionals*. However, they are beyond the scope of this thesis and not considered here.

The following sections give an overview on the most frequently used functionals for audio analysis and describe all the functionals used in this thesis in detail. A univariate time series will be denoted by $x(n)$. Each time series has N elements $n = 0 \ldots N-1$.

2.4.2.1 Means

Various types of mean values are implemented.[56] The most common one, often referred to simply as *mean* μ, is the arithmetic mean μ_a:

---

[55] In openSMILE univariate functionals are accessible via the `cFunctionals` component.

[56] Implementations of mean value related functionals are contained in the `cFunctionalMeans` component in openSMILE, which can be activated by setting `functionalsEnabled = Means` in the configuration of `cFunctionals`.

$$\mu_a = \frac{1}{N}\sum_{n=0}^{N-1} x(n) \tag{2.249}$$

Variations are the arithmetic mean of absolute values $\mu_{|a|}$:

$$\mu_{|a|} = \frac{1}{N}\sum_{n=0}^{N-1} |x(n)|\,, \tag{2.250}$$

and the arithmetic mean of positive (μ_{a+}) or negative (μ_{a-}) values:

$$\mu_{a+} = \frac{1}{N_+}\sum_{n=0}^{N-1} x(n)\cdot s_p\left(x(n)\right), \tag{2.251}$$

$$\mu_{a-} = \frac{1}{N_-}\sum_{n=0}^{N-1} x(n)\cdot s_n\left(x(n)\right), \tag{2.252}$$

where N_+ and N_- are the counts of positive and negative values in $x(n)$, respectively. s_p and s_n mask out non-positive and non-negative values, respectively:

$$s_p\left(x(n)\right) = \begin{cases} 0 & x(n) \le 0 \\ 1 & x(n) > 0 \end{cases}, \tag{2.253}$$

$$s_n\left(x(n)\right) = \begin{cases} 0 & x(n) \ge 0 \\ 1 & x(n) < 0 \end{cases}. \tag{2.254}$$

Other types of means implemented, are the quadratic mean μ_q:

$$\mu_q = \frac{1}{N}\sum_{n=0}^{N-1} x(n)^2, \tag{2.255}$$

the root-quadratic mean μ_{rq}:

$$\mu_{rq} = \sqrt{\frac{1}{N}\sum_{n=0}^{N-1} x(n)^2}, \tag{2.256}$$

and the geometric mean μ_g:

$$\mu_g = \sqrt[N]{\prod_{n=0}^{N-1} |x(n)|}. \tag{2.257}$$

The above equation for the geometric mean is numerically inefficient for large N. A more efficient method—which is used in this thesis[57]—operates in the logarithmic domain:

$$\mu_g = \exp\left(\frac{1}{N}\sum_{n=0}^{N-1} \log|x(n)|\right). \tag{2.258}$$

The geometric mean is only defined, if all $x(n) \stackrel{!}{=} 0$. Thus, zero-values are excluded in the computation of the geometric mean, i.e., μ_g in this thesis always refers to the geometric mean of all $x(n) \stackrel{!}{=} 0$.

Further, for some LLDs (e.g., pitch) it might be of high relevance to compute other types of mean values only from non-zero values. Therefore, the superscript $^{(nz)}$ shall denote a non-zero mean, i.e., a mean value computed only from values $x(n) \stackrel{!}{=} 0$. For instance, $\mu_a^{(nz)}$, $\mu_{|a|}^{(nz)}$, $\mu_q^{(nz)}$, and $\mu_{rq}^{(nz)}$. By the definition in the last paragraph, the geometric mean is always computed from non-zero values only, thus $\mu_g = \mu_g^{(nz)}$. In the same way as for the arithmetic mean, the quadratic mean and the root-quadratic mean can be computed only for positive and negative values, denoted by $\mu_{(r)q+}$ and $\mu_{(r)q-}$.

Not a mean value, but related to the non-zero mean values[58] (thus listed in this section), is the number of non-zero values N_{nz} in $x(n)$. This number can be normalised to the total number of values in the series (N)[59]: $N_{nz,rel} = N_{nz}/N$.

Further, the *flatness* μ_f of a time series is described by the ratio of the geometric mean to the arithmetic mean (of absolute values in both cases):

$$\mu_f = \begin{cases} \frac{\mu_g}{\mu_{|a|}} & \mu_{|a|} > 0 \\ 1 & \mu_{|a|} = 0 \end{cases} \tag{2.259}$$

This measure has originated as spectral flatness LLD, where it was applied to power spectra (cf. Sect. 2.2.4.5). Here it is generalised as a mean related functional which can be applied to any time series $x(n)$.

2.4.2.2 Moments

The arithmetic mean, as described in the previous section is also known as the first order statistical moment. Higher order statistical moments[60] are also very important functionals, in particular the variance and standard deviation (Ververidis and Kotropoulos 2006; Schuller 2006, 2013).

---

[57] And is the implementation used in openSMILE.

[58] And also implemented in the `cFunctionalMeans` component.

[59] In openSMILE the `norm` option of `cFunctionalMeans` can be set to *segment* to normalise counts and times etc. by N.

[60] Implemented in openSMILE in the `cFunctionalMoments` component.

The second order statistical moment, also called *variance* σ^2, is defined as:

$$m_2 = \sigma^2 = \frac{1}{N}\sum_{n=0}^{N-1}(x(n) - \mu_a)^2 . \tag{2.260}$$

The standard deviation σ is defined as the root of the variance:

$$\sigma = \sqrt{\frac{1}{N}\sum_{n=0}^{N-1}(x(n) - \mu_a)^2}. \tag{2.261}$$

A variation of the standard deviation which neutralises the influence of the mean, is the Coefficient of Variation (CV) (cf. Reed et al. 2002, for example), or normalised standard deviation:

$$\overline{\sigma} = \frac{\sigma}{\mu_a}. \tag{2.262}$$

Using the CV is only recommended for variables with an expected mean way greater than zero (this is not true, however, if σ is also very small). It is undefined for $\mu_a = 0$. To avoid invalid values in large-scale data-analysis in this thesis, in the case of $\mu_a = 0$ the following substitution is used: $\overline{\sigma} = \sigma$.

Higher order statistical moments are of less relevance, but for the completeness of the presented set of features, order three and four are considered nonetheless. The third central moment m_3 is defined as:

$$m_3 = \frac{1}{N}\sum_{n=0}^{N-1}(x(n) - \mu_a)^3 . \tag{2.263}$$

Due to the third power in the sum, the values of m_3 can be very large (or close to zero). Thus, instead of the central moment m_3, the third standardised moment $\overline{m}_3$ is used here (commonly known as *skewness*):

$$\overline{m}_3 = \frac{m_3}{\sigma^3}. \tag{2.264}$$

In the same way the fourth central and fourth standardised moment are defined:

$$m_4 = \frac{1}{N}\sum_{n=0}^{N-1}(x(n) - \mu_a)^4 , \tag{2.265}$$

$$\overline{m}_4 = \frac{m_4}{\sigma^4}. \tag{2.266}$$

The fourth standardised moment is also known as *kurtosis*.

2.4.2.3 Extreme Values

Extreme values[61] can be important markers in a signal, i.e., the maximum pitch value or the maximum energy value. In this thesis the following functionals related to global extrema have been considered: global maximum and minimum value of $x(n)$ (x_{max}, x_{min}), the positions (indices) of the global maximum and minimum value (n_{max}, n_{min}), and the range R_x of the signal: $R_x = x_{max} - x_{min}$.

Two additional functionals are formulated in this thesis, based on the maximum and minimum values: the difference $d_{max,\mu}$ between the maximum value and the arithmetic mean ($d_{max,\mu} = x_{max} - \mu_a$) and the difference $d_{min,\mu}$ between the arithmetic mean and the minimum value ($d_{\mu,\,min} = \mu_a - x_{min}$).

2.4.2.4 Percentiles

The arithmetic mean, and especially the global extreme values (maximum and minimum) are all sensitive to outliers, i.e., single values which are well out of the typical range of the majority of values. If there is one single very large value which has been caused by noise or corruption of the input, the maximum value (and its position) will relate to this value (erroneously). Instead, for more robustness against outliers, percentiles should be considered.[62] The jth percentile P_j is defined as the value x below which j percent of all the values $x(n)$ are, i.e., for j percent of the values in $x(n)$ the following is true: $x(n) < P_j$.

For computational efficiency when computing multiple percentiles of the same time series $x(n)$, the series is sorted in ascending order (denoted here as $s(n)$). The jth percentile can the be found in the sorted set of values at the index n_j, i.e., $P_j = s(n_j)$ with

$$n_j = \lfloor \frac{j}{100}(N-1) + 0.5 \rfloor. \tag{2.267}$$

This equation is inprecise if N is small and N is not evenly dividable by j, i.e., when the actual percentile value lies between $s(n_j)$ and either $s(n_j - 1)$ or $s(n_j + 1)$. For these cases a linear interpolation method to improve the precision of the percentile values has been implemented. With the the real valued (exact) index of the percentile P_j location n'_j, and the lower and upper bound integer indices ($n_{j,l}$ and $n_{j,u}$, respectively), the weights $w_{j,l}$ and $w_{j,u}$ for the values $s(n_{j,l})$ and $s(n_{j,u})$ can be computed:

$$n'_j = \frac{j}{100}(N-1), \tag{2.268}$$

$$n_{j,l} = \lfloor n'_j \rfloor, \tag{2.269}$$

[61] In openSMILE extreme values can be extracted with the `cFunctionalExtremes` component.

[62] Percentiles are implemented in openSMILE in the `cFunctionalPercentiles` component.

$$n_{j,u} = \lceil n'_j \rceil, \tag{2.270}$$
$$w_{j,l} = n'_j - n_{j,l}, \tag{2.271}$$
$$w_{j,u} = n_{j,u} - n_j. \tag{2.272}$$

With these weights, the linearly interpolated jth percentile value is expressed as:

$$P'_j = \begin{cases} w_{j,l}s(n_{j,l}) + w_{j,u}s(n_{j,u}) & n_{j,l} \stackrel{!}{=} n_{j,u} \\ s(n_{j,l}) & n_{j,l} = n_{j,u} \end{cases}. \tag{2.273}$$

The median of a set of values (e.g., a time series $x(n)$) is equivalent to the 50th percentile of that set of values. Other well known percentiles are the quartiles I–III, which correspond to the 25th, 50th, 75th percentile and the quintiles, which correspond to the 20th, 40th, 60th, and 80th percentile.

Further, percentile ranges have been often used in related work (Schuller 2006, 2013). The range between percentile j and percentile i is defined as the difference between the respective percentile values: $P^r_{j,i} = P_j - P_i$ for $j > i$. Special cases of percentile ranges are the three Inter-Quartile Ranges (IQRs): IQR1-2 is $P_{50} - P_{25}$, IQR2-3 is $P_{75} - P_{50}$ and IQR1-3 is $P_{75} - P_{25}$.

In order to robustly express maximum and minimum signal values for noisy signals, it is suggested here to use the 95th and 05th percentile instead of the maximum and minimum value, respectively. Alternatively, for longer segments ($\geq \approx 2$ s), the 99th and 01st percentile can be used.

2.4.2.5 Temporal Centroid

Similar to the spectral centroid (Sect. 2.2.4.6), the temporal centroid $t_{centroid}$ of the signal $x(n)$ can be computed as[63]:

$$x_{centroid} = \frac{\sum_{n=0}^{N-1} t(n)x(n)}{\sum_{n=0}^{N-1} x(n)}, \tag{2.274}$$

where $t(n)$ is a time-scaling function that can be used to modify the time units. For a time scale in seconds $t(n) = nT_f$, for a time-scale in frames $t(n) = n$, and for a time scale in relative percentage of the segment $t(n) = n/N$.

[63] In openSMILE the temporal centroid is implemented by the `cFunctionalRegression` component, as the sums are shared with the regression equations, thus computing both descriptors in the same component increases the efficiency.

2.4.2.6 Regression

In order to describe the general shape of the time series $x(n)$, linear and quadratic regression coefficients can be used.[64]

For the linear regression the values of the series $y = x(n)$ are approximated by a line $\hat{y} = mn + o$ in a way that the quadratic error between the line and the actual series is minimised. The regression coefficients m and o[65] are sometimes also known as m (slope) and t (offset), respectively, or plainly a and b, respectively. The derivation for the linear regression coefficients is the same as for the spectral slope (Sect. 2.2.4.2). It is thus, not repeated here. A generalisation for different time-scales (other than frame index n) can be easily obtained by substituting a general time-scale function $t(n)$ for the frame index n in all equations and $L' = t(N)$ for the length of the sequence $x(n)$ on the new time scale. The solutions for m and o for the general case of $y = x(t(n))$ are as follows:

$$m = \frac{N\Sigma_{ty} - \Sigma_t \Sigma_y}{N\Sigma_{t^2} - \Sigma_t^2}, \tag{2.275}$$

$$o = \frac{1}{N-1}\left(\Sigma_y - \frac{\Sigma_{ty}}{\Sigma_t}\right), \tag{2.276}$$

where

$$\Sigma_y = \sum_{n=0}^{N-1} x(n), \tag{2.277}$$

$$\Sigma_t = \sum_{n=0}^{N-1} t(n), \tag{2.278}$$

$$\Sigma_{t^2} = \sum_{n=0}^{N-1} t(n)^2, \tag{2.279}$$

$$\Sigma_{ty} = \sum_{n=0}^{N-1} t(n)x(n). \tag{2.280}$$

For a linear time scale $t(n) = gn$ with a general time-stretch factor g, the following simplifications to speed up the computations can be made (using exponential series expansions from Rade et al. 2000, p. 189):

$$\Sigma_t = g\frac{1}{2}N(N-1), \tag{2.281}$$

---

[64] In openSMILE the `cFunctionalRegression` component computes linear and quadratic regression coefficients.

[65] As used in this thesis, in order to avoid a name conflict with the quadratic regression coefficients a and b and time t.

$$\Sigma_{t^2} = g^2 \frac{1}{6} N(N-1)(2N-1). \tag{2.282}$$

The unit of the slope m is the unit of $x(n)$ (e.g., an amplitude, or power unit, or Hz for a frequency) per time-unit (unit of $t(n)$, e.g., frames, seconds, or relative segment percentage). The unit of the offset b is the same as the unit of $x(n)$.

For the quadratic regression the values of the series $y = x(t(n))$ are approximated by a parabola $\hat{y} = \hat{x}(t(n)) = at(n)^2 + bt(n) + c$. The derivation of the least-squares fit of the parabola to the series follows the same pattern as for the linear regression: The minimisation of the quadratic error e^2 between the quadratic approximation of the function and the function itself is expressed as:

$$\begin{aligned} e^2 &= \sum_{n=0}^{N-1} \left(x\left(t\left(n\right)\right) - \hat{x}\left(t\left(n\right)\right)\right)^2 = \\ &= \sum_{n=0}^{N-1} \left(x\left(t\left(n\right)\right) - at\left(n\right)^2 - bt\left(n\right) - c\right)^2 \stackrel{!}{=} min. \end{aligned} \tag{2.283}$$

From this, the following three differential equations for a, b, and c for the points $\left(t\left(n\right), x\left(t\left(n\right)\right)\right)$ with $n = 0 \ldots N-1$ are obtained:

$$\frac{\delta}{\delta a} e^2 = \sum_{i=0}^{N-1} -2t(n)^2 \left(x(t(n)) - at(n)^2 - bt(n) - c\right) \stackrel{!}{=} 0, \tag{2.284}$$

$$\frac{\delta}{\delta b} e^2 = \sum_{i=0}^{N-1} -2t(n) \left(x(t(n)) - at(n)^2 - bt(n) - c\right) \stackrel{!}{=} 0, \tag{2.285}$$

$$\frac{\delta}{\delta c} e^2 = \sum_{i=0}^{N-1} -2 \left(x(t(n)) - at(n)^2 - bt(n) - c\right) \stackrel{!}{=} 0. \tag{2.286}$$

Solving these three equations for a, b, and c yields solutions for these parameters in terms of the given points $\left(t(n), x(t(n))\right)$. With

$$\Sigma_{t^2 y} = \sum_{n=0}^{N-1} t(n)^2 y(n), \tag{2.287}$$

$$\Sigma_{t^3} = \sum_{n=0}^{N-1} t(n)^3, \tag{2.288}$$

$$\Sigma_{t^4} = \sum_{n=0}^{N-1} t(n)^4, \tag{2.289}$$

the above system of equations can be re-written in matrix form:

$$\begin{pmatrix} \Sigma_{t^4} & \Sigma_{t^3} & \Sigma_{t^2} \\ \Sigma_{t^3} & \Sigma_{t^2} & \Sigma_t \\ \Sigma_{t^2} & \Sigma_t & N \end{pmatrix} \begin{pmatrix} a \\ b \\ c \end{pmatrix} = \begin{pmatrix} \Sigma_{t^2 y} \\ \Sigma_{ty} \\ \Sigma_y \end{pmatrix}. \tag{2.290}$$

Again, for a linear time scale $t(n) = gn$, the following substitutions can be applied to speed up the computation (using exponential series expansions from Rade et al. 2000, p. 189):

$$\Sigma_{t^3} = \frac{g^3}{4} N^2 (N-1)^2, \tag{2.291}$$

$$\Sigma_{t^4} = \frac{g^4}{5} (N-1)^5 + \frac{1}{4} (N-1)^4 + \frac{1}{3} (N-1)^3 - \frac{1}{30} (N-1). \tag{2.292}$$

The solution is then given by Cramer's rule (Rade et al. 2000, p. 93) as:

$$a = \frac{1}{d} \begin{vmatrix} \Sigma_{t^2 y} & \Sigma_{t^3} & \Sigma_{t^2} \\ \Sigma_{ty} & \Sigma_{t^2} & \Sigma_t \\ \Sigma_y & \Sigma_t & N \end{vmatrix}, \tag{2.293}$$

$$b = \frac{1}{d} \begin{vmatrix} \Sigma_{t^4} & \Sigma_{t^2 y} & \Sigma_{t^2} \\ \Sigma_{t^3} & \Sigma_{ty} & \Sigma_t \\ \Sigma_{t^2} & \Sigma_y & N \end{vmatrix}, \tag{2.294}$$

$$c = \frac{1}{d} \begin{vmatrix} \Sigma_{t^4} & \Sigma_{t^3} & \Sigma_{t^2 y} \\ \Sigma_{t^3} & \Sigma_{t^2} & \Sigma_{ty} \\ \Sigma_{t^2} & \Sigma_t & \Sigma_y \end{vmatrix}, \tag{2.295}$$

$$d = \begin{vmatrix} \Sigma_{t^4} & \Sigma_{t^3} & \Sigma_{t^2} \\ \Sigma_{t^3} & \Sigma_{t^2} & \Sigma_t \\ \Sigma_{t^2} & \Sigma_t & N \end{vmatrix}. \tag{2.296}$$

Regression coefficients obtained with the above equations, where $t(n) = n$ (frame index time scale) can be converted to another linear time-scale $t(n) = gn$ with the linear scale factor g by the following transformations:

$$m' = g^{-1} m, \tag{2.297}$$

$$a' = g^{-2} a, \tag{2.298}$$

$$b' = g^{-1} b. \tag{2.299}$$

The scale factor g can be set to $g = T_f$ for an absolute time scale in seconds, or to $g = 1/N$ for a percentage time scale relative to the segment length N.[66] The derivation

[66] In openSMILE, the time scaling feature is enabled by the `normRegCoeff` option in `cFunctionalRegression` component. Setting it to 1 enables the relative time scale $g = 1/N$ and setting it to 2 enables the absolute time scale in seconds.

of these transformations is obtained by setting $t(n) = gn$ in Eqs. (2.278)–(2.280) and in Eqs. (2.287)–(2.289) and re-evaluating Eqs. (2.275)–(2.276) and Eqs. (2.293)–(2.295). In this way it is also shown that linear scaling of the time scale has no influence on the linear regression coefficients o (linear offset), and the quadratic regression offset c.

If the coefficients must be independent of the range R_x of values in $x(n)$, the coefficients m, a, and b must be divided by R_x:

$$m^{(norm)} = \frac{m}{R_x}. \tag{2.300}$$

Equation (2.300) also applies to a and b. The offset coefficients o and c must be transformed according to (also applies for c)[67]:

$$o^{(norm)} = \frac{o - x_{min}}{R_x}. \tag{2.301}$$

The derivations are made by setting

$$x^{(norm)}(t(n)) = \frac{1}{R_x}\left(x(t(n)) - x_{min}\right) \tag{2.302}$$

in the regression coefficient equations.

Besides the linear or quadratic regression coefficients, an informative measure is the regression error e, i.e., the remaining (minimised) difference between the computed regression line or curve $\hat{x}(t(n))$ and the actual time series $x(t(n))$ (Eq. (2.283)). The error is normalised by the length of the series to obtain a measure $\bar{e}$ which is independent of the sequence length:

$$\bar{e} = \frac{1}{N}e. \tag{2.303}$$

The error is either called *linear regression error* when it is computed as difference between $x(n)$ and a line, or *quadratic regression error*, when it is a difference between $x(n)$ and a parabola. As alternative to the normalised sum-square error $\bar{e}$ (Eq. (2.283)), the normalised absolute error $\bar{e}_a$ is additionally used:

$$\bar{e}_a = \frac{1}{N}\sum_{n=0}^{N-1}\left|x(t(n)) - \hat{x}(t(n))\right|. \tag{2.304}$$

[67]Option `normInputs` in openSMILE component `cFunctionalRegression`—also affects linear and quadratic error.

The regression errors can be normalised to the range R of $x(n)$ in order to have comparable errors across segments with different magnitudes[68]:

$$\overline{e}_a^{(norm)} = \frac{1}{R}\overline{e}_a, \tag{2.305}$$

$$\overline{e}^{(norm)} = \frac{1}{R^2}\overline{e}. \tag{2.306}$$

For quadratic regression, also the (relative) temporal location n_v of the parabola vertex and its corresponding amplitude $\hat{x}(n_v)$ are of interest. If the vertex is within $n = 0 \ldots N-1$, then also the slope of the line connecting the point at $n = 0$ with the vertex (left slope) and the slope of the line connecting the vertex with the last point at $n = N-1$ (right slope) can be computed. The vertex coordinates n_v and $\hat{x}(n_v)$ are computed according to the following equations (cf. Sect. 2.2.11.4—quadratic interpolation):

$$n_v = \frac{b}{-2a}, \tag{2.307}$$

$$\hat{x}(n_v) = c - \frac{b^2}{4a}. \tag{2.308}$$

To avoid outliers, the range of n_v is limited to $[0; N-1]$, thus the corresponding $\hat{x}(n_v)$ is actually computed as:

$$\hat{x}(n_v) = an_v^2 + bn_v + c. \tag{2.309}$$

The parabola vertex coordinates can be normalised to the scale $n = 0 \ldots 1$ (for n_v) and to the range R of the values in $x(n)$ (for $\hat{x}(n_v)$) by:

$$n_v^{(norm)} = \frac{n_v}{N-1}, \tag{2.310}$$

$$\hat{x}(n_v)^{(norm)} = \frac{\hat{x}(n_v) - min}{R}. \tag{2.311}$$

For the computation of linearised left and right slopes of the parabola, the parabola is evaluated for $n = 0$, yielding $\hat{x}(0) = c$, and for $N-1$, yielding $\hat{x}(N-1) = a(N-1)^2 + b(N-1) + c$. The left slope m_{left} is then formulated as:

$$m_{left} = \begin{cases} \dfrac{\hat{x}(n_v) - c}{n_v} & n_v > 0 \\ 0 & n_v \leq 0 \end{cases}, \tag{2.312}$$

and the right slope m_{right} is given as:

[68]Option `normInputs` in the openSMILE component `cFunctionalRegression`—note that this option also affects the regression coefficients as it effectively normalises the input range.

$$m_{right} = \begin{cases} \dfrac{\hat{x}(N-1) - \hat{x}(n_v)}{N-1-n_v} & n_v < N-1 \\ 0 & n_v \geq N-1 \end{cases}. \tag{2.313}$$

The slopes normalised to the value range, or the scale $n = 0 \ldots 1$, or both, are given by using the respective normalised values in Eqs. (2.312) and (2.313):

$$m_{left}^{(norm)} = \begin{cases} \dfrac{\hat{x}(n_v)^{(norm)} - c^{(norm)}}{n_v^{(norm)}} & n_v > 0 \\ 0 & n_v \leq 0 \end{cases}, \tag{2.314}$$

$$m_{right}^{(norm)} = \begin{cases} \dfrac{\hat{x}(N-1)^{(norm)} - \hat{x}(n_v)^{(norm)}}{1 - n_v^{(norm)}} & n_v < 1 \\ 0 & n_v \geq 1 \end{cases}. \tag{2.315}$$

An important property of the regression coefficient functionals is the *irreversibility*, i.e., that if the input sequence is reversed, a different value is obtained (except for the rare case of exactly symmetric input series and quadratic regression coefficients). Most other functionals, such as the means, moments, extreme values (excluding positions), temporal distributions, peak amplitudes, and some of the modulation features, will return the same value even if the input sequence is reversed in time.

2.4.2.7 Temporal Distribution

The temporal distribution of the signal is described by statistics which give the percentage of values $x(n)$ in the time series which are above or below a certain threshold x_θ. Hence, these functionals are assigned the name *up-/down-level times*, respectively.[69]

In this thesis a range-relative threshold x_θ^{rel} is used to ensure independence from variations of the signal dynamics. x_θ^{rel} can be in the range [0; 1]. The absolute threshold x_θ is computed dynamically from the range (R_x) and minimum value (x_{min}) of $x(n)$ and the relative threshold:

$$x_\theta = x_{min} + x_\theta^{rel} R_x. \tag{2.316}$$

Two kinds of threshold-based temporal descriptors are implemented: the *up-level* and *down-level times*. The *up-level time* is defined as the time or number of frames (absolute or relative)[70] the signal is above the threshold ($x(n) > x_\theta$). The *down-level time* is defined as the time or number of frames (again, absolute or relative) the signal is below or equal to the threshold ($x(n) \leq x_\theta$). Typical relative thresholds which are considered in standard feature sets are 0.25, 0.5, 0.75, and 0.9. The corresponding

[69] In openSMILE these functionals are implemented in the component `cFunctionalTimes`.

[70] Configurable with the `norm` option in openSMILE.

descriptors are then referred to as the 25, 50, 75, and 90 % up- or down-level times, respectively. The up- and down-level times with the same threshold are always complementary, i.e., their sum is 1 (for relative times) or the length of $x(n)$ for absolute times. Due to this redundancy, it is sufficient to use one or the other (up or down) in a feature set.

Additionally, *rise time* and *fall time* are considered. These are defined as the time or number of frames (absolute or relative) the signal is rising or falling, i.e., $x(n-1) < x(n)$ or $x(n-1) > x(n)$, respectively. Similarly, the *left-curvature time* and *right-curvature time* are defined as the time or number of frames the signal has a left or right curvature, i.e., $x(n) - x(n-1) < x(n+1) - x(n)$ (left) or $x(n) - x(n-1) > x(n+1) - x(n)$, respectively. In contrast to the up- and down-level times the rise-/fall times and the curvature times are not complementary because the case of a flat signal ($x(n-1) = x(n)$ and $x(n+1) = x(n)$) is not considered in the definition, i.e., for a constant signal all four of these descriptors will be 0, while for a signal which represents a line, the curvature times will be 0.

2.4.2.8 Peaks and Valleys

Peaks (maximum values) and valleys (minimum values) can be important markers in a signal. Both the amplitude and the position of the peaks are valuable. Various peak based descriptors have been implemented for this thesis.[71] These descriptors are described in the following, assuming a given set of I peaks $(x_{p,i}; n_{p,i})$ with $i = 0 \ldots I-1$ and J valleys $(x_{v,j}; n_{v,j})$ with $j = 0 \ldots J-1$:

- Number of peaks (I)—optionally normalised to peaks per second,
- Arithmetic mean of the peak amplitudes $x_{\mu}^{(peaks)}$:

$$x_{\mu}^{(peaks)} = \frac{1}{I} \sum_{i=0}^{I-1} x_{p,i}, \tag{2.317}$$

- Absolute peak amplitude range (maximum peak amplitude - minimum peak amplitude),
- Peak amplitude range normalised to the input contour's arithmetic mean,
- Difference of the arithmetic mean of the peak amplitudes to the arithmetic mean of $x(n)$,
- Ratio of the arithmetic mean of the peak amplitudes to the arithmetic mean of $x(n)$ (relative peak mean),
- Mean of peak to peak amplitude differences,
- Mean of peak to peak amplitude differences normalised by range R_x,
- Standard deviation of peak to peak amplitude differences,
- Standard deviation of peak to peak amplitude differences normalised by range R_x,

[71] In openSMILE these functionals can be applied with the `cFunctionalPeaks2` component; the `cFunctionalPeaks` component contains an older, obsolete peak picking algorithm.

- Mean distance between peaks $d_{\mu}^{(peaks)}$:

$$d_{\mu}^{(peaks)} = \frac{1}{I}\sum_{i=1}^{I-1}\left(n_{p,i} - n_{p,i-1}\right), \tag{2.318}$$

- Mean of difference of consecutive peak distances,
- Standard deviation of inter-peak distances $d_{\sigma}^{(peaks)}$:

$$d_{\sigma}^{(peaks)} = \frac{1}{I}\sum_{i=1}^{I-1}\left(n_{p,i} - n_{p,i-1} - x_{\mu}^{(peaks)}\right)^2, \tag{2.319}$$

- Arithmetic mean of the valley amplitudes $x_{\mu}^{(minima)}$:

$$x_{\mu}^{(minima)} = \frac{1}{J}\sum_{j=0}^{J-1} x_{m,j}, \tag{2.320}$$

- Absolute valley amplitude range,
- Valley amplitude range normalised to the input contour's arithmetic mean,
- Difference of the arithmetic mean of $x(n)$ to the arithmetic mean of the valley amplitudes,
- Ratio of the arithmetic mean of the valley amplitudes to the arithmetic mean of $x(n)$,
- Mean of valley to valley amplitude differences
- Mean of valley to valley amplitude differences normalised by range R_x,
- Standard deviation of valley to valley amplitude differences,
- Standard deviation of valley to valley amplitude differences normalised range R_x,
- Arithmetic mean of rising slopes, i.e., the slopes of the lines connecting a valley with the following peak. The rising slope $m_{j,i}$ for valley j and peak i, where $i = j$ (if $x(n)$ starts with a valley) or $i = j + 1$ (if $x(n)$ starts with a peak), is given by:

$$m_{j,i} = \frac{x_{p,i} - x_{v,j}}{T\left(n_{p,i} - n_{v_i}\right)}, \tag{2.321}$$

  where T is a normalisation factor. T is equivalent to the frame period length (in seconds) for all experiments in this thesis[72] which results in a unit for the slope as *amplitude difference per second*. T can, however, also be set to $T = 1$ to obtain a slope in terms of amplitude difference per frame[73] or $T = 1/N$ for a slope in terms of amplitude difference per relative segment length.[74] The first ($x(0)$) and

[72]In openSMILE in `cFunctionalPeaks2` `norm=second` has to be set for this behaviour (default).

[73]`norm=frame` in openSMILE.

[74]`norm=segment` in openSMILE.

last ($x(N-1)$) value of the signal x are treated as first/last valley/peak (the latter depending on whether the respective value is followed/preceded by a peak/valley, respectively) in order to compute at least one slope, even if 0 or only 1 peak (or valley) is present in the signal.
- Standard deviation of rising slopes,
- Maximum rising slope,
- Minimum rising slope,
- Arithmetic mean of (positive) falling slopes, i.e., the slopes of the lines which connect a peak with the following valley. The positive falling slope $m_{i,j}$ for peak i and valley j, where $j = i$ (if $x(n)$ starts with a peak) or $j = i + 1$ (if $x(n)$ starts with a valley), is given by:

$$m_{i,j} = \frac{x_{p,i} - x_{v,j}}{T\left(n_{v,j} - n_{p_i}\right) \cdot}, \tag{2.322}$$

  where T is a time-domain normalisation factor (see *rising slope* above for details),
- Standard deviation of falling slopes,
- Maximum falling slope,
- Minimum falling slope.

The inter peak and valley distances measured in frames, and their standard deviations, can be converted to either distances in seconds by multiplying with the frame period length T (in seconds), or to distances measured as proportions of the segment length N by dividing by N.[75] Normalising parameters to a time scale of seconds is recommended for all peak and valley related functionals because this makes the descriptors independent of the input length and the underlying frame rate.

To find the peaks, a peak picking algorithm must be applied. The most simple algorithm is the extreme value search for maxima ($x(n-1) < x(n) > x(n+1)$), which finds all local maximum values. However, in speech and music signals small local extreme values might occur due to windowing artefacts or noise. These values are insignificant and have no global meaning. Thus, a threshold should be used to discard extreme values with a low local/relative amplitude. The algorithm implemented for this thesis consists of three steps. It

1. finds all local maxima and minima (extrema) in $x(n)$,
2. discards extrema with a low relative amplitude (based on an absolute threshold which is a configurable fraction of the range of $x(n)$—see Eq. (2.316)),[76] and
3. enforces a constraint of alternating maxima/minima, i.e., it discards two adjacent maxima with no valid (!) minimum in between and vice versa.

[75] In openSMILE the `norm` option controls this behaviour (frames, seconds, segment—respectively).

[76] See the `absThresh` and `relThresh` options in the openSMILE component `cFunctionalPeaks2`.

2.4.2.9 Segments

Some signals can be divided into meaningful, continuous segments, such as voiced and unvoiced parts of the pitch contour, or high and low signal energy regions. The shape and structure of these segments contains valuable paralinguistic information, in particular temporal and rhythmic information. Based on unsupervised statistical segmentation algorithms, segments based on statistics can be found in any arbitrary signal (e.g., MFCC, etc.).[77] Depending on the segmentation algorithm and the meaning of the segments when considering the underlying LLD, segments are either defined as adjacent, i.e., the end of the first segment is identical to the beginning of the second segment, or as separated by gaps, i.e., between segment one and two there is a gap of at least one sample/frame length. The following segmentation algorithms which return adjacent segments are considered in this thesis:

delta threshold A segment boundary is placed wherever $x(n) > \overline{x}(n) + \delta_x$, with $\overline{x}(n)$ being the sliding window moving average over w frames (Schuller 2006). w is estimated from the maximum number of allowed segments $N_{seg,\,max}$ as $w = 2N/N_{seg,\,max}$. If required, w can also be set to a fixed custom value[78]—however, this has not been applied in the experiments in this thesis. δ_x is the threshold that has to be exceeded in order for a new segment to begin. Is is computed from a range-relative threshold θ_R as $\delta_x = R_x\theta_R$.

relative threshold A segment boundary is placed wherever $x(n)$ or a running average $\overline{x}$ (sliding window, size $w = 3$ frames) of $x(n)$ crosses or touches a pre-defined threshold a, i.e., the boundary is placed at frame n, when $\overline{x}(n-1) \leq a < \overline{x}(n)$ or $\overline{x}(n-1) \geq a > \overline{x}(n)$. The absolute threshold a can be computed from a range-relative threshold θ_R:

$$a = x_{min} + R_x\theta_R, \tag{2.323}$$

where R_x is the range of $x(n)$, or from a mean-relative threshold θ_μ:

$$a = \mu_x\theta_\mu, \tag{2.324}$$

where μ_x is the arithmetic mean of $x(n)$.

change from/to constant A segment boundary is placed wherever $x(n)$ changes from $x(n) = a$ to $x(n) \neq a$ or vice versa.

Further, the following algorithms which return segments of interest which are separated by gap segments are implemented:

unequal to Segments are separated by continuous regions in which $x(n) = a$. Within a segment, $x(n) \neq a$ must be true for the samples within the segment (except for small gaps, see below). a is a configurable threshold parameter and depends on the type of input (LLD) the functional is applied to. In this thesis, this

[77] In openSMILE segment-based temporal functionals can be computed with the component `cFunctionalSegments`.

[78] Use the `ravgLng` option of the `cFunctionalSegments` component in openSMILE.

functional is applied to the F_0 LLD, for example, and $a = 0$. Thus, the segments considered correspond to voiced segments where by definition $F_0 > 0$ is true. In order to increase noise robustness (e.g., robustness to F_0 extraction errors), a minimum gap (pause) duration of 2 frames is enforced.[79] If a gap is shorter or equal to the minimum pause duration, the segments left and right of the short gap are considered as a single segment, including the gap.

equal to In analogy to the *unequal to* condition, segments are separated by continuous regions where $x(n) \neq a$, and $x(n) = a$ is true within a segment.

Additionally, for all algorithms, a minimum segment length constraint is enforced to avoid segments of one frame length due to noise. The minimum segment length $l_{seg,min}$ is estimated from the maximum number of allowed segments $N_{seg,max}$: $l_{seg,min} = N/N_{seg,max} - 1$.

From the segment boundaries, the following temporal descriptors are computed:

- Number of segments,
- Arithmetic mean of segment lengths,
- Standard deviation of segment lengths,
- Maximum segment length,
- Minimum segment length,

If the segments are not defined as adjacent segments, but as segments separated by gaps, the following functionals can be additionally computed:

- Arithmetic mean of the gaps between segments,
- Standard deviation of the gaps between segments,
- Maximum length of gap between segments,
- Minimum length of gap between segments.

2.4.2.10 Onsets

An onset is generally defined as a sudden, steep amplitude rise in $x(n)$, e.g., the beginning/onset of an acoustic event. Here, onsets as a functional follow a more simple definition, similar to the threshold definition for segment based functionals (Sect. 2.4.2.9): An onset is marked at each position n where $x(n)$ rises above a given absolute threshold a: $x(n-1) \leq a < x(n)$. An "offset" marks the inverse, i.e., the case when $x(n)$ falls below a: $x(n-1) > a \geq x(n)$. For some LLDs it might make sense to use the absolute value of $x(n)$ to define onsets only by the magnitude of the descriptor. While this has not been used in this thesis, as onset functionals have been applied to F_0 contours only, the equations for the conditions are given here as: $|x(n-1)| \leq a < |x(n)|$ for an onset and $|x(n-1)| > a \geq |x(n)|$ for an "offset".

[79]This length can be changed via the `pauseMinLng` option of the `cFunctionalSegments` component.

The following functionals are computed from the onset and "offset" positions[80]:

- The position of the first onset, either as frame number n, relative to the segment length $n_{rel} = n/N$, or as a time in seconds $t = n \cdot T$,
- The position of the last "offset", either as frame number n, relative to the segment length $n_{rel} = n/N$, or as a time in seconds $t = n \cdot T$,
- The absolute number of onsets N_{onsets},
- The absolute number of "offsets" $N_{offsets}$ (only provided for completeness, but not used, as it is highly redundant with the number of onsets—a more informative feature for future work would be the difference between onsets and "offsets", which can be either $-1, 0$, or $+1$, depending on whether $x(n)$ starts/ends with a value above/below the threshold a),
- The onset rate (frequency) in Hz: $f_{onsets} = \frac{N_{onsets}}{N \cdot T}$.

2.4.2.11 Crossings

With the definition of ZCR from Sect. 2.2.1.1, ZCR can be applied as a functional.[81] Since—in contrast to time domain audio signals—most LLDs are not symmetric around zero, the MCR is favoured. The MCR is defined like the ZCR but computed from $x(n)$ after mean removal as given by Eq. (2.37).

2.4.2.12 Sample Values

In order to capture sample values of $x(n)$ at given temporal positions independent of the length of $x(n)$, values are sampled at given relative time instants n_{rel}, where n_{rel} is in the range [0; 1] and gives the sample time as percentage of the length of $x(n)$. Thereby the absolute position n is computed as $n = N \cdot n_{rel}$. This functional is recommended for rather short segments, such as isolated phonemes, words, beats or bars. In this case the relative positions have meanings, such as the beginning, middle, and end of the segment (for 3 sample values placed accordingly).[82]

2.4.3 Modulation Functionals

Modulation features have recently gained attention in speech related tasks (cf. Mubarak et al. 2006; Wu et al. 2011; Chi et al. 2012) and have been inherently used for rhythmic analysis in Music Information Retrieval (Mubarak et al. 2006; Schuller et al. 2007).

---

[80] Computed in openSMILE by the `cFunctionalOnset` component.

[81] Provided by the `cFunctionalCrossings` component in openSMILE.

[82] Sample-based functionals are provided by the `cFunctionalSamples` component in openSMILE.

The goal of modulation features is to qualitatively describe the modulations and periodicities of a signal $x(n)$, as well as the shapes thereof with a set of numerical descriptors. For this thesis the following types of modulation functionals applicable to LLD signals have been considered: Discrete Cosine Transformation coefficients (Sect. 2.4.3.1), LPC coefficients (Sect. 2.4.3.2), and modulation spectrum coefficients (Sect. 2.4.3.3).

2.4.3.1 Discrete Cosine Transformation Coefficients

For the DCT coefficient functional,[83] the DCT base functions from order o_0 to order o_l are computed and applied to the signal to obtain the $o_l - o_0 + 1$ DCT coefficients from order o_0 to o_l. The base functions are expressed relative to the length N of the input $x(n)$. This leads to functionals which describe the overall contour of each input segment as if all segments were normalised to a common length. Thus, the DCT coefficients do not correspond to absolute frequencies, but rather frequencies relative to the segment length. For modulation coefficients corresponding to absolute frequencies, independent of the segment length, the modulation spectrum coefficients (Sect. 2.4.3.3) were designed.

The same DCT-II as used in Sect. 2.2.10.1 for MFCC is applied here. The oth DCT base function $b_{DCT}^{(o)}(n)$ is thus given as $(n = 0 \dots N-1)$:

$$b_{DCT}^{(o)}(n) = \cos\left(\pi \frac{n}{N}(o + 0.5)\right). \tag{2.325}$$

The oth DCT coefficient is then computed as:

$$DCT(o) = \sqrt{\frac{2}{N}} \sum_{n=0}^{N-1} x(n) b_{DCT}^{(o)}(n). \tag{2.326}$$

Typically DCT coefficients 1–5 are used as functionals because most affective and paralinguistic information is hidden in the lower frequency modulations, which correspond to the lower order DCT coefficients. If the length of segments varies strongly, DCT coefficient functionals should not be used, due to reasons discussed above.

2.4.3.2 Linear Predictor Coefficients

Although Linear Predictive Coding has been derived from the source-filter model of speech production, linear prediction is a general method to model linear dependencies present among subsequent samples of an arbitrary signal $x(n)$. LP analysis can therefore be used as a functional on LLD signals.[84]

[83] In openSMILE the `cFunctionalDCT` component computes DCT coefficient functionals.

[84] In openSMILE the `cFunctionalLpc` component computes LP-analysis functionals.

LPC coefficients are computed from a signal $x(n)$ (length N) with the autocorrelation method which is described in Sect. 2.2.7.1. In most experiments reported on in this thesis the coefficients a_i with $i = 1 \dots 6$ are used, as well as the gain of LP analysis (the energy of the remaining error signal). A low gain indicates a signal which can be well approximated by LP analysis, i.e., a deterministic signal, while a high gain indicates a stochastic signal.

As most LLDs are sampled at a constant rate of 100 Hz, LP coefficients are independent of the segment length and can be used when the segment length varies strongly—in contrast to the DCT coefficients. I.e., if a signal of $N/2$ frames is appended to itself to obtain a sequence of length N, the resulting LP coefficient functionals will be very similar to those LP coefficients computed from only the $N/2$ length signal. The resulting DCT coefficients, however, will differ significantly across both cases. More precisely, they will be shifted by one order, i.e., $DCT(1)$ of the $N/2$ length signal will be roughly equal to $DCT(2)$ of the full (N) length signal.

2.4.3.3 Modulation Spectrum

The modulation spectrum[85] of a (LLD) signal $x(n)$ of length N is computed via a Short-Time Fourier Transform (STFT) approach (short-time analysis, cf. Sect. 2.1.3). Thereby the signal $x(n)$ is divided into I frames (each N_f samples long), a windowing function (cf. Sect. 2.1.3) is applied to the frames, zero-padding is performed, and a FFT is applied to each windowed signal $x_i(n)$ ($i = 1 \dots I$) and the linear magnitudes of the resulting complex spectrum are computed (cf. Sect. 2.2.3.1). The magnitude spectra X_i of each short-time frame are averaged over all I frames, resulting in the raw modulation spectrum $X^{(mod,raw)}$. With a constant window size N_f, this method enables the efficient computation of the modulation spectrum on arbitrary length signals in linear time with respect to the signal length N.

As a last step, the spectrum $X^{(mod,raw)}$ is scaled and interpolated to a fixed number of bins M', a number which is independent of the window size N_f. This way it is ensured that modulation spectrum features computed with arbitrary (N_f, N) combinations are compatible. Spline interpolation (Steffensen 2012) is used to map $X^{(mod,raw)}(m)$ to $X^{(mod)}(m')$ ($m' = 1 \dots M'$). The spline function is constructed with the points $(F(m), X^{(mod)}(m))$, where $F(m)$ maps a bin index m to a linear frequency (Hz) f (cf. Sect. 2.2.3.1). The spline function is then evaluated for the frequencies $f' = F'(m')$ corresponding to the bins of the spectrum $X^{(mod)}$ to obtain the respective (interpolated) magnitudes.

Phase information of the modulation spectrum is not considered here. In future work, however, this could be addressed when segments with meaningful start and end points are analysed, i.e., the segments are not selected as fixed length segments at a fixed rate from a stream. Such segments could be for speech: whole words, phrases,

[85] In openSMILE the `cFunctionalModulation` component computes modulation spectrum functionals.

parts of sentences, or full sentences, for example, and for music: beats, one or more bars, or the chorus, for example.

As a default recommendation, the frequency resolution δ_f for $X^{(mod)}$ was chosen between 0.1 to 0.5 Hz and a range from 0.25 to 30 Hz was considered by the author of this thesis, which results in $N_f = 298$ bins ($\delta_f = 0.1$) and $N_f = 60$ bins ($\delta_f = 0.5$).

The modulation spectrum can either be used as a feature vector directly, or further statistics can be computed from this spectrum. All statistics applicable to normal magnitude and power spectra can be applied. Most important, however, seem to be the frequency(-ies) related to the peak(s) of the modulation spectrum and the flatness of the modulation spectrum.[86]

2.4.3.4 Rhythmic Features

For music analysis, rhythmic features are of high relevance. While for speech, modulation spectrum, and LPC based modulation features might be sufficient, the highly structured rhythm of music requires features developed specifically for music rhythm analysis. Such features have been proposed by the author of this thesis in (Schuller et al. 2007), originally.[87] The features are derived by a 2-step comb-filterbank analysis of Mel-scale auditory band spectra. A so-called *Tatum vector* (resembling a modulation spectrum in a range which covers very fast tempi) and a *meter vector*, which is a modulation spectrum computed at multiples of the Tatum tempo, are computed for a music segment or a whole piece of music. The Tatum tempo is computed by finding the most prominent peak in the Tatum vector. For details, the reader is referred to Schuller et al. (2007) and Eyben and Schuller (2010b). By definition in (Schuller et al. 2007), the Tatum vector has 57 elements, and the meter vector has 19 elements, which covers 19 metrical levels, i.e., 19 multiples of the Tatum tempo.

2.5 Modelling

In order to evaluate the acoustic features and especially the standard feature sets introduced in the previous chapters, classification experiments were performed. This section gives a brief overview over the classification and modelling methods used in this thesis and a theoretical introduction to the two categories of modelling: static (Sect. 2.5.1) and dynamic (Sect. 2.5.2).

---

[86] In openSMILE, the statistics can be applied to the modulation spectrum with the `cSpectral` component. Also other components which expect magnitude spectra (e.g., ACF in `cAcf`) can read from the output of `cFunctionalModulation`.

[87] These features are not part of openSMILE (yet). It is planned to include them in future releases. C code is available from the author of this thesis upon request.

In the field of affect recognition and Computational Paralinguistics, most methods in the past have dealt with classification on an utterance level, e.g., Ververidis and Kotropoulos (2006); Vlasenko et al. (2007); Schuller et al. (2009b, 2013b). Thereby each utterance or a part of an utterance is associated with one set of affective labels and the classifier/regressor assigns exactly one set of predictions to one utterance. As shown by Vlasenko et al. (2007), for example, either a static classifier, such as a Support Vector Machine (SVM) (e.g., Schuller et al. 2006) or a dynamic classifier, such as a Hidden Markov Models (HMMs) can be used (e.g., Schuller et al. (2003)). The SVM estimates a classification label from a single high dimensional feature vector, which summarises the utterance. The HMM computes a likelihood score for every frame of low-level features and from the most likely path over the utterance estimates the most likely class. Recently, databases with dimensional affect ratings have emerged: the Sensitive Artificial Listener (SAL) set in the HUMAINE database (Douglas-Cowie et al. 2007), the SEMAINE database (Schröder et al. 2012), the RECOLA database (Ringeval et al. 2013), and a set of continuous music mood annotations (Soleymani et al. 2013). Tools like Feeltrace (Cowie et al. 2000) have been used for continuous rating, both in time and value. Such databases have caused a shift in methods, first of all moving from classification to regression to be able to model continuous affective dimensions (Grimm et al. 2007; Wöllmer et al. 2008), and next moving from utterance or segment level labels to quasi time-continuous labels (Eyben et al. 2010c, 2012; Schröder et al. 2012; Schuller et al. 2012b; Weninger et al. 2013, 2014)—creating a need for research on dynamic, context aware modelling methods. This thesis proposes and evaluates such a modelling method based on combining supra-segmental features with dynamic modelling by Long Short-Term Memory Recurrent Neural Networks (LSTM-RNNs) in Sect. 6.4.

In the field of MIR, static modelling has also been applied for several tasks such as dance-style recognition (Schuller et al. 2007), genre identification (e.g., Tzanetakis and Cook 2002; Eyben and Schuller 2010a), and chord and key transcription (e.g., Lee and Slaney 2008; Schuller et al. 2008). Music fingerprinting and identification of artist and title based on a fingerprint can also be seen as a static classification task (e.g., Wang 2003).

As the main aim and novelty of this thesis is to explore efficient large-scale, online feature extraction and propose ground-breaking standard feature sets, only the two most popular modelling methods Support-Vector Machines for the static case, and (Long Short-Term Memory) Recurrent Neural Networks for the dynamic case, are discussed in this section.

2.5.1 Static Modelling with Support Vector Machines

In static modelling a segment of audio is represented by a single, fixed dimensional vector of parameters (supra-segmental features, see Sect. 2.4) and modelled with a static classifier—usually a distance based, i.e., nearest neighbour or polynomial classifier, or a statistical classifier, such as a Bayes classifier (Kroschel et al. 2011). Due to the great success of SVMs in speech analysis for emotion recognition and Computational Paralinguistics (Schuller and Batliner 2013), this distance based polynomial classifier is the favoured classifier for static classification in this thesis.

SVMs—originally introduced by Cortes and Vapnik (1995)—are probably the most frequently used classifier for paralinguistic speech and music analysis tasks (Schuller 2013; Schuller and Batliner 2013) at present. This fact can be attributed to a handful of convenient properties—mainly their ability to handle high dimensional feature spaces (e.g., the baseline acoustic feature sets created for this thesis—cf. Chap. 3), noisy and sparse features (e.g., features which are almost always zero, such as frequently encountered in vector space modelling of linguistic features, cf. Schuller and Batliner 2013), and the robustness of their training algorithms to over-fitting. These properties have been discussed and pointed out by Joachims (1998) in the context of classification of texts with linguistic features, and they have subsequently been exploited for acoustic classification, too. Another advantage is that SVMs can be easily extended to continuous class labels (regression tasks) by Support Vector Regression (SVR) as introduced by Cortes and Vapnik (1995). In order to avoid repetition of material, which has been already described many times elsewhere, only a very coarse summary of the concept of linear SVM (as used in this thesis) is given at this point. For more details the reader is referred to the excellent summary of SVMs and SVR by Schuller and Batliner (2013) and the original paper by Cortes and Vapnik (1995).

The core idea of SVM is built around the concept of binary linear classifiers and is optimised towards providing the best possible separation between classes in the given feature space—which is the core difference to other classifiers with linear decision boundaries, such as the nearest neighbour classifiers. When training a SVM a hyperplane which separates the two classes in the feature space is constructed. In order to improve generalisation and limit over-fitting, a margin between the two classes which should be free of feature vectors is enforced and maximised during the construction of the hyperplane (Cortes and Vapnik 1995). Thereby a trade-off between margin maximisation and data points which then fall into the margin region must be found.

This optimisation criterion leads to a description of the hyperplane based on so called 'support vectors' which lie in between the centres of gravity of the classes and define the decision boundary (the hyperplane including a margin between the classes). The concept of the hyperplane and maximum margin is illustrated in Fig. 2.9.

The support vectors are chosen by solving a quadratic optimisation problem, for which efficient algorithms are available (cf. Cortes and Vapnik 1995; Platt 1998). As a result, the classification is based on a small subset of the training set of data points, effectively reducing the risk of over-fitting and improving generalisation performance. In order to solve non-linear decision tasks, i.e., tasks where the two classes cannot be separated by a linear decision border (hyperplane) in the feature space, the 'kernel trick' (Schölkopf and Smola 2002) is applied to map the non-linear problem into a higher dimensional space where it can be solved linearly in order to retain the low complexity of the (linear) support vector principle. For the

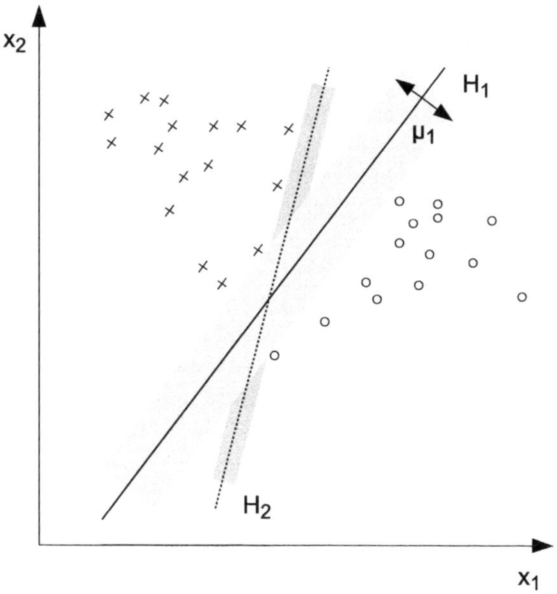

Fig. 2.9 Example of an (optimal) hyper plane H_1 (*solid line*) with maximum margin (μ_1) and a sub-optimal hype plane with smaller margin H_2 in a 2d feature space. The symbols "x" and "o" represent instances (data points) of the two classes, respectively (Schuller and Batliner 2013)

mathematics behind the training algorithms of support vector machines and support vector regression the summary by Schuller and Batliner (2013) can be consulted.

The SVMs as described so far are capable of discriminating between two classes only. For multi-class problems extensions must be built on top of the binary SVM. This could be, e.g., by building SVMs for each pair of classes, summing up the 'votes' for each class during recognition and then choosing the class with the most votes as winner, or by forming a binary decision tree (cf. Schuller and Batliner 2013) with each binary decision being performed by a SVM classification.

For linear kernel SVM, the hyperplane can be expressed in a compact representation by its normal vector. This makes linear kernel SVM highly suitable for real-time, on-line recognition tasks, as the decision function for an unknown instance can be computed as the scalar product of the normal vector and the feature vector of the unknown instance.

2.5.2 *Dynamic Modelling*

Dynamic modelling involves modelling of signal dynamics and context. In contrast to static modelling, where a single vector is mapped to a class or regression label, in dynamic modelling a sequence of feature vectors is mapped to a label or a sequence of labels. The modelling framework must be capable of modelling not only the associations of feature vectors to labels or label probabilities but also the dependencies between the labels in relation to the inputs (feature vectors) and the labels.

The most well known dynamic modelling technique in the field of ASR is HMMs (Rabiner and Juang 1986; Rabiner 1989). There, a statistical framework models the transitions between frames, while a probabilistic classifier models the data for each feature vector. For the latter, most commonly Gaussian Mixture Models (GMMs) are used. However, any classifier which is capable of returning an observation probability for a given feature vector could be used. Common examples are Neural Networks (NNs) (Stadermann and Rigoll 2006) and SVMs (Stadermann and Rigoll 2004). As HMMs are not a core topic of this thesis, the author refers to Rabiner (1989), Young et al. (2006), and Schuller and Batliner (2013) for more details on the concepts of HMMs for speech and music analysis.

An alternative dynamic modelling approach is based on neural networks. A standard Feed-Forward Neural Network (FFNN) can be extended to have access to data from previous timesteps. This extension is called Recurrent Neural Network (RNN) and will be described in the following section.

2.5.2.1 Recurrent Neural Networks

A standard FFNN is given by one or more hidden layers of sigmoid units. Each sigmoid unit (also referred to as *neuron*) consists of a weighted summation of inputs followed by a nonlinearity (Fig. 2.10, left). The output y of a sigmoid unit is described by the following equation:

$$y = g\left(b + \underline{\boldsymbol{w}} \cdot \underline{\boldsymbol{x}}\right) = g\left(b + \sum_{i=0}^{N-1} w_i x_i\right), \tag{2.327}$$

where N is the dimensionality of the input vector $\underline{\boldsymbol{x}}$ and the weight vector $\underline{\boldsymbol{w}}$ (both dimensions must match), and w_i and x_i are the ith elements of these vectors.

The function $g(.)$ can be any non-linear differentiable function in theory. Practically, however, the *sigmoid* function is used most often (Schuller 2013):

$$g_{sigmoid}(x) = \frac{1}{1 + e^{-\alpha x}}, \tag{2.328}$$

in its special case where $\alpha = 1$, which is known as the *logistic* function (Verhulst 1945). Alternatively, the hyperbolic tangent function tanh (Rade et al. 2000):

$$g_{tanh}(x) = \frac{1 - e^{-2x}}{1 + e^{-2x}}, \tag{2.329}$$

is used.

The purpose of the non-linearity is to **(a)** limit, i.e., compress, the output values and ensure stable outputs for a wide range of inputs, and—most important—**(b)** to enable the network to approximate any arbitrary non-linear function. For more details on the fundamentals of FFNNs, such as the training algorithms, the reader is referred to Bishop (1995).

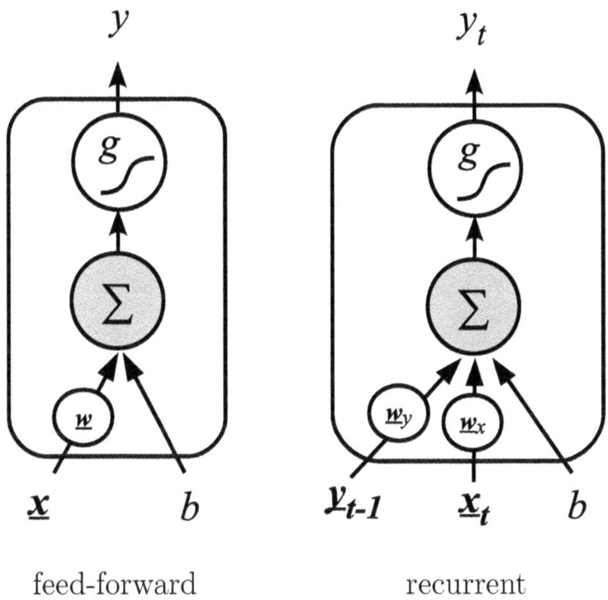

Fig. 2.10 Sigmoid neuron in a feedforward neural network (*left*) and sigmoid neuron in a recurrent neural network (*right*). Input vector $\underline{x}$, previous output (from all cells in the current layer) in vector $\underline{y}_{t-1}$, output activation y (scalar), constant bias b, and non-linearity $g(.)$

FFNNs with the standard sigmoid units have no memory, i.e., they have no knowledge of other inputs than those of the current timestep. A logical extension is to make the network recurrent, i.e., add a feedback from the output to the input with a delay of one timestep. Such networks are known as RNNs. In a RNN typically the outputs of all neurons of a recurrent hidden layer are connected back to the inputs of each of the neurons in that layer through a recurrent connection which has a time delay of one timestep. A single recurrent sigmoid unit is shown in Fig. 2.10 (right).

2.5.2.2 Long Short-Term Memory Recurrent Neural Networks

RNNs as introduced in the previous section, however, suffer from the Vanishing Gradient Problem (Hochreiter et al. 2001). This means that the activations and the back-propagated error on the recurrent connections (with weights ranging from $0 \ldots 1$) decay exponentially. This severely limits the amount of temporal context which is accessible to the networks effectively to approximately 10 frames. To overcome this problem, LSTM-RNNs have been proposed originally by Hochreiter and Schmidhuber (1997) and extended to the version used in this thesis by Graves and Schmidhuber (2005). The main difference between the original version and the version used here (and by Graves and Schmidhuber 2005) is the use of *peep-hole* connections (and according weights) from the internal cell memory state c_t to the input, output, and

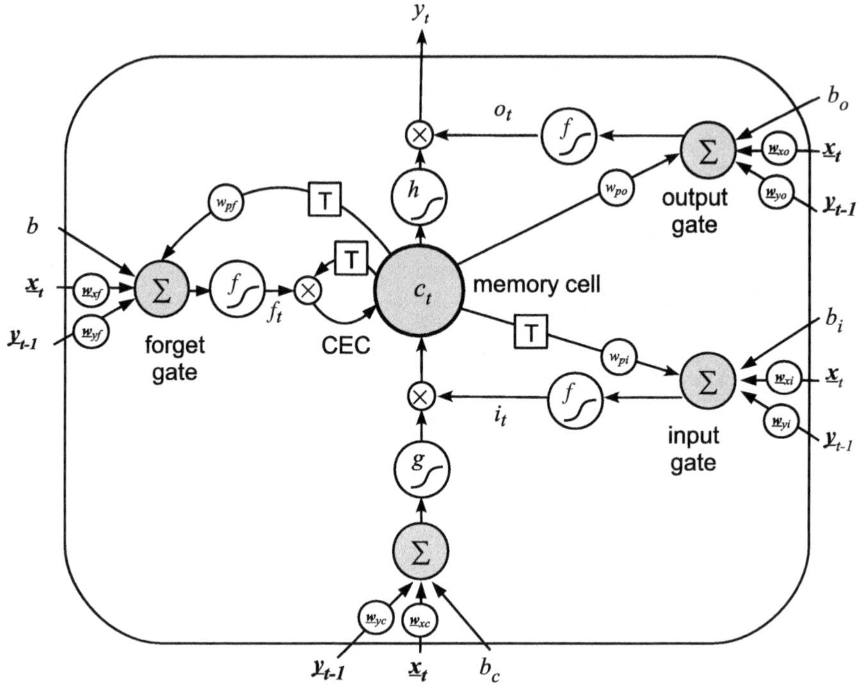

Fig. 2.11 Long Short-Term Memory block, with one LSTM cell and the input (i), output (o) and forget (f) gates. The block state is shown at timestep t. Input data vector ($\underline{\boldsymbol{x}}$), connection weights w_{ab} (multiplicative), bias values b, block output y. Vectors are indicated by *underlined bold face font*, all other variables are scalars. The vector containing all cell outputs y_t of the current hidden layer at timestep t is denoted as $\underline{\boldsymbol{y}}_t$. T denotes a time delay unit of one timestep/frame. X in a *circle* denotes a multiplicative unit. Σ denotes a summation unit. $f()$, $g()$, and $h()$ are non-linear activation functions (squashing functions)

forget gate summation units (cf. $w_{p..}$ in Fig. 2.11). Compared to a conventional RNN, the sigmoid summation units in the hidden layers are replaced by so-called Long Short-Term Memory (LSTM) blocks. LSTM blocks can theoretically store information in the cell variable c_t for an infinite amount of time due to the Constant Error Carousel (CEC) in which the previous cell state c_{t-1} is connected to the current state via a recurrent connection with a constant weight of 1 (excluding the multiplicative influence of the forget gate)—see Fig. 2.11. In this way, the network can dynamically exploit long-range temporal context present in the input data. In practice this ability has successfully been demonstrated on many speech and music analysis tasks, e.g., Wöllmer et al. (2008, 2010, 2013), Eyben et al. (2009b, 2010b), Böck and Schedl (2012), and Weninger et al. (2014).

Each LSTM block consists of a memory cell and three multiplicative gates: the input gate, the output gate, and the forget gate, as shown in Fig. 2.11. These gates control the access to the block's internal memory cell c_t. According to Fig. 2.11, the input, output and forget gate activation values i_t, o_t, and f_t are computed, respectively, as:

$$i_t = f(\underline{\boldsymbol{w}}_{xi}\underline{\boldsymbol{x}}_t + \underline{\boldsymbol{w}}_{yi}\underline{\boldsymbol{y}}_{t-1} + w_{pi}c_{t-1} + b_i), \tag{2.330}$$
$$o_t = f(\underline{\boldsymbol{w}}_{xo}\underline{\boldsymbol{x}}_t + \underline{\boldsymbol{w}}_{yo}\underline{\boldsymbol{y}}_{t-1} + w_{po}c_t + b_o), \tag{2.331}$$
$$f_t = f(\underline{\boldsymbol{w}}_{xf}\underline{\boldsymbol{x}}_t + \underline{\boldsymbol{w}}_{yf}\underline{\boldsymbol{y}}_{t-1} + w_{pf}c_{t-1} + b_f), \tag{2.332}$$

where $\underline{\boldsymbol{w}}_x$ and $\underline{\boldsymbol{w}}_y$ are weight vectors (row vectors) matching the dimensionality of $\underline{\boldsymbol{x}}$ or $\underline{\boldsymbol{y}}$, respectively. $\underline{\boldsymbol{x}}_t$ is the input vector at timestep t, $\underline{\boldsymbol{y}}_{t-1}$ is the vector of hidden layer activations (outputs of all N cells in the hidden layer) at the previous timestep, and $b_{i,o,f}$ denotes the respective input, output, or forget gate bias value for the cell.

The forget gate controls the decay of the stored input c_t. If $f_t = 0$, the previous cell state c_{t-1} is fully erased. The input and output gates are responsible for dynamically weighting the cell input and output, respectively. The cell state c_t at timestep t is expressed as:

$$c_t = f_t c_{t-1} + i_t g(\underline{\boldsymbol{w}}_{xc}\underline{\boldsymbol{x}}_t + \underline{\boldsymbol{w}}_{yc}\underline{\boldsymbol{y}}_{t-1} + b_c), \tag{2.333}$$

and the cell output is given as:

$$\boldsymbol{y}_t = o_t h(c_t). \tag{2.334}$$

The activation functions f (for the gates), g (for the input), and h (for the output) are non-linear squashing functions like those in normal sigmoid neurons. Common choices are the *logistic* function or the tanh function (Sect. 2.5.2.1). Recently, also Rectified Linear Units (ReLUs) have been proposed for standard (deep) NNs and for LSTM (Krizhevsky et al. 2012; Pham et al. 2013). While they give performance gains for (deep) NNs (Krizhevsky et al. 2012; Dahl et al. 2013), this was not reported for LSTM (Pham et al. 2013). Thus, the standard configuration as used by Graves and Schmidhuber (2005), for example, and most other related work[88] which uses LSTM and/or Bidirectional Long Short-Term Memory (BLSTM), is applied for this thesis: tanh activation functions for the cell input and output (functions g and h) and *logistic* activation functions for the gates (functions f).

In a LSTM-RNN a LSTM hidden layer consists of N LSTM blocks (Fig. 2.11) which are fully connected to all inputs and recurrently to all outputs of this layer as shown in Fig. 2.12.

In addition to LSTM-RNN, Bidirectional Long Short-Term Memory Recurrent Neural Networks (BLSTM-RNNs) (Schuster and Paliwal 1997) are employed in this thesis. A bidirectional recurrent network can access context from both past and future inputs, which makes it very suitable for processing data with de-synchronised inputs and outputs, or where the outputs (targets) have been centred at the middle (time-wise) of an input event. The bidirectional context is made possible by processing the data in both directions with two separate hidden layers, one processing the data sequence forward, the other backward. The output activations from both hidden layers are then fed to the same output layer, where they are fused. The combination of the

---

[88] E.g., as is also implemented in the CURRENNT toolkit (http://sourceforge.net/projects/currennt) and the RNNLIB (http://sourceforge.net/projects/rnnl/).

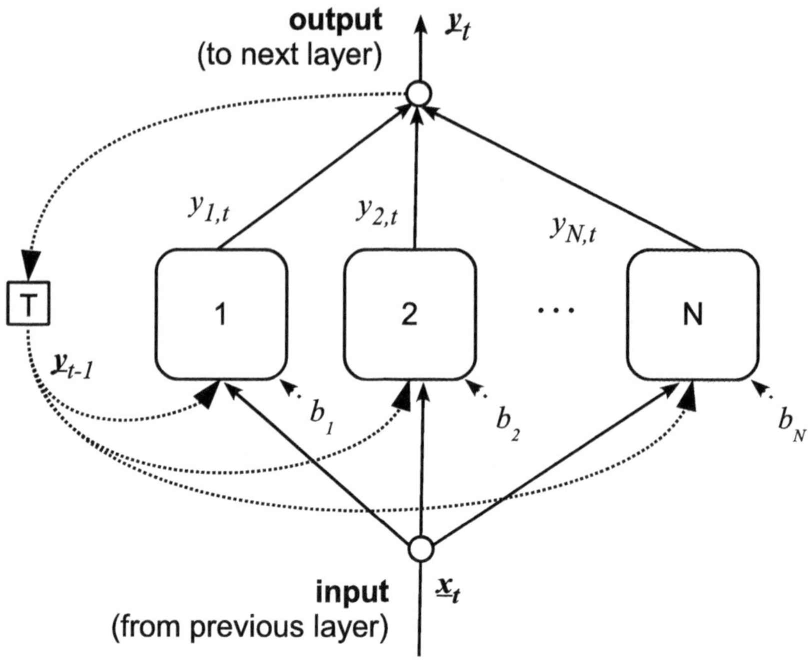

Fig. 2.12 Long Short-Term Memory hidden layer with N LSTM blocks, showing the fully connected recurrent connections of the outputs y

concept of bidirectional RNNs and LSTM blocks leads to BLSTM-RNNs (Graves and Schmidhuber 2005; Graves 2008).

LSTM-RNNs and BLSTM-RNNs as well as standard RNNs can be trained with a gradient descent method where the weights are iteratively updated, known from the backpropagation algorithm for FFNNs and extended to Backpropagation Through Time (BPTT) by Werbos (1990) for recurrent networks. A variation of the backpropagation algorithm is Resilient Propagation (rProp) introduced by Riedmiller and Braun (1993). There, only the sign of the error gradient is backpropagated and used for weight updates instead of the absolute value of the error weighted by the learn rate. Resilient propagation produces more stable convergence (e.g., Eyben et al. 2010c) and thus can outperform standard backpropagation, especially with respect to the number of training epochs required.

A more detailed summary of BLSTM-RNNs and the training algorithms is found in Schuller (2013) and an extensive discussion is found in the PhD thesis of Graves (2008).

References

R.G. Bachu, S. Kopparthi, B. Adapa, B.D. Barkana, Voiced/unvoiced decision for speech signals based on zero-crossing rate and energy, in *Advanced Techniques in Computing Sciences and Software Engineering*, ed. by K. Elleithy (Springer, Netherlands, 2010), pp. 279–282. doi:10.1007/978-90-481-3660-5_47. ISBN 978-90-481-3659-9

A. Batliner, S. Steidl, B. Schuller, D. Seppi, T. Vogt, L. Devillers, L. Vidrascu, N. Amir, L. Kessous, V. Aharonson, The impact of F0 extraction errors on the classification of prominence and emotion, in *Proceedings of 16-th ICPhS* (Saarbrücken, Germany, 2007), pp. 2201–2204

L.L. Beranek, *Acoustic Measurements* (Wiley, New York, 1949)

C.M. Bishop, *Neural Networks for Pattern Recognition* (Oxford University Press, New York, 1995)

R.B. Blackman, J. Tukey, Particular pairs of windows, *The Measurement of Power Spectra, from the Point of View of Communications Engineering* (Dover, New York, 1959)

S. Böck, M. Schedl, Polyphonic piano note transcription with recurrent neural networks, in *Proceedings of ICASSP 2012* (Kyoto, 2012), pp. 121–124

P. Boersma, Accurate short-term analysis of the fundamental frequency and the harmonics-to-noise ratio of a sampled sound. IFA Proc. **17**, 97–110 (1993)

P. Boersma, Praat, a system for doing phonetics by computer. Glot Int. **5**(9/10), 341–345 (2001)

B.P. Bogert, M.J.R. Healy, J.W. Tukey, The quefrency alanysis of time series for echoes: cepstrum, pseudo-autocovariance, cross-cepstrum, and saphe cracking, in *Proceedings of the Symposium on Time Series Analysis, chapter 15*, ed. by M. Rosenblatt (Wiley, New York, 1963), pp. 209–243

C.H. Chen, *Signal Processing Handbook*. Electrical Computer Engineering, vol. 51 (CRC Press, New York, 1988), 840 p. ISBN 978-0824779566

A. Cheveigne, H. Kawahara, YIN, a fundamental frequency estimator for speech and music. J. Acoust. Soc. Am. (JASA) **111**(4), 1917–1930 (2002)

T.-S. Chi, L.-Y. Yeh, C.-C. Hsu, Robust emotion recognition by spectro-temporal modulation statistic features. J. Ambient Intell. Humaniz. Comput. **3**, 47–60 (2012). doi:10.1007/s12652-011-0088-5

J. Cooley, P. Lewis, P. Welch, The finite fourier transform. IEEE Trans. Audio Electroacoust. **17**(2), 77–85 (1969)

C. Cortes, V. Vapnik, Support-vector networks. Mach. Learn. **20**(3), 273–297 (1995)

R. Cowie, E. Douglas-Cowie, S. Savvidou, E. McMahon, M. Sawey, M. Schröder, Feeltrace: an instrument for recording perceived emotion in real time, in *Proceedings of the ISCA Workshop on Speech and Emotion* (Newcastle, Northern Ireland, 2000), pp. 19–24

G. Dahl, T. Sainath, G. Hinton, Improving deep neural networks for LVCSR using rectified linear units and dropout, in *Proceedings of ICASSP 2013* (IEEE, Vancouver, 2013), pp. 8609–8613

G. Dalquist, A. Björk, N. Anderson, *Numerical Methods* (Prentice Hall, Englewood Cliffs, 1974)

S. Damelin, W. Miller, *The Mathematics of Signal Processing* (Cambridge University Press, Cambridge, 2011). ISBN 978-1107601048

G. de Krom, A cepstrum-based technique for determining a harmonics-to-noise ratio in speech signals. J. Speech Hear. Res. **36**, 254–266 (1993)

J.R. Deller, J.G. Proakis, J.H.L. Hansen, *Discrete-Time Processing of Speech Signals*, University of Michigan, Macmillan Publishing Company (1993)

P. Deuflhard, *Newton Methods For Nonlinear Problems: Affine Invariance and Adaptive Algorithms*. Springer Series in Computational Mathematics, vol. 35 (Springer, Berlin, 2011), 440 p

E. Douglas-Cowie, R. Cowie, I. Sneddon, C. Cox, O. Lowry, M. McRorie, J.C. Martin, L. Devillers, S. Abrilian, A. Batliner, N. Amir, K. Karpouzis, *The HUMAINE Database*. Lecture Notes in Computer Science, vol. 4738 (Springer, Berlin, 2007), pp. 488–500

J. Durbin, The fitting of time series models. Revue de l'Institut International de Statistique (Review of the International Statistical Institute) **28**(3), 233–243 (1960)

C. Duxbury, M. Sandler, M. Davies, A hybrid approach to musical note onset detection, in *Proceedings of the Digital Audio Effect Conference (DAFX'02)* (Hamburg, Germany, 2002), pp. 33–38

L.D. Enochson, R.K. Otnes, *Programming and Analysis for Digital Time Series Data*, 1st edn. U.S. Department of Defense, Shock and Vibration Information Center (1968)

F. Eyben, B. Schuller, Music classification with the Munich openSMILE toolkit, in *Proceedings of the Annual Meeting of the MIREX 2010 community as part of the 11th International Conference on Music Information Retrieval (ISMIR)* (ISMIR, Utrecht, 2010a). http://www.music-ir.org/mirex/abstracts/2010/FE1.pdf

F. Eyben, B. Schuller, Tempo estimation from tatum and meter vectors, in *Proceedings of the Annual Meeting of the MIREX 2010 community as part of the 11th International Conference on Music Information Retrieval (ISMIR)* (ISMIR, Utrecht, 2010b). www.music-ir.org/mirex/abstracts/2010/ES1.pdf

F. Eyben, M. Wöllmer, B. Schuller, openEAR—introducing the Munich open-source emotion and affect recognition toolkit, in *Proceedings of the 3rd International Conference on Affective Computing and Intelligent Interaction (ACII 2009)*, vol. I (IEEE, Amsterdam, 2009a), pp. 576–581

F. Eyben, M. Wöllmer, B. Schuller, A. Graves, From speech to letters—using a novel neural network architecture for grapheme based ASR, in *Proceedings of the IEEE Workshop on Automatic Speech Recognition and Understanding (ASRU) 2009* (IEEE, Merano, 2009b), pp. 376–380

F. Eyben, M. Wöllmer, B. Schuller, openSMILE—The Munich versatile and fast open-source audio feature extractor, in *Proceedings of ACM Multimedia 2010* (ACM, Florence, 2010a), pp. 1459–1462

F. Eyben, S. Böck, B. Schuller, A. Graves, Universal onset detection with bidirectional long-short term memory neural networks, in *Proceedings of ISMIR 2010* (ISMIR, Utrecht, The Netherlands, 2010b), pp. 589–594

F. Eyben, M. Wöllmer, A. Graves, B. Schuller, E. Douglas-Cowie, R. Cowie, On-line emotion recognition in a 3-D activation-valence-time continuum using acoustic and linguistic cues. J. Multimodal User Interfaces (JMUI) **3**(1–2), 7–19 (2010c). doi:10.1007/s12193-009-0032-6

F. Eyben, M. Wöllmer, B. Schuller, A multi-task approach to continuous five-dimensional affect sensing in natural speech, ACM Trans. Interact. Intell. Syst. **2**(1), Article No. 6, 29 p. Special Issue on Affective Interaction in Natural Environments (2012)

G. Fant, *Speech Sounds and Features* (MIT press, Cambridge, 1973), p. 227

H.G. Feichtinger, T. Strohmer, *Gabor Analysis and Algorithms* (Birkhäuser, Boston, 1998). ISBN 0-8176-3959-4

J.-B.-J. Fourier, Théorie analytique de la chaleur, University of Lausanne, Switzerland (1822)

T. Fujishima, Realtime chord recognition of musical sound: a system using common lisp music, in *Proceedings of the International Computer Music Conference (ICMC) 1999* (Bejing, China, 1999), pp. 464–467

S. Furui, *Digital Speech Processing: Synthesis, and Recognition*. Signal Processing and Communications, 2nd edn. (Marcel Denker Inc., New York, 1996)

C. Glaser, M. Heckmann, F. Joublin, C. Goerick, Combining auditory preprocessing and bayesian estimation for robust formant tracking. IEEE Trans. Audio Speech Lang. Process. **18**(2), 224–236 (2010)

E. Gómez, Tonal description of polyphonic audio for music content processing. INFORMS J. Comput. **18**(3), 294–304 (2006). doi:10.1287/ijoc.1040.0126

F. Gouyon, F. Pachet, O. Delerue. Classifying percussive sounds: a matter of zero-crossing rate? in *Proceedings of the COST G-6 Conference on Digital Audio Effects (DAFX-00)* (Verona, Italy, 2000)

A. Graves, Supervised sequence labelling with recurrent neural networks. Doctoral thesis, Technische Universität München, Munich, Germany (2008)

A. Graves, J. Schmidhuber, Framewise phoneme classification with bidirectional LSTM and other neural network architectures. Neural Netw. **18**(5–6), 602–610 (2005)

W.D. Gregg, *Analog & Digital Communication* (Wiley, New York, 1977). ISBN 978-0-471-32661-8

M. Grimm, K. Kroschel, S. Narayanan, Support vector regression for automatic recognition of spontaneous emotions in speech, in *Proceedings of ICASSP 2007*, vol. 4 (IEEE, Honolulu, 2007), pp. 1085–1088

B. Hammarberg, B. Fritzell, J. Gauffin, J. Sundberg, L. Wedin, Perceptual and acoustic correlates of abnormal voice qualities. Acta Otolaryngol. **90**, 441–451 (1980)

H. Hanson, Glottal characteristics of female speakers: acoustic correlates. J. Acoust. Soc. Am. (JASA) **101**, 466–481 (1997)

H. Hanson, E.S. Chuang, Glottal characteristics of male speakers: acoustic correlates and comparison with female data. J. Acoust. Soc. Am. (JASA) **106**, 1064–1077 (1999)

F.J. Harris, On the use of windows for harmonic analysis with the discrete fourier transform. Proc. IEEE **66**, 51–83 (1978)

H. Hermansky, Perceptual linear predictive (PLP) analysis for speech. J. Acoust. Soc. Am. (JASA) **87**, 1738–1752 (1990)

H. Hermansky, N. Morgan, A. Bayya, P. Kohn, RASTA-PLP speech analysis technique, in *Proceedings of ICASSP 1992*, vol. 1 (IEEE, San Francisco, 1992), pp. 121–124

D.J. Hermes, Measurement of pitch by subharmonic summation. J. Acoust. Soc. Am. (JASA) **83**(1), 257–264 (1988)

W. Hess, *Pitch Determination of Speech Signals: Algorithms and Devices* (Springer, Berlin, 1983)

S. Hochreiter, J. Schmidhuber, Long short-term memory. Neural Comput. **9**(8), 1735–1780 (1997)

S. Hochreiter, Y. Bengio, P. Frasconi, J. Schmidhuber, Gradient flow in recurrent nets: the difficulty of learning long-term dependencies, in *A Field Guide to Dynamical Recurrent Neural Networks*, ed. by S.C. Kremer, J.F. Kolen (IEEE Press, New York, 2001)

ISO16:1975. ISO Standard 16:1975 Acoustics: Standard tuning frequency (Standard musical pitch). International Organization for Standardization (ISO) (1975)

T. Joachims, Text categorization with support vector machines: learning with many relevant features, in *Proceedings of the 10th European Conference on Machine Learning (ECML-98)*, ed. by C. Nédellec, C. Rouveirol (Springer, Chemnitz, 1998), pp. 137–142

J.D. Johnston, Transform coding of audio signals using perceptual noise criteria. IEEE J. Sel. Areas Commun. **6**(2), 314–332 (1988)

P. Kabal, R.P. Ramachandran, The computation of line spectral frequencies using Chebyshev polynomials. IEEE Trans. Acoust. Speech Signal Process. **34**(6), 1419–1426 (1986)

J.F. Kaiser, Some useful properties of teager's energy operators, in *Proceedings of ICASSP 1993*, vol. 3, pp. 149–152, (IEEE, Minneapolis, 1993). doi:10.1109/ICASSP.1993.319457

G.S. Kang, L.J. Fransen, Application of line spectrum pairs to low bit rate speech encoders, in *Proceedings of ICASSP 1985*, vol.10 (IEEE, Tampa, 1985), pp. 244–247. doi:10.1109/ICASSP.1985.1168526

R. Kendall, E. Carterette, Difference thresholds for timbre related to spectral centroid, in *Proceedings of the 4-th International Conference on Music Perception and Cognition (ICMPC)* (Montreal, Canada, 1996), pp. 91–95

J.F. Kenney, E.S. Keeping, Root mean square, *Mathematics of Statistics*, vol. 1, 3rd edn. (Van Nostrand, Princeton, 1962), pp. 59–60

A. Khintchine, Korrelationstheorie der stationären stochastischen prozesse. Math. Ann. **109**, 604–615 (1934)

A. Kießling, *Extraktion und Klassifikation prosodischer Merkmale in der automatischen Sprachverarbeitung* (Shaker, Aachen, 1997). ISBN 978-3-8265-2245-1

A. Krizhevsky, I. Sutskever, G.E. Hinton, Imagenet classification with deep convolutional neural networks, in *Advances in Neural Information Processing Systems*, vol. 25, ed. by F. Pereira, C.J.C. Burges, L. Bottou, K.Q. Weinberger (Curran Associates, Inc., 2012), pp. 1097–1105

K. Kroschel, G. Rigoll, B. Schuller, *Statistische Informationstechnik*, 5th edn. (Springer, Berlin, 2011)

K. Lee, M. Slaney, Acoustic chord transcription and key extraction from audio using key-dependent HMMs trained on synthesized audio. IEEE Trans. Audio Speech Lang. Process. **16**(2), 291–301 (2008). doi:10.1109/TASL.2007.914399. ISSN 1558-7916

P. Lejeune-Dirichlet, Sur la convergence des séries trigonométriques qui servent à représenter une fonction arbitraire entre des limites données. Journal für die reine und angewandte Mathematik **4**, 157–169 (1829)

N. Levinson, A heuristic exposition of wiener's mathematical theory of prediction and filtering. J. Math. Phys. **25**, 110–119 (1947a)
N. Levinson, The Wiener RMS error criterion in filter design and prediction. J. Math. Phys. **25**(4), 261–278 (1947b)
P.I. Lizorkin, Fourier transform, in *Encyclopaedia of Mathematics*, ed. by M. Hazewinkel (Springer, Berlin, 2002). ISBN 1-4020-0609-8
I. Luengo, Evaluation of pitch detection algorithms under real conditions, in *Proceedings of ICASSP 2007*, vol. 4 (IEEE, Honolulu, 2007), pp. 1057–1060
J. Makhoul, Linear prediction: a tutorial review. Proc. IEEE **63**(5), 561–580 (1975)
J. Makhoul, L. Cosell, LPCW: an LPC vocoder with linear predictive spectral warping, in *Proceedings of ICASSP 1976* (IEEE, Philadelphia, 1976), pp. 466–469
B.S. Manjunath, P. Salembier, T. Sikoraa (eds.), *Introduction to MPEG-7: Multimedia Content Description Interface* (Wiley, Berlin, 2002), 396 p. ISBN 978-0-471-48678-7
P. Martin, Détection de f_0 par intercorrelation avec une fonction peigne. J. Etude Parole **12**, 221–232 (1981)
P. Martin, Comparison of pitch detection by cepstrum and spectral comb analysis, in *Proceedings of ICASSP 1982* (IEEE, Paris, 1982), pp. 180–183
J. Martinez, H. Perez, E. Escamilla, M.M. Suzuki, Speaker recognition using mel frequency cepstral coefficients (MFCC) and vector quantization (VQ) techniques, in *Proceedings of the 22nd International Conference on Electrical Communications and Computers (CONIELECOMP)* (Cholula, Puebla, 2012), pp. 248–251. doi:10.1109/CONIELECOMP.2012.6189918
P. Masri, Computer modelling of sound for transformation and synthesis of musical signal. Doctoral thesis, University of Bristol, Bristol (1996)
S. McCandless, An algorithm for automatic formant extraction using linear prediction spectra. IEEE Trans. Acoust. Speech Signal Process. **22**, 134–141 (1974)
D.D. Mehta, D. Rudoy, P.K. Wolfe, Kalman-based autoregressive moving average modeling and inference for formant and antiformant tracking. J. Acoust. Soc. Am. (JASA) **132**(3), 1732–1746 (2012)
H. Misra, S. Ikbal, H. Bourlard, H. Hermansky, Spectral entropy based feature for robust ASR, in *Proceedings of ICASSP 2004*, vol. 1 (IEEE, Montreal, Canada, 2004), pp. I–193–6. doi:10.1109/ICASSP.2004.1325955
O. Mubarak, E. Ambikairajah, J. Epps, T. Gunawan, Modulation features for speech and music classification, in *Proceedings of the 10th IEEE Singapore International Conference on Communication systems (ICCS) 2006* (IEEE, 2006), pp. 1–5. doi:10.1109/ICCS.2006.301515
M. Müller, *Information Retrieval for Music and Motion* (Springer, Berlin, 2007)
M. Müller, F. Kurth, M. Clausen, Audio matching via chroma-based statistical features, in *Proceedings of the 6th International Conference on Music Information Retrieval (ISMIR)* (London, 2005a), pp. 288–295
M. Müller, F. Kurth, M. Clausen, Chroma-based statistical audio features for audio matching, in *Proceedings of the IEEE Workshop on Applications of Signal Processing to Audio and Acoustics* (IEEE, 2005b), pp. 275–278
N.J. Nalini, S. Palanivel, Emotion recognition in music signal using AANN and SVM. Int. J. Comput. Appl. **77**(2), 7–14 (2013)
A.M. Noll, Cepstrum pitch determination. J. Acoust. Soc. Am. (JASA) **41**(2), 293–309 (1967)
A.M. Noll, Pitch determination of human speech by the harmonic product spectrum, the harmonic sum spectrum, and a maximum likelihood estimate, in *Symposium on Computer Processing in Communication*, vol. 19 (University of Brooklyn, New York, 1970), pp. 779–797, edited by the Microwave Institute
A.H. Nuttal, Some windows with very good sidelobe behavior. IEEE Trans. Acoust. Speech Signal Process. ASSP **29**, 84–91 (1981)
A.V. Oppenheim, R.W. Schafer, *Digital Signal Processing* (Prentice-Hall, Englewood Cliffs, 1975)
A.V. Oppenheim, A.S. Willsky, S. Hamid, *Signals and Systems*, 2nd edn. (Prentice Hall, Upper Saddle River, 1996)

A.V. Oppenheim, R.W. Schafer, J.R. Buck, *Discrete-Time Signal Processing* (Prentice Hall, Upper Saddle River, 1999)

T.W. Parsons, *Voice and Speech Processing*. Electrical and Computer Engineering (University of Michigan, McGraw-Hill, 1987)

S. Patel, K.R. Scherer, J. Sundberg, E. Björkner, Acoustic markers of emotions based on voice physiology, in *Proceedings of Speech Prosody 2010* (ISCA, Chicago, 2010), pp. 100865:1–4

G. Peeters, A large set of audio features for sound description. Technical report, IRCAM, Switzerland (2004). http://recherche.ircam.fr/equipes/analyse-synthese/peeters/ARTICLES/Peeters_2003_cuidadoaudiofeatures.pdf. Accessed 3 Sept. 2013

V. Pham, C. Kermorvant, J. Louradour, Dropout improves recurrent neural networks for handwriting recognition, in *CoRR* (2013) (online), arXiv:1312.4569

J. Platt, Sequential minimal optimization: a fast algorithm for training support vector machines, Technical report MSR-98-14, Microsoft Research (1998)

L.R. Rabiner, On the use of autocorrelation analysis for pitch detection. IEEE Trans. Acoust. Speech Signal Process. **25**(1), 24–33 (1977). doi:10.1109/TASSP.1977.1162905

L.R. Rabiner, A tutorial on hidden markov models and selected applications in speech recognition. Proc. IEEE **77**(2), 257–286 (1989)

L.R. Rabiner, B.H. Juang, An introduction to hidden markov models. IEEE ASSP Mag. **3**(1), 4–16 (1986)

L. Rabiner, B.-H. Juang, *Fundamentals of Speech Recognition*, 1st edn. (Prentice Hall, Englewood Cliffs, 1993)

L. Rade, B. Westergren, *Springers Mathematische Formeln (German translation by P. Vachenauer)*, 3rd edn. (Springer, Berlin, 2000). ISBN 3-540-67505-1

J.F. Reed, F. Lynn, B.D. Meade, Use of coefficient of variation in assessing variability of quantitative assays. Clin. Diagn. Lab. Immunol. **9**(6), 1235–1239 (2002)

M. Riedmiller, H. Braun, A direct adaptive method for faster backpropagation learning: the RPROP algorithm, in *Proceedings of the IEEE International Conference on Neural Networks*, vol. 1 (IEEE, San Francisco, 1993), pp. 586–591. doi:10.1109/icnn.1993.298623

F. Ringeval, A. Sonderegger, J. Sauer, D. Lalanne, Introducing the RECOLA multimodal corpus of remote collaborative and affective interactions, in *Proceedings of the 2nd International Workshop on Emotion Representation, Analysis and Synthesis in Continuous Time and Space (EmoSPACE), held in conjunction with FG 2013* (IEEE, Shanghai, 2013), pp. 1–8

S. Rosen, P. Howell, The vocal tract as a linear system, *Signals and Systems for Speech and Hearing*, 1st edn. (Emerald Group, 1991), pp. 92–99. ISBN 978-0125972314

G. Ruske, *Automatische Spracherkennung*. Methoden der Klassifikation und Merkmalsextraktion, 2nd edn. (Oldenbourg, Munich, 1993)

K.R. Scherer, J. Sundberg, L. Tamarit, G.L. Salomão, Comparing the acoustic expression of emotion in the speaking and the singing voice. Comput. Speech Lang. **29**(1), 218–235 (2015). doi:10.1016/j.csl.2013.10.002

B. Schölkopf, A. Smola, *Learning with Kernels: Support Vector Machines, Regularization, Optimization, and Beyond (Adaptive Computation and Machine Learning)* (MIT Press, Cambridge, 2002)

M. Schröder, E. Bevacqua, R. Cowie, F. Eyben, H. Gunes, D. Heylen, M. ter Maat, G. McKeown, S. Pammi, M. Pantic, C. Pelachaud, B. Schuller, E. de Sevin, M. Valstar, M. Wöllmer, Building autonomous sensitive artificial listeners. IEEE Trans. Affect. Comput. **3**(2), 165–183 (2012)

M.R. Schroeder, Period histogram and product spectrum: new methods for fundamental-frequency measurement. J. Acoust. Soc. Am. (JASA) **43**, 829–834 (1968)

M.R. Schroeder, Recognition of complex acoustic signals, in *Life Sciences Research Reports*, vol. 5, ed. by T.H. Bullock (Abakon Verlag, Berlin, 1977), 324 p

B. Schuller, Automatische Emotionserkennung aus sprachlicher und manueller Interaktion. Doctoral thesis, Technische Universität München, Munich, Germany (2006)

B. Schuller, *Intelligent Audio Analysis*. Signals and Communication Technology (Springer, Berlin, 2013)

B. Schuller, A. Batliner, *Computational Paralinguistics: Emotion, Affect and Personality in Speech and Language Processing* (Wiley, Hoboken, 2013), 344 p. ISBN 978-1119971368
B. Schuller, G. Rigoll, M. Lang, Hidden Markov model-based speech emotion recognition, in *Proceedings of ICASSP 2003*, vol. 2 (IEEE, Hong Kong, 2003), pp. II 1–4
B. Schuller, D. Arsić, F. Wallhoff, G. Rigoll, Emotion recognition in the noise applying large acoustic feature sets, in *Proceedings of the 3rd International Conference on Speech Prosody (SP) 2006* (ISCA, Dresden, 2006), pp. 276–289
B. Schuller, F. Eyben, G. Rigoll, Fast and robust meter and tempo recognition for the automatic discrimination of ballroom dance styles, in *Proceedings of ICASSP 2007*, vol. I (IEEE, Honolulu, 2007), pp. 217–220
B. Schuller, F. Eyben, G. Rigoll, Beat-synchronous data-driven automatic chord labeling, in *Proceedings 34. Jahrestagung für Akustik (DAGA) 2008* (DEGA, Dresden, 2008), pp. 555–556
B. Schuller, S. Steidl, A. Batliner, F. Jurcicek, The INTERSPEECH 2009 emotion challenge, in *Proceedings of INTERSPEECH 2009* (Brighton, 2009a), pp. 312–315
B. Schuller, B. Vlasenko, F. Eyben, G. Rigoll, A. Wendemuth, Acoustic emotion recognition: A benchmark comparison of performances, in *Proceedings of the IEEE Workshop on Automatic Speech Recognition and Understanding (ASRU) 2009* (IEEE, Merano, 2009b), pp. 552–557
B. Schuller, S. Steidl, A. Batliner, F. Burkhardt, L. Devillers, C. Müller, S. Narayanan, The INTERSPEECH 2010 paralinguistic challenge, in *Proceedings of INTERSPEECH 2010* (ISCA, Makuhari, 2010), pp. 2794–2797
B. Schuller, A. Batliner, S. Steidl, F. Schiel, J. Krajewski, The INTERSPEECH 2011 speaker state challenge, in *Proceedings of INTERSPEECH 2011* (ISCA, Florence, 2011), pp. 3201–3204
B. Schuller, S. Steidl, A. Batliner, E. Nöth, A. Vinciarelli, F. Burkhardt, R. van Son, F. Weninger, F. Eyben, T. Bocklet, G. Mohammadi, B. Weiss, The INTERSPEECH 2012 speaker trait challenge, in *Proceedings of INTERSPEECH 2012* (ISCA, Portland, 2012a)
B. Schuller, M. Valstar, R. Cowie, M. Pantic, AVEC 2012: the continuous audio/visual emotion challenge—an introduction, in *Proceedings of the 14th ACM International Conference on Multimodal Interaction (ICMI) 2012*, ed. by L.-P. Morency, D. Bohus, H.K. Aghajan, J. Cassell, A. Nijholt, J. Epps (ACM, Santa Monica, 2012b), pp. 361–362. October
B. Schuller, F. Pokorny, S. Ladstätter, M. Fellner, F. Graf, L. Paletta. Acoustic geo-sensing: recognising cyclists' route, route direction, and route progress from cell-phone audio, in *Proceedings of ICASSP 2013* (IEEE, Vancouver, 2013a), pp. 453–457
B. Schuller, S. Steidl, A. Batliner, A. Vinciarelli, K. Scherer, F. Ringeval, M. Chetouani, et al., The INTERSPEECH 2013 computational paralinguistics challenge: social signals, conflict, emotion, autism, in *Proceedings of INTERSPEECH 2013* (ISCA, Lyon, 2013b), pp. 148–152
M. Schuster, K.K. Paliwal, Bidirectional recurrent neural networks. IEEE Trans. Signal Process. **45**(11), 2673–2681 (1997)
C.E. Shannon, A mathematical theory of communication. Bell Syst. Tech. J. **27**, 379–423, 623–656 (1948). (Reprint with corrections in: ACM SIGMOBILE Mobile Computing and Communications Review **5**(1), 3–55 (2001))
M. Slaney, An efficient implementation of the patterson-holdsworth auditory filter bank. Technical Report 35, Apple Computer Inc. (1993)
M. Soleymani, M.N. Caro, E.M. Schmidt, Y.-H. Yang, The MediaEval 2013 brave new task: emotion in music, in *Proceedings of the MediaEval 2013 Workshop* (CEUR-WS.org, Barcelona, 2013)
F.K. Soong, B.-W. Juang, Line spectrum pair (LSP) and speech data compression, in *Proceedings of ICASSP 1984* (IEEE, San Diego, 1984), pp. 1.10.1–1.10.4
A. Spanias, T. Painter, V. Atti, *Audio Signal Processing and Coding* (Wiley, Hoboken, 2007), 464 p. ISBN 978-0-471-79147-8
J. Stadermann, G. Rigoll, A hybrid SVM/HMM acoustic modeling approach to automatic speech recognition, in *Proceedings of INTERSPEECH 2004* (ISCA, Jeju, 2004), pp. 661–664
J. Stadermann, G. Rigoll, Hybrid NN/HMM acoustic modeling techniques for distributed speech recognition. Speech Commun. **48**(8), 1037–1046 (2006)

J.F. Steffensen, *Interpolation*, 2nd edn. (Dover Publications, New York, 2012), 256 p. ISBN 978-0486154831

P. Suman, S. Karan, V. Singh, R. Maringanti, Algorithm for gunshot detection using mel-frequency cepstrum coefficients (MFCC), in *Proceedings of the Ninth International Conference on Wireless Communication and Sensor Networks*, ed. by R. Maringanti, M. Tiwari, A. Arora. Lecture Notes in Electrical Engineering, vol. 299 (Springer, India, 2014), pp. 155–166. doi:10.1007/978-81-322-1823-4_15. ISBN 978-81-322-1822-7

J. Sundberg, *The Science of the Singing Voice* (Northern Illinois University Press, Dekalb, 1987), p. 226. ISBN 978-0-87580-542-9

D. Talkin, A robust algorithm for pitch tracking (RAPT), in *Speech Coding and Synthesis*, ed. by W.B. Kleijn, K.K. Paliwal (Elsevier, New York, 1995), pp. 495–518. ISBN 0444821694

L. Tamarit, M. Goudbeek, K.R. Scherer, Spectral slope measurements in emotionally expressive speech, in *Proceedings of SPKD-2008* (ISCA, 2008), paper 007

H.M. Teager, S.M. Teager, Evidence for nonlinear sound production mechanisms in the vocal tract, in *Proceedings of Speech Production and Speech Modelling, Bonas, France*, ed. by W.J. Hardcastle, A. Marchal. NATO Advanced Study Institute Series D, vol. 55 (Kluwer Academic Publishers, Boston, 1990), pp. 241–261

E. Terhardt, Pitch, consonance, and harmony. J. Acoust. Soc. Am. (JASA) **55**, 1061–1069 (1974)

E. Terhardt, Calculating virtual pitch. Hear. Res. **1**, 155–182 (1979)

H. Traunmueller, Analytical expressions for the tonotoc sensory scale. J. Acoust. Soc. Am. (JASA) **88**, 97–100 (1990)

K. Turkowski, S. Gabriel, Filters for common resampling tasks, in *Graphics Gems*, ed. by A.S. Glassner (Academic Press, New York, 1990), pp. 147–165. ISBN 978-0-12-286165-9

G. Tzanetakis, P. Cook, Musical genre classification of audio signals. IEEE Trans. Speech Audio Process. **10**(5), 293–302 (2002). doi:10.1109/TSA.2002.800560. ISSN 1063-6676

P.-F. Verhulst, Recherches mathématiques sur la loi d'accroissement de la population (mathematical researches into the law of population growth increase). Nouveaux Mémoires de l'Académie Royale des Sciences et Belles-Lettres de Bruxelles **18**, 1–42 (1945)

D. Ververidis, C. Kotropoulos, Emotional speech recognition: resources, features, and methods. Speech Commun. **48**(9), 1162–1181 (2006)

A.J. Viterbi, Error bounds for convolutional codes and an asymptotically optimum decoding algorithm. IEEE Trans. Inf. Theory **13**(2), 260–269 (1967)

B. Vlasenko, B. Schuller, A. Wendemuth, G. Rigoll., Frame vs. turn-level: emotion recognition from speech considering static and dynamic processing, in *Proceedings of the 2nd International Conference on Affective Computing and Intelligent Interaction (ACII) 2007*, ed. by A. Paiva, R. Prada, R.W. Picard. Lecture Notes in Computer Science, Lisbon, Portugal, vol. 4738 (Springer, Berlin, 2007), pp. 139–147

A.L. Wang, An industrial-strength audio search algorithm, in *Proceedings of ISMIR* (Baltimore, 2003)

F. Weninger, F. Eyben, B. Schuller, The TUM approach to the mediaeval music emotion task using generic affective audio features, in *Proceedings of the MediaEval 2013 Workshop* (CEUR-WS.org, Barcelona, 2013)

F. Weninger, F. Eyben, B. Schuller, On-line continuous-time music mood regression with deep recurrent neural networks, in *Proceedings of ICASSP 2014* (IEEE, Florence, 2014), pp. 5449–5453

P. Werbos, Backpropagation through time: what it does and how to do it. Proc. IEEE **78**, 1550–1560 (1990)

N. Wiener, Generalized harmonic analysis. Acta Math. **55**(1), 117–258 (1930)

N. Wiener, *Extrapolation, Intrapolation and Smoothing of Stationary Time Series*, M.I.T. Press Paperback Series (Book 9) (MIT Press, Cambridge, 1964), 163 p

M. Wöllmer, F. Eyben, S. Reiter, B. Schuller, C. Cox, E. Douglas-Cowie, R. Cowie, Abandoning emotion classes—towards continuous emotion recognition with modelling of long-range dependencies, in *Proceedings of INTERSPEECH 2008* (ISCA, Brisbane, 2008), pp. 597–600

M. Wöllmer, F. Eyben, A. Graves, B. Schuller, G. Rigoll, Improving keyword spotting with a tandem BLSTM-DBN architecture, in *Advances in Non-linear Speech Processing: Revised selected papers of the International Conference on Nonlinear Speech Processing (NOLISP) 2009*, ed. by J. Sole-Casals, V. Zaiats. Lecture Notes on Computer Science (LNCS), vol. 5933/2010 (Springer, Vic, 2010), pp. 68–75

M. Wöllmer, M. Kaiser, F. Eyben, B. Schuller, G. Rigoll, LSTM-Modeling of Continuous Emotions in an Audiovisual Affect Recognition Framework. Image Vis. Comput. (IMAVIS) **31**(2), 153–163. Special Issue on Affect Analysis in Continuous Input (2013)

S. Wu, T.H. Falk, W.-Y. Chan, Automatic speech emotion recognition using modulation spectral features. Speech Commun. **53**(5), 768–785 (2011). doi:10.1016/j.specom.2010.08.013. ISSN 0167-6393 (Perceptual and Statistical Audition)

Q. Yan, S. Vaseghi, E. Zavarehei, B. Milner, J. Darch, P. White, I. Andrianakis, Formant-tracking linear prediction model using hmms and kalman filters for noisy speech processing. Comput. Speech Lang. **21**(3), 543–561 (2007). doi:10.1016/j.csl.2006.11.001

S. Young, G. Evermann, M. Gales, T. Hain, D. Kershaw, X. Liu, G. Moore, J. Odell, D. Ollason, D. Povey, V. Valtchev, P. Woodland, *The HTK Book*, Cambridge University Engineering Department, for HTK version 3.4 edition (2006)

E. Yumoto, W.J. Gould, Harmonics-to-noise ratio as an index of the degree of hoarseness. J. Acoust. Soc. Am. (JASA) **71**(6), 1544–1549 (1981)

G. Zhou, J.H.L. Hansen, J.F. Kaiser, Nonlinear feature based classification of speech under stress. IEEE Trans. Speech Audio Process. **9**(3), 201–216 (2001). doi:10.1109/89.905995

X. Zuo, P. Fung, A cross gender and cross lingual study of stress recognition in speech without linguistic features, in *Proceedings of the 17th ICPhS* (Hong Kong, China, 2011)

E. Zwicker, Subdivision of the audible frequency range into critical bands. J. Acoust. Soc. Am. (JASA) **33**(2), 248–248 (1961)

E. Zwicker, Masking and psychological excitation as consequences of ear's frequency analysis, in *Frequency Analysis and Periodicity Detection in Hearing*, ed. by R. Plomp, G.F. Smoorenburg (Sijthoff, Leyden, 1970)

E. Zwicker, E. Terhardt, Analytical expressions for critical-band rate and critical bandwidth as a function of frequency. J. Acoust. Soc. Am. (JASA) **68**, 1523–1525 (1980)

E. Zwicker, H. Fastl, *Psychoacoustics—Facts and Models*, 2nd edn. (Springer, Berlin, 1999), 417 p. ISBN 978-3540650638

Chapter 3
Standard Baseline Feature Sets

A central aim of this thesis was to define standard acoustic feature sets for both speech and music, which contain a large and comprehensive list of descriptors, based on previous efforts to combine features, e.g., by Batliner et al. (2011).

Standard sets have been defined by the author of this thesis and have been evaluated as baseline feature sets of international research competitions held at the INTERSPEECH conference. The first such challenge was the INTERSPEECH 2009 Emotion Challenge (Schuller et al. 2009) for which the IS09 feature set was defined (Sect. 3.1). It was followed by the INTERSPEECH 2010 Paralinguistics Challenge (Schuller et al. 2010) with the IS10 feature set (Sect. 3.2) and the INTERSPEECH 2011 Speaker State Challenge (Schuller et al. 2011a) and the INTERSPEECH 2012 Speaker Trait Challenge (Schuller et al. 2012a) with the respective feature sets IS11 (Sect. 3.3) and IS12 (Sect. 3.4). In 2013 the next Challenge was launched under the unified name Computational Paralinguistics ChallengE (ComParE) (Schuller et al. 2013). A high-impact baseline feature set, known as the ComParE 2013 set has been designed for this challenge (Sect. 3.5). The same set has been adopted for the follow-up challenge ComParE 2014 by Schuller et al. (2014).

Two feature sets with a reduced number of acoustic features have been introduced for the first and second Audio-visual Emotion Challenge (AVEC), namely AVEC 2011 (Schuller et al. 2011b), and AVEC 2012 (Schuller et al. 2012b). The need for lower dimensional acoustic feature vectors was given by the larger data-sets and the inclusion of visual features. Based on experience with the INTERSPEECH feature sets, only the most important and most comprehensive Low-level Descriptors (LLDs) were included and the set of functionals was reduced. An extended version of the AVEC 2012 set was proposed for the first Audiovisual Emotion and Depression Challenge (Valstar et al. 2013). The AVEC sets are described in Sect. 3.7.

Next to the challenge baseline feature sets, a minimalistic standard parameter set for speech analysis (Geneva Minimalistic Acoustic Parameter Set (GeMAPS)—Sect. 3.8), and a standard music feature set, for tempo, dance-style, and genre recognition (cf. Eyben and Schuller (2010); Sect. 3.9) have been designed for this thesis.

F. Eyben, *Real-time Speech and Music Classification by Large Audio Feature Space Extraction*, Springer Theses, DOI 10.1007/978-3-319-27299-3_3

The Geneva Minimalistic Acoustic Parameter Set is the result of close collaborations of engineers, psychologists, and linguists.[1]

This chapter contains brief descriptions of all the baseline feature sets that were developed during the course of this thesis. An in-depth description of the parametrisation of the feature extraction processes for each of the feature sets is found in Appendix A.1. A compact summary of all the feature sets up to IS12 is given by Schuller and Batliner (2013). An evaluation of all the feature sets on a large pool of audio databases and paralinguistic tasks is presented in Sect. 6.3.

3.1 INTERSPEECH 2009 Emotion Challenge Set

This feature set was the first of its kind, a baseline feature set provided for an international research challenge on speech emotion recognition—the first INTERSPEECH Emotion Challenge (Schuller et al. 2009). Basing on the state-of-the-art and prior work by Schuller et al. (2003) and Ververidis and Kotropoulos (2006), it was decided to design a minimalistic feature set for static modelling (cf. Sect. 2.5.1) with widely used and well-known acoustic LLDs, especially Mel-Frequency Cepstral Coefficients (MFCCs). In order to keep the size of the resulting feature vectors small, a low number of functionals was chosen. A larger feature vector would have caused very large feature files—given the size of the FAU-AIBO database (cf. Sect. 6.1.2), which was used for the challenge—and thus would have caused problems for the challenge participants during their experiments with the standard computer hardware available in 2008/2009.

The INTERSPEECH 2009 Emotion Challenge (IS09) feature set contains 16 LLD (see Table 3.1) and 16 corresponding delta regression coefficients (Sect. 2.3.2). To those 32 descriptors, 12 functionals (see Table 3.1) are applied, which yields a total of **384 features**. Due to the success of MFCCs for speech emotion recognition in prior work (e.g., Schuller et al. 2003), and the importance of prosodic features (e.g., Batliner et al. 1999; Batliner and Möbius 2005), the 16 LLDs contain 12 cepstral parameters (MFCCs 1–12). More details on this set are found in Appendix A.1.1.

3.2 INTERSPEECH 2010 Paralinguistics Challenge Set

This set was designed for the INTERSPEECH 2010 Paralinguistics Challenge, where one task for age recognition and one task for level of interest recognition were given (Schuller et al. 2010). Since these tasks were new paradigms, compared to the 2009 Emotion Recognition Challenge, a larger set of features was created and existing features updated, in order to enable participants to perform feature (type) selection and gain a deeper insight into which features are of importance for the tasks. Compared to

[1]An article is to appear in IEEE Transactions on Affective Computing (Eyben et al. 2015).

Table 3.1 INTERSPEECH 2009 Emotion Challenge (IS09) set

LLD	Functionals
MFCC 1–12,	Arithmetic mean,
F_0 via ACF,	Standard deviation, skewness, kurtosis,
Prob. of voicing,	Maximum and minimum value, range,
RMS energy.	Relative position of max. and min. value,
	Linear regression slope, offset, quadratic error.

Overview of Low-level Descriptor (LLD) contained in the set and functionals applied to these LLDs. Details in Appendix A.1.1

the IS09 set, Zero-Crossing Rate (ZCR) was removed and a more robust pitch extraction algorithm (Subharmonic Summation (SHS) algorithm instead of Autocorrelation function (ACF)/Cepstrum—Sect. 2.2.11.4) was chosen. Logarithmic Mel-Frequency Bands (MFBs) and Line Spectral Frequencies (LSFs) were included, as well as the voice quality descriptors Jitter and Shimmer, which potentially carry information about speaker age (cf. e.g., Sadeghi Naini and Homayounpour 2006).

The INTERSPEECH 2010 Paralinguistics Challenge (IS10) set contains 38 LLDs (see Table 3.2) and 38 corresponding delta regression coefficients. 21 functionals (see Table 3.2) are applied to 34 of these LLDs (group A in the ongoing) and their corresponding delta coefficients, yielding $34 \cdot 2 \cdot 21 = 1{,}428$ features. To the remaining 4 LLDs (group B in the ongoing) and the corresponding 4 delta regression coefficients, 19 functionals (see Table 3.2) are applied, yielding another $4 \cdot 2 \cdot 19 = 152$ features. Two more descriptors are included to reflect speaking rate and the segment length, respectively: the number of F_0 onsets (N_{onsets}, cf. onset functionals, Sect. 2.4.2.10) and the total duration of the segment in seconds. Overall, the IS10 set contains **1,582 features**. More details on this set are found in Appendix A.1.2.

Table 3.2 INTERSPEECH 2010 Paralinguistics Challenge (IS10) set

LLD	Functionals
Group A: (34)	Arithmetic mean,
Loudness,	Standard deviation, skewness, kurtosis,
MFCC 0–14,	Relative position of max. and min. value,
log. MFB 1–8, LSP 0–7,	99th and 1-st percentile^A, range of these^A,
F_0 envelope, Prob. of voicing.	Linear regression slope, offset,
Group B: (4)	Linear regression linear and quadratic error,
	Quartiles 1–3,
F_0 via SHS,	Inter-quartile ranges 1–2, 2–3, 1–3,
Jitter (local and delta), Shimmer.	Up-level time 75 % and 90 %.

Overview of Low-level Descriptors (LLDs) contained in the set and functionals applied to these LLDs. Functionals marked with ^{A} are only applied to group A LLDs. Details in Appendix A.1.2

Table 3.3 INTERSPEECH 2011 Speaker State Challenge (IS11) set

LLD	Functionals
Group A: (54)	Arithmetic and quadratic^B mean,
Loudness,	Standard deviation, skewness, kurtosis,
Modulation loudness,	Quartiles 1–3,
RMS energy,	Inter-quartile ranges 1–2, 2–3, 1–3,
ZCR,	99th and 1-st percentile, range of these,
RASTA auditory bands 1–26,	Linear regression slope a, quadratic error,
MFCC 1–12,	Quadratic regression a and b and quad. error,
Bands 25–650 Hz and 1–4 kHz,	Temporal centroid,
Spectral RoP .25, .50, .75, .90,	Peak mean and peak mean to arith. mean distance,
Spectral flux and entropy,	Mean and std. dev. of peak to peak distances,
Spectral variance,	Segment length mean^A, min.^{A}, max.^{A}, std. dev.^{A},
Spectral skewness and kurtosis,	Up-level time 90 %,
Spectral slope.	Down-level time 75 %,
Group B: (5)	Rise/fall time, left/right curvature time^B,
	Linear Prediction gain,
F_0 via SHS,	Linear Prediction Coefficients 1–5.
Prob. of voicing,	
Jitter (local and delta),	
Shimmer.	

Overview of Low-level Descriptors (LLDs) contained in the set and functionals applied to these LLDs. Functionals marked with ^{A} are only applied to group A LLDs; functionals marked with ^{B} are only applied to group B LLDs. Details in Appendix A.1.3

At this point it shall be noted, that in the 2010 Paralinguistic Challenge, two versions of this feature set were used. The full-sized set—as described above—was used for the interest recognition sub-challenge, and a down-sized version was used for the age recognition sub-challenge to reduce the size of training data due to the high number of instances in the age data set (for details, cf. Schuller et al. 2010).

3.3 INTERSPEECH 2011 Speaker State Challenge Set

To explore the full potential of large-scale brute-force feature extraction and reach out to novel speaker state recognition tasks such as alcohol intoxication and sleepiness, an even larger, more comprehensive set (IS11) was proposed for the INTERSPEECH 2011 Speaker State Challenge baseline (cf. Schuller et al. 2011a).

Similar to the IS10 set, the INTERSPEECH 2011 Speaker State Challenge (IS11) set contains two groups of LLDs: group A with 54 energy, spectral, and cepstral LLDs (see Table 3.3), and group B with 5 source/excitation signal related LLDs (see Table 3.3). To group A, 37 functionals are applied, and to group B, 36 functionals (see Table 3.3). This results in $54 \cdot 2 \cdot 37 = 3{,}996$ features from group A and $5 \cdot 2 \cdot 36 = 360$ features from group B, giving a total of 4,356 features. In addition to this brute-force combination of LLDs and functionals, the IS11 set contains temporal statistic descriptors based on voiced/unvoiced segments which are computed by applying the following functionals to the F_0 LLD and its first order delta coefficients: *Ratio of non-zero values* (ratio of non-zero values, i.e., percentage of voiced frames of total frames) and *segment length statistics* (minimum, mean, maximum, and standard deviation of the lengths of voiced segments, i.e., $F_0 > 0$ (cf. Sect. 2.4.2.9)). Additionally, as in the IS10 set, the total duration of the segment (in seconds) is included as a feature. All together this adds 11 more features, and thus, in total the IS11 set contains **4,367 features**.[2] More details on this set are found in Appendix A.1.3.

3.4 INTERSPEECH 2012 Speaker Trait Challenge Set

For the INTERSPEECH 2012 Speaker Trait Challenge a yet larger set of baseline acoustic features has been proposed by the author of this thesis (Schuller et al. 2012a). At the same time, the systematic brute-forcing, i.e., applying all functionals to all LLDs, has been reduced. This means, some functionals, which—when applied to delta LLDs—do not describe meaningful signal statistics, have been removed from the set.

As the IS11 set, the INTERSPEECH 2012 Speaker Trait Challenge (IS12) set (cf. Table 3.4) contains two group of LLDs: group A with 58 energy, spectral, and cepstral LLDs, and group B with 6 source/excitation signal related LLDs. To the 58 LLDs of group A, 58 functionals are applied. To the delta LLDs of group A, 38 functionals are applied. To the LLDs of group B, 56 functionals are applied, and to the delta LLDs of group B, 36 functionals are applied. This results in a total of 6,120 features.

In addition to this (semi) brute-force combination of LLDs and functionals, the IS12 set contains a part of the temporal statistic descriptors from the IS11 set. These are based on voiced/unvoiced segments which are computed by applying the following functionals to the F_0 LLD (not to δ as in IS11): *Ratio of non-zero values* (ratio of non-zero values, i.e., percentage of voiced frames of total frames) and *segment length statistics* (minimum, mean, maximum, and standard deviation of the lengths of voiced segments, i.e., $F_0 > 0$ (cf. Sect. 2.4.2.9)). In contrast to the IS11 set, the

[2]According to Schuller et al. (2011a)—and the openSMILE configuration file—the IS11 set contains 4,368 features in total. This is also the size of the baseline feature vectors provided for the challenge. However, the duration of the segment is counted twice there, due to the way it was implemented in the openSMILE configuration file. Thus, the correct number of unique features in IS11 is 4,367.

Table 3.4 INTERSPEECH 2012 Speaker Trait Challenge (IS12) set

LLD	Functionals
Group A: (58)	Arithmetic* or positive arithmetic$^{\delta}$ mean,
Loudness,	Root-quadratic mean, flatness,
Modulation loudness,	Standard deviation, skewness, kurtosis,
RMS energy,	Quartiles 1–3,
ZCR,	Inter-quartile ranges 1–2, 2–3, 1–3,
RASTA auditory bands 1–26,	99th and 1-st percentile, range of these,
MFCC 1–14,	Relative position of max. and min. value,
Energy 250–650 Hz,	Range (maximum to minimum value),
Energy 1–4 kHz,	Linear regression slope* a, offset* b,
Spectral RoP .25, .50, .75, .90,	Linear regression quadratic error*,
Spectral flux and entropy,	Quadratic regression coeff.* a, b, c,
Spectral variance,	Quadratic regression quadratic error*,
Spectral skewness and kurtosis,	Temporal centroid*,
Spectral slope,	Peak mean value* and dist. to arithm. mean*,
Spectral harmonicity,	Mean* and std. dev.* of peak to peak distances,
Spectral sharpness (auditory).	Peak and valley range* (absolute and relative),
Group B: (6)	Peak-valley-peak slopes mean* and std. dev.*,
	Segment length mean^{A}, min.^{A}, max.^{A}, std. dev.^{A},
F_0 via SHS, Prob. of voicing,	Up-level time 25 %, 50 %, 75 %, 90 %,
Jitter (local and delta),	Down-level time 25 %, 50 %, 75 %, 90 %,
Shimmer,	Rise/fall time, left/right curvature time^{B},
logHNR (time domain).	Linear Prediction gain and coefficients 1–5.

Overview of Low-level Descriptors (LLDs) and functionals applied to these LLDs. Functionals marked with ^{A} and ^{B} are only applied to group A or B LLDs, respectively; functionals marked with * or $^{\delta}$ are not or only (respectively) applied to the delta LLDs. Details in Appendix A.1.4

total duration of the segment is *not* included as a feature. The decision to remove the duration has been made because the tasks in the challenge involve speaker trait recognition (Schuller et al. 2012a). Speaker trait is a long-term phenomenon, and thus the duration of a single utterance is irrelevant if the data-set contains mixed utterances with random lengths. With these 5 additional descriptors, the IS12 set totals **6,125 features**. More details on this set are found in Appendix A.1.4.

3.5 INTERSPEECH 2013 ComParE Set

At INTERSPEECH 2013, the first ComParE has been held (Schuller et al. 2013). It is an attempt to extend the idea of the previous challenges to manifold tasks of Computational Paralinguistics and continue the annual series of challenges at INTERSPEECH in a unified way under a common name (ComParE).

For the ComParE baseline feature set, the IS12 feature set has been slightly enlarged by approximately 200 features. Irregularites and redundancies have been removed, parameters have been tuned, and updates to some feature extraction algorithms have been made. As for IS12, two groups of LLDs are defined. To the 59 LLDs in group A, 54 functionals are applied and to the 59 delta LLDs of group A, 46 functionals are applied (see Table 3.5). To the 6 LLDs in group B, 39 functionals are applied, and to the 6 delta LLDs of set B, 39 functionals are applied. This results in a total of 6,368 features.

In addition to this (semi) brute-force combination of LLDs and functionals, the ComParE set contains the same 5 global temporal statistics as the IS12 set. These are based on voiced/unvoiced segments which are computed by applying the following functionals to the F_0 LLD: *Ratio of non-zero values* (ratio of non-zero values, i.e., percentage of voiced frames of total frames) and *segment length statistics* (minimum,

Table 3.5 INTERSPEECH 2013 Computational Paralinguistics ChallengE (Com- ParE) set

LLD	Functionals
Group A: (59)	Arithmetic^{A*,B} or positive arithmetic$^{A\delta,B}$ mean,
Loudness,	Root-quadratic mean, flatness,
Modulation loudness,	Standard deviation, skewness, kurtosis,
RMS energy, ZCR,	Quartiles 1–3,
RASTA auditory bands 1–26,	Inter-quartile ranges 1–2, 2–3, 1–3,
MFCC 1–14,	99th and 1-st percentile, range of these,
Energy 250–650 Hz,	Relative position of max. and min. value,
Energy 1–4 kHz,	Range (maximum to minimum value),
Spectral RoP .25, .50, .75, .90,	Linear regression slope^{A*,B} a, offset^{A*,B} b,
Spectral flux, entropy, variance,	Linear regression quadratic error^{A*,B},
Spectral skewness and kurtosis,	Quadratic regression coeff.^{A*,B} a, b, c,
Spectral slope,	Quadratic regression quadratic error^{A*,B},
Spectral harmonicity,	Temporal centroid^{A*,B},
Spectral sharpness (auditory),	Peak mean value^{A} and dist. to arithm. mean^{A},
Spectral centroid (linear).	Mean^{A} and std. dev.^{A} of peak to peak distances,
Group B: (6)	Peak and valley range^{A} (absolute and relative),
	Peak-valley-peak slopes mean^{A} and std. dev.^{A},
F_0 via SHS, Prob. of voicing,	Segment length mean^{A}, min.^{A}, max.^{A}, std. dev.^{A},
Jitter (local and delta),	Up-level time 25 %, 50 %, 75 %, 90 %,
Shimmer,	Rise time, left curvature time,
logHNR (time domain).	Linear Prediction gain and coefficients 1–5.

Overview of Low-level Descriptors (LLDs) and functionals applied to these LLDs. Functionals marked with ^{A} and ^{B} are only applied to group A or B LLDs (and deltas), respectively; functionals marked with * or $^{\delta}$ are not or only (respectively) applied to the delta LLDs. Details in Appendix A.1.5

mean, maximum, and standard deviation of the lengths of voiced segments, i.e., $F_0 > 0$ (cf. Sect. 2.4.2.9)). In total, the ComParE set contains **6,373 features**. More details on this set are found in Appendix A.1.5.

3.6 INTERSPEECH 2014 ComParE Set

The second ComParE is held at INTERSPEECH 2014. Due to the excellent universal performance of the 2013 ComParE feature set on various speech and music tasks (e.g., Weninger et al. 2013b) it was decided not to propose a new feature set for the follow up evaluation. Therefore, the INTERSPEECH 2014 ComParE baseline feature set is identical to the IS13 ComParE feature set (cf. previous section). In the 2014 ComParE, the feature set is evaluated on a cognitive load and a physical load task, i.e., classification of the respective load levels from voice recordings (Schuller et al. 2014).

3.7 Audio-Visual Emotion Challenge Sets

Next to the INTERSPEECH series of emotion and paralinguistics challenges, an Audiovisual Emotion Challenge (AVEC) has been established. The first challenge was held in 2011 (Schuller et al. 2011b). Audio-visual recordings from the SEMAINE database were used in the evaluations, divided into training, development, and a non-public test set. The same data-set was used again in 2012 where fully continuous dimensional emotion recognition on the data-set was performed (Schuller et al. 2012b). In 2013 a new data-set, featuring depression annotations and continuous emotion labels, was introduced (Valstar et al. 2013).

The feature set has remained nearly the same for all three challenges all based on the systematically brute-forced original 2011 feature set with 1,941 features (Schuller et al. 2011b). The feature set of the AVEC 2012 was reduced by 100 features that were found to carry very little information, as they were zero or close to zero most of the time, leaving 1,841 features (Schuller et al. 2012b). In the AVEC 2013 feature set the extraction of jitter and shimmer were improved, the spectral flatness was added to the set of spectral LLDs and the MFCCs 11–16 were included in the set in order for the features to better reflect speaker characteristics. As AVEC 2013 is the most complete set, the description of the set here and the details in Appendix A.1.6 are limited to this set.

As in the IS11, IS12, and ComParE sets, the AVEC 2013 feature set contain two groups of LLDs: group A with spectral, cepstral, and energy related LLDs, and group B with voicing related LLDs. Group A contains 32 LLDs, and group B contains 6 LLDs (see Table 3.6). A total of 43 functionals (see Table 3.6) is found in the set,

Table 3.6 2013 Audio-visual Emotion Challenge (AVEC) feature set

LLD	Functionals
Group A: (32)	Arithmetic* or positive arithmetic$^{\delta}$ mean,
Loudness,	Root-quadratic mean, flatness,
ZCR,	Standard deviation, skewness, kurtosis,
MFCC 1–16,	Quartiles 1–3,
Energy 250–650 Hz,	Inter-quartile ranges 1–2, 2–3, 1–3,
Energy 1–4 kHz,	99th and 1-st percentile, range of these,
Spectral RoP 0.25, 0.50, 0.75, 0.90,	Linear regression slope* a,
Spectral flux, entropy, flatness,	Linear regression linear error*,
Spectral variance, skewness, kurtosis,	Quadratic regression coeff.* a,
Spectral harmonicity,	Quadratic regression linear error*,
Spectral sharpness (auditory).	Peak mean value relative to arithm. mean*,
Group B: (6)	Mean* and std. dev.* of peak to peak dists.,
	Peak and valley range* (relative to mean),
F_0 via SHS, Prob. of voicing,	Peak-valley-peak slopes mean* and std. dev.*,
Jitter (local and delta),	Segment len. mean^{A}*, min/max.^{A}*, std. dev.^{A}*,
Shimmer,	Up-level time 25 %, 50 %, 90 %,
log$\dot{\text{H}}$NR (time domain).	Rise time,
	Linear Prediction gain^{A}* and coefficients 1–5^{A}*.

Overview of Low-level Descriptors (LLDs) and functionals applied to these LLDs. Functionals marked with ^{A} are only applied to group A LLDs (and deltas); functionals marked with * or $^{\delta}$ are not or only (respectively) applied to the delta LLDs. Details in Appendix A.1.6

however, only 42 are applied to the LLDs in group A, 19 to the delta LLDs of group A, 32 to the LLDs in group B, and 19 to the delta LLDs of group B. This combination results in 2,258 features. In addition to these, 10 temporal statistics are included: *Ratio of non-zero values* (ratio of non-zero values, i.e., percentage of voiced frames of total frames), mean, min., max., standard deviation of the voiced segment lengths (as in the IS12 and ComParE sets), mean, min., max., standard deviation of unvoiced segment lengths, and the duration of the input (in seconds). In total the AVEC 2013 audio baseline feature set contains **2,268 features**. More details on this set are found in Appendix A.1.6.

3.8 Geneva Minimalistic Acoustic Parameter Set

Very recently—in contrast to the large-scale, brute-forced feature sets, there have been efforts to propose a minimalistic feature set for affect processing and automatic voice analysis, which utilises knowledge from early analysis of emotional speech by psychologists (cf. Banse and Scherer 1996; Scherer 1986). While the large scale features sets such as the ComParE feature set (Sect. 3.5), contain a vast number of features and thus capture many acoustic properties reliably, interpreting the results in terms of relevant features is non-intuitive and sometimes impossible. Applications such as the fine grained control of emotionality in speech synthesis (cf. Schröder 2004; Schröder et al. 2010) require a deeper understanding of the mechanism of production and perception of emotion in humans. To reach this understanding, finding and interpreting relevant acoustic parameters is crucial.

Thus, based on previous findings in the area of speech analysis (cf. Banse and Scherer 1996; Juslin and Laukka 2003; Patel and Scherer 2013; Sundberg et al. 2011), a recommendation for a minimalistic standard parameter set for acoustic analysis of speech is proposed for this thesis. *Note:* The recommendation published and evaluated here is a preliminary version (0.9) of the final recommendation (1.0) as published in (Eyben et al. 2015).

The Geneva Minimalistic Acoustic Parameter Set (GeMAPS) (0.9) contains a compact set of 18 LLDs: **Pitch**, **Jitter**, **Shimmer**, **Loudness**, **Harmonics to Noise Ratio (HNR)**, **Spectral Slope**, **Alpha Ratio**, **Hammarberg Index**, **Formant 1–3 frequency** and **relative level**, **Formant 1 bandwidth**, and **Harmonic ratios** (H1–H2, H1–A3). In contrast to previous, brute-forced feature sets, no delta regression coefficients are computed from the LLDs in this set. More details on the LLDs in this set are found in Sect. A.1.7.

Arithmetic mean and **Coefficient of Variation** (Sect. 2.4.2.2) are applied as functionals to all 18 LLDs, yielding 36 parameters. To pitch and loudness the following 8 functionals are additionally applied: **percentiles 20, 50, and 80**, the **range of percentiles 20–80**, and the **mean and standard deviation of the slope of rising/falling signal parts**. This leads to 52 parameters. Additionally, 6 temporal features are included: the rate of loudness peaks per second (see Sect. 2.4.2.8), mean length and standard deviation of continuous voiced segments ($F_0 > 0.0$, see *Pitch* above), mean length and standard deviation of unvoiced segments ($F_0 = 0$; approximating pauses), and the rate of voiced segments per time (approximating the pseudo syllable rate). Overall, **58 features** are contained in this minimalistic parameter set.

This minimalistic set does not contain any cepstral parameters and only very few dynamic parameters (i.e., no delta regression coefficients and no difference features; only the percentage of rising and falling F_0 and loudness frames encapsulate some dynamic information). Dynamic parameters and especially cepstral parameters have proven highly successful in the modelling of affective states, e.g., by Schuller et al. (2007a), Schuller and Rigoll (2009), Marchi et al. (2012). Therefore, an *extension* to the minimalistic set is proposed which contains the following 10 extra LLDs: **Spectral Energy Proportions** (0–500 Hz and 0–1000 Hz), **MFCC 1–4**, **linear pitch**, **Formant 2–3 bandwidth**, and **spectral flux**. Arithmetic mean and standard deviation as functionals are applied to all of these 10 additional LLDs resulting in 20 additional features. To the linear scaled pitch the 8 additional functionals, which are also applied to the semitone scale pitch, are applied. Further, the equivalent sound level (LEq) is included. It is computed as the average Root Mean Square (RMS) energy ($\overline{E}_{rms}$), converted to dB by $LEq = 10 \log_{10}(\overline{E}_{rms})$. In total the extended Geneva Minimalistic Acoustic Parameter Set (eGeMAPS) (0.9) consists of **87 features**. More details on the features in both sets are found in Appendix A.1.7.

Since the GeMAPS parameter set is intented to become a widely adopted standard in the community, researchers are encouraged to shape the future of the parameter set recommendation together. Discussions, updates, and latest parameter recommendations can always be found on-line at http://www.audeering.com/research. A detailed description and evaluation of the final set (1.0) is published in (Eyben et al. 2015). In contrast to the preliminary set (0.9) as evaluated in this thesis, the final set differentiates between voiced and unvoiced parts when applying certain functionals to LLD.

3.9 Music Genre Sets

A feature set for Music Genre recognition has been proposed by the author of this thesis in (Eyben and Schuller 2010). It contains rhythmic features introduced by Schuller et al. (2007b) and a basic set of static supra-segmental features (functionals applied to LLDs). In total the set contains 1,351 acoustic descriptors, optimised for music signal type discrimination. 87 of these features are the rhythmic descriptors computed from Mel-frequency band envelopes as described by Schuller et al. (2007b) (and also summarised in Appendix A.1.8.1 in this thesis). The remaining 1,264 features are constructed by applying 14 statistical functionals to 40 generic LLDs and their deltas, and 12 functionals to 12 CHROMA Energy-distribution Normalised Statistics (CENS) (cf. Sect. 2.2.14.3 and Müller et al. (2005) for details) descriptors (see Table 3.7). More details on the statistic descriptors are found in Appendix A.1.8.1.

Two versions of this set are evaluated in this thesis: The *Genre Set*, which is the full set as described above, and the *Genre Set without rhythmic descriptors*, which consists only of the 1,264 statistical features.

3.10 Summary

12 acoustic feature sets were proposed for various tasks and challenges during the writing of this thesis. Table 3.8 shows the number of features in each of the 12 sets. Excluding the AVEC 2011 and AVEC 2012 sets, which are highly redundant with the AVEC 2013 set, 10 sets remain, which are systematically evaluated on 10 speech/singing voice tasks and 2 music style identification tasks in Chap. 6.

Table 3.7 Music Genre set, statistical descriptor sub-set

LLD	Functionals
Group A: (40)	Arithmetic mean,
Loudness,	Standard deviation, skewness, kurtosis,
F_0 envelope,	Quartiles 1–3 ^{A},
ZCR,	Inter-quartile ranges 1–2, 2–3, 1–3,
MFCC 1–12,	99th and 1-st percentile, range of these,
12 auditory spectral band energies,	Down-level time 25 %,
LSP 0–7,	Up-level time 75 %,
Spectral RoP .95,	Mean of peak amplitudes,
Spectral variance,	Number of peaks,
Spectral entropy, and centroid.	Distance of arithmetic mean to min. value^B.
Group B: (12)	
CENS 1–12.	

Overview of Low-level Descriptors (LLDs) and functionals applied to these LLDs. Functionals marked with ^{A} or ^{B} are only applied to group A or B LLDs (and deltas), respectively. Details in Appendix A.1.8

Table 3.8 Dimensionalities of proposed acoustic feature sets and Real-time factors (RTFs) for the extraction of a single feature vector of a 10 s audio segment and a 100 s audio segment on a standard desktop PC

Set	# features	Real-time factors	
		10 s	100 s
GeMAPS	58	0.047	0.044
eGeMAPS	87	0.049	0.046
IS09	384	0.015	0.012
IS10	1,582	0.044	0.041
IS11	4,367	0.040	0.037
IS12	6,125	0.039	0.035
ComParE	6,373	0.045	0.042
AVEC 2011	1,941	0.041	0.037
AVEC 2012	1,841	0.041	0.037
AVEC 2013	2,268	0.042	0.039
Genre (full)	1,351	–	–
Genre (w/o rhythmic)	1,264	0.048	0.043

Further, Table 3.8 shows the Real-time factors (RTFs) at which these sets[3] can be extracted on standard desktop PC (AMD64 (Phenom), 3.1 GHz, using 1 CPU core). The RTFs were computed by measuring the total execution time of the openSMILE toolkit for 100 extraction runs of 10 second speech files (16 kHz, mono) and 10

[3]The RTF for the rhythmic features was not evaluated, as they are not implemented in the open SMILE framework.

extraction runs of 100 s music files (16 kHz, mono). The execution time was then divided by the length of the audio multiplied by the number of runs to obtain the RTF. A RTF of 1.0 means that the extraction of the features takes as long as it takes to play the audio sample (i.e., processing requires the "real" time of the signal). A RTF smaller than 1.0 indicates that the processing is faster than the playback time. All feature sets proposed for this thesis can be extracted with a RTF of smaller than 0.05, which means that the feature extraction of up to 6 k features runs over 20 times faster than real-time. Most notably, the IS09 set is the fastest to extract with a RTF of 0.015. This is due to the absence of Voice Quality features and the simpler ACF based pitch algorithm in contrast to the SHS algorithm. Processing of the smaller segments is slower because it comes with more overhead (mainly parsing the configuration files and initialising all components in the current implementation).

References

R. Banse, K.R. Scherer, Acoustic profiles in vocal emotion expression. J. Personal. Soc. Psychol. **70**(3), 614–636 (1996)

A. Batliner, J. Buckow, R. Huber, V. Warnke, E. Nöth, H. Niemann, Prosodic Feature Evaluation: Brute Force or Well Designed? In *Proceedings of the 14th ICPhS*, vol 3, San Francisco, CA, USA, pp. 2315–2318 (1999)

A. Batliner, S. Steidl, B. Schuller, D. Seppi, T. Vogt, J. Wagner, L. Devillers, L. Vidrascu, V. Aharonson, N. Amir, Whodunnit—Searching for the most important feature types signalling emotional user states in speech. Comput. Speech Lang. **25**(1), 4–28 (2011)

A. Batliner, B. Möbius, Prosodic models, automatic speech understanding, and speech synthesis: towards the common ground?, in *The Integration of Phonetic Knowledge in Speech Technology*, ed. by W. Barry, W. Dommelen (Springer, Dordrecht, 2005), pp. 21–44

F. Eyben, K. Scherer, B. Schuller, J. Sundberg, E. André, C. Busso, L. Devillers, J. Epps, P. Laukka, S. Narayanan, K. Truong, The Geneva Minimalistic Acoustic Parameter Set (GeMAPS) for Voice Research and Affective Computing. IEEE Trans. Affect. Comput. doi:10.1109/TAFFC.2015.2457417

F. Eyben, B. Schuller, Music Classification with the Munich openSMILE Toolkit. In *Proceedings of the Annual Meeting of the MIREX 2010 community as part of the 11th International Conference on Music Information Retrieval (ISMIR)*, Utrecht, The Netherlands, August 2010. ISMIR. http://www.music-ir.org/mirex/abstracts/2010/FE1.pdf

P.N. Juslin, P. Laukka, Communication of emotions in vocal expression and music performance: Different channels, same code? Psychol. Bull. **129**(5), 770–814 (2003)

E. Marchi, A. Batliner, B. Schuller, S. Fridenzon, S. Tal, O. Golan, Speech, Emotion, Age, Language, Task, and Typicality: Trying to Disentangle Performance and Feature Relevance. In *Proceedings of the First International Workshop on Wide Spectrum Social Signal Processing (WS^{3}P 2012), held in conjunction with the ASE/IEEE International Conference on Social Computing (SocialCom 2012)*, IEEE Computer Society. pp. 961–968, Amsterdam, The Netherlands, September 2012

M. Müller, F. Kurth, M. Clausen. Audio matching via chroma-based statistical features. In *Proceedings of the 6th International Conference on Music Information Retrieval (ISMIR)*, pp. 288–295, London, UK, (2005)

S. Patel, K.R. Scherer, Vocal behaviour, in *Handbook of Nonverbal Communication*, ed. by J.A. Hall, M.L. Knapp (Mouton-DeGruyter, Berlin, 2013), pp. 167–204

A. Sadeghi Naini, M. Homayounpour, Speaker age interval and sex identification based on jitters, shimmers and mean MFCC using supervised and unsupervised discriminative classification

methods. In *Proceedings of the 8th International Conference on Signal Processing (ICSP)*, vol 1, Beijing, China, 2006. doi:10.1109/ICOSP.2006.345516
K.R. Scherer, Vocal affect expression: A review and a model for future research. Psychol. Bull. **99**, 143–165 (1986)
M. Schröder, *Speech and Emotion Research: An Overview of Research Frameworks and a Dimensional Approach to Emotional Speech Synthesis*, volume PHONUS 7 of *Research Report of the Institute of Phonetics, Saarland University*. Ph.D thesis, Institute for Phonetics, University of Saarbrücken, 2004
M. Schröder, F. Burkhardt, S. Krstulovic, Synthesis of emotional speech, in *Blueprint for Affective Computing*, ed. by K.R. Scherer, T. Bänziger, E. Roesch (Oxford University Press, Oxford, 2010), pp. 222–231
B. Schuller, A. Batliner, D. Seppi, S. Steidl, T. Vogt, J. Wagner, L. Devillers, L. Vidrascu, N. Amir, L. Kessous, V. Aharonson, The Relevance of Feature Type for the Automatic Classification of Emotional User States: Low Level Descriptors and Functionals. In *Proceedings of INTERSPEECH 2007*, ISCA .pp. 2253–2256, Antwerp, Belgium, August 2007a
B. Schuller, F. Eyben, G. Rigoll, Fast and Robust Meter and Tempo Recognition for the Automatic Discrimination of Ballroom Dance Styles. In *Proceedings of the ICASSP 2007*, IEEE. vol I, pp 217–220, Honolulu, HI, USA, April 2007b
B. Schuller, A. Batliner, S. Steidl, F. Schiel, J. Krajewski, The INTERSPEECH 2011 Speaker State Challenge. In *Proceedings of INTERSPEECH 2011*, ISCA. Florence, Italy, pp. 3201–3204 August 2011a
B. Schuller, M. Valstar, F. Eyben, G. McKeown, R. Cowie, M. Pantic, AVEC 2011—The First International Audio/Visual Emotion Challenge, in *Proceedings of the First International Audio/Visual Emotion Challenge and Workshop, AVEC 2011, held in conjunction with the International HUMAINE Association Conference on Affective Computing and Intelligent Interaction (ACII) 2011*, vol. II, ed. by B. Schuller, M. Valstar, R. Cowie, M. Pantic (Springer, Memphis, TN, USA, October 2011b), pp. 415–424
B. Schuller, G. Rigoll, Recognising Interest in Conversational Speech—Comparing Bag of Frames and Supra-segmental Features. In *Proceedings of INTERSPEECH 2009*, ISCA pp. 1999–2002, Brighton, UK, September 2009
B. Schuller, G. Rigoll, M. Lang, Hidden Markov Model-based Speech Emotion Recognition. In *Proceedings of the ICASSP 2003*, IEEE. vol 2, pp. II 1–4, Hong Kong, China, April 2003
B. Schuller, S. Steidl, A. Batliner, F. Burkhardt, L. Devillers, C. Müller, S. Narayanan, The INTERSPEECH 2010 Paralinguistic Challenge. In *Proceedings of INTERSPEECH 2010*, ISCA. Makuhari, Japan, pp. 2794–2797 September 2010
B. Schuller, S. Steidl, A. Batliner, J. Epps, F. Eyben, F. Ringeval, E. Marchi, Y. Zhang, The INTERSPEECH 2014 computational paralinguistics challenge: Cognitive and physical load. In *Proceedings of the INTERSPEECH 2014*, ISCA. Singapore, 2014. (to appear)
B. Schuller, S. Steidl, A. Batliner, F. Jurcicek, The INTERSPEECH 2009 Emotion Challenge. In *Proceedings of INTERSPEECH 2009*, Brighton, UK, pp. 312–315 September 2009
B. Schuller, S. Steidl, A. Batliner, E. Nöth, A. Vinciarelli, F. Burkhardt, R. van Son, F. Weninger, F. Eyben, T. Bocklet, G. Mohammadi, B. Weiss, The INTERSPEECH 2012 Speaker Trait Challenge. In *Proceedings of INTERSPEECH 2012*, ISCA. Portland, OR, USA, September 2012a
B. Schuller, M. Valstar, R. Cowie, M. Pantic, AVEC 2012: the continuous audio/visual emotion challenge—an introduction, in *Proceedings of the 14th ACM International Conference on Multimodal Interaction (ICMI) 2012*, ed. by L.-P. Morency, D. Bohus, H.K. Aghajan, J. Cassell, A. Nijholt, J. Epps (ACM, Santa Monica, CA, USA, October 2012b), pp. 361–362
B. Schuller, S. Steidl, A. Batliner, A. Vinciarelli, K. Scherer, F. Ringeval, M. Chetouani, et al., The INTERSPEECH 2013 Computational Paralinguistics Challenge: Social Signals, Conflict, Emotion, Autism. In *Proceedings of the INTERSPEECH 2013*, ISCA. Lyon, France, pp. 148–152 2013
B. Schuller, A. Batliner, *Computational Paralinguistics: Emotion, Affect and Personality in Speech and Language Processing* (Wiley, Hoboken, 2013), p. 344. ISBN 978-1119971368

J. Sundberg, S. Patel, E. Bjorkner, K .R. Scherer, Interdependencies among voice source parameters in emotional speech. IEEE Trans. Affect. Comput. **2**(3), 162–174 (2011). doi:10.1109/T-AFFC.2011.14. ISSN 1949-3045

M. Valstar, B. Schuller, K. Smith, F. Eyben, B. Jiang, S. Bilakhia, S. Schnieder, R. Cowie, M. Pantic, AVEC 2013—The Continuous Audio/Visual Emotion and Depression Recognition Challenge. In *Proceedings of the ACM Multimedia 2013*, CM. Barcelona, Spain, October 2013

D. Ververidis, C. Kotropoulos, Emotional speech recognition: Resources, features, and methods. Speech Commun. **9**, 1162–1181 (2006)

F. Weninger, F. Eyben, B. W. Schuller, M. Mortillaro, K. R. Scherer, On the Acoustics of Emotion in Audio: What Speech, Music and Sound have in Common. *Frontiers in Psychology*, 4(Article ID 292): 1–12, May 2013b. doi:10.3389/fpsyg.2013.00292

Chapter 4
Real-time Incremental Processing

In Chap. 2 a large set of acoustic features based on Low-level Descriptors (LLDs) and statistical functionals has been introduced and the most common modelling methods were summarised. The features and the modelling methods used in this thesis have been selected with the goal of on-line processing in mind, however, most of them are all general methods that are suitable both for on-line and off-line processing. This section deals specifically with the issues encountered in on-line processing. Instead of on-line processing, the terms real-time processing or incremental processing are sometimes used synonymously, although they are not the same. Thus, in order to avoid confusions, these terms are defined here to clarify the differences between them:

Real-time processing refers to processing of data in real-time. This implies two things: (a) the processing of the data requires less time than the duration of the data, i.e., the time it will take to record or play-back the data at normal speed, and (b) the processing lag is as low as possible, ideally zero. In other fields, such as computer operating systems, real-time also refers to scheduling requirements, e.g., that events are processed exactly at the time they occur or within a certain deadline (Ben-Ari 1990, p. 164). Here, the definition of real-time will be restricted to points (a) and (b), where (a) *must* be enforced, and for (b) the lag must have a fixed upper bound and should be as low as possible, which is roughly between 100 ms and 5–10 s for speech and music analysis tasks.

On-line processing refers to processing of a continuous live stream of data while data are recorded or transmitted, i.e., computing and returning results continuously before the end of the data stream has been seen. This is in contrast to off-line processing, where a whole segment is recorded (e.g., a speech utterance or a song) before it is processed. In on-line processing only single pass processing algorithms can be used, while off-line processing allows for multiple processing passes. Algorithms with multiple processing steps or algorithms which use context (past or future data) are not suitable for on-line processing without major modifications and, even then,

F. Eyben, *Real-time Speech and Music Classification by Large Audio Feature Space Extraction*, Springer Theses, DOI 10.1007/978-3-319-27299-3_4

they are (in most cases) not able to give the same results as in off-line processing. For on-line processing, real-time processing is implicitly required. If the processing of the incoming data is too slow, the lag between data input and output will constantly increase, soon making the system unusable. To implement on-line processing, generally data are aggregated in some (rather short) buffer and a result is generated every time the buffer is full. Depending on the application only the data in the buffer is used to produce the result, or potentially also more past data or the previous results can be used. The lag between inputs and outputs is determined by the buffer size. For affect recognition, for example, a buffer size of $\approx$1–4 s is suitable (e.g., Eyben et al. 2012). I.e., in the most naive on-line processing approach, audio data would be recorded for, e.g., 2 s, then features would be extracted and a classification would be performed, returning the result. The total lag of the system then is given by the buffer size (2 s) and the time required for feature extraction and classification.

Incremental processing is a more optimised implementation of on-line processing, which tries to minimise the lag between data input and analysis result output. The trick is to process incoming data as soon as possible, while waiting for new data to arrive. In the above example of affect processing, the LLDs can be computed from the signal already before the 2 second buffer of audio data is filled because for computing a LLD vector only 20–60 ms (depending on the frame size) of audio data are required. Thus, all LLDs are already computed when the 2 second audio segment has been recorded and the lag is reduced by the time required to compute the LLDs. If, for instance, the computation of the LLDs requires 0.2 s, the lag is reduced by this amount. Moreover, a preliminary result can be computed from a 1 second buffer, for example, and another, final result would be then computed from the full 2 second buffer. This reduces the lag further, but possibly at the cost of accuracy of the preliminary result.

The remaining parts of this section show important issues of on-line processing in speech and music analysis in detail, such as input segmentation issues (Sect. 4.1) and feature extraction issues (Sect. 4.2). Next, the implementation and architecture of the incremental on-line processing which has been implemented for this thesis in the open-source feature extraction toolkit openSMILE (Sect. 4.3) is described. Finally, as an example, a fully continuous and incremental approach to automatic speech emotion recognition is described in Sect. 4.4.

4.1 Segmentation Issues

The specific issues encountered with incremental processing in various speech and music analysis tasks depend on the task set-up and its requirements, especially the temporal level the analysed phenomenon occurs on. In most applications multiple hierarchical (temporal) levels of processing are used.

For speech, the most important first step is to identify segments in a continuous audio stream, which actually contain speech. This is well-known as Voice Activity

Detection (VAD) or Speech Activity Detection (SAD). A robust VAD method for neutral and emotional speech, which was developed for this thesis, is described in Sect. 5.1. It shows excellent results in high noise and reverberation levels at low latency. The low lag required is a challenge for the VAD algorithm in an incremental on-line processing framework. Some algorithms, such as simple energy based VAD, require thresholds to be tuned over one or more utterances. They cannot be used for on-line processing, for example. For more details see Sect. 5.1.

Once the speech segments and the speech pauses have been identified, segments need to be either combined (if too short) or split into sub-segments (if too long) in order to match the optimal analysis unit length. The choice of this length is a highly debated topic (cf. Schuller and Rigoll 2006; Batliner et al. 2010; Eyben et al. 2012), and the answer depends strongly on the type of data and labels. For instance, some pathological vocal chord disorders may be evident from only single vowels (cf. Parsa and Jamieson 2001), while speaker traits like age, or depression are only evident from multiple utterances and sometimes are also reflected in the linguistic content rather than only in the acoustic content. Emotion lies somewhat in between—very strong and prototypical emotions can be expressed in single words, while others are only evident in utterances which contain a handful of words (cf. Batliner et al. 2010). Finding the optimal unit of analysis is still an active area of research, especially for affect recognition (Schuller and Rigoll 2006; Schuller et al. 2007a; Busso et al. 2007; Mower and Narayanan 2011). As stated by Zeng et al. (2009), the audio segmentation is one of the most important issues for real-life use-cases, but it has been "largely unexplored so far". This thesis does not claim to solve this issue, but shall highlight various constraints that arise in the context of real-time and incremental processing, and propose a suitable segmentation algorithm for these scenarios.

4.1.1 On-Line Segmentation

For fast on-line processing, linguistic analysis units such as words, phrases, etc. are unsuitable as long as an Automatic Speech Recognition (ASR) system has to be used to perform the necessary segmentation. The delay introduced is too high for most on-line applications and the computing resources required by the ASR reduce the resources available for the paralinguistic analysis of speech, for example. Thus, for on-line systems, a low-resource, probabilistic pause based segmentation method is suggested in this thesis: A range for the length of speech units is defined, ranging from 1–10 s for affect recognition, for example. The algorithm starts in the 'pause' state. If the VAD function (a generic speech activity function returning a probability value is assumed here) crosses the voicing threshold for at least N_{pre} consecutive frames, the start of a speech segment is assumed at the first frame above the threshold. The end of the speech segment is detected when the VAD function falls below the threshold for N_{post} consecutive frames. An end is not detected if the current segment length is smaller than the minimum allowed segment length L_{min}. If the current segment length exceeds the maximum allowed segment length L_{max}, N_{post} is linearly reduced down to 0 up to a final maximum length $L_{max,grace}$. Thus, between L_{max} and $L_{max,grace}$ splitting at short pauses becomes more and more likely, and at $L_{max,grace}$, an end of the current speech segment is always defined, even if the VAD function is above the threshold. In that case, the next segment will start at exactly the next frame, if the current segment is followed by at least N_{pre} consecutive frames above the threshold. The length of the ith speech segment detected with this method is denoted by $N_{seg}^{(i)}$ in the following.[1]

4.1.2 Incremental Segmentation

To avoid problems which occur when long utterances are force cut to a maximum length as described above and to decrease the maximum lag (which is the maximum segment size in the above method), the above on-line segmentation has been extended by an incremental segmentation algorithm (published by the author of this thesis in Eyben et al. 2012): A window length N_{win} and a window step $N_{step} \leq N_{win}$ (both in frames) are defined in addition to the minimum and maximum length parameters above. The corresponding variables, which express these lengths in seconds are denoted as L_{step} and L_{win}, respectively. Speech segments are detected with the same on-line method as above, but split into analysis sub-segments. Each sub-segment is labelled with an index pair (i, j), where i is the index of the master segment, and j is the index of the sub-segment within the master segment. The first sub-segment $(j = 0)$ starts at the beginning of the ith master segment and is of maximum length N_{win} or $N_{seg}^{(i)}$, whichever is lower. The following sub-segments start at $j \cdot N_{step}$ as long as $j \cdot N_{step} < N_{seg}^{(i)}$ and have a corresponding length of N_{win} or $N_{seg}^{(i)} - j \cdot N_{step}$, whichever is lower. The proposed method was used in the SEMAINE system (Schröder et al. 2012) and for the first two Audio-visual Emotion Challenge (AVEC) baseline evaluations (Schuller et al. 2011a, 2012a). A typical choice for L_{step} is one second; for L_{win} typical values are between 2 and 5 s.

Two *variations* of this incremental approach are possible, which can be both combined arbitrarily:

1. to reduce the initial response lag, the window length N'_{win} grows incrementally up to the maximum length N_{win} with the step size N_{step}; thus, the sub-segment at $j = 0$ has the length N_{step}, at $j = 1 \ldots \lfloor \frac{N_{win}}{N_{step}} \rfloor$ the length $(j + 1)\, N_{step}$ and at $j > \frac{N_{win}}{N_{step}}$ the length N_{win},
2. and/or the maximum window length N_{win} can be set to infinity, i.e., the fixed maximum length constraint removed—the maximum sub-segment length will then correspond to the current segment length $N_{seg}^{(i)}$, i.e., the last sub-segment contains the complete ith segment.

[1] In openSMILE this behaviour is implemented in the `cTurnDetector` component.

With the above approach the segment length can be controlled, however, it remains variable within the specified bounds and excludes long speech pause segments. For some applications a constant segment length might be required, or for music, for example, a segmentation based on beats or bars should be chosen.[2]

4.2 Feature Issues

In an incremental processing framework, certain constraints are put on feature extraction algorithms. Most important, the processing delay (lag) must be as low as possible, ideally only one audio frame. Consequently, multi-pass processing is not possible, i.e., all audio features used in an incremental system must be extractable in a single processing pass. For instance, features, which are centred to zero mean for one or more utterances cannot be used because these would require the mean value of the feature to be computed in a first pass, and the mean subtraction to be performed in a second pass. Such a normalisation would be applied to pitch, for example, to adapt to speaker specific pitch ranges. For incremental processing a different strategy has to be applied for such normalisations, i.e., the mean values have to be computed on-the-fly and updated with every new incoming value, and then immediately applied. These normalisation issues are discussed in more detail in Sect. 5.2.

Another issue is post-processing of formant and pitch contours (cf. Sects. 2.2.8 and 2.2.11.5, respectively). Thereby typically the best path over the full input is found – given some side constraints such as smoothness or the most likely range—and chosen as the final contour. In incremental processing, it is not possible to wait for all input to be complete. Neither is it feasible in this case of Viterbi based post-smoothing, for example, to update the best path decision with every frame, as in the case of an incrementally updated mean value. This would on the one side involve a memory overhead, as N best paths, which are constantly growing, need to be stored, and on the other side would also introduce long delays, as it cannot be predicted at which point in the past the paths will merge to a single path (and thus a decision on the best path can be made for a frame at time $(t - \tau)$—cf. Sect. 2.2.11.5 for details).

The features used and described in this thesis (Sect. 2.2) have all been designed to be extractable in incremental processing, or suitable modifications have been made to limit the processing lag to an acceptably small, constant maximum lag. As LLDs are extracted frame by frame anyhow, virtually all LLDs can be extracted incrementally—except for global post-processing and smoothing applied to the raw LLDs as discussed above. Supra-segmental features (cf. Sect. 2.4), however, are computed from LLDs on a higher temporal level. For on-line processing, this level is the most important factor which determines the lag of the overall processing. Segments

---

[2] Incremental segmentation of music into beats and bars is not part of this thesis. An on-line segmentation approach for music has not been investigated in this thesis, however, an off-line segmentation method basing on the beat tracker presented by Schuller et al. (2007b) and Eyben et al. (2007) has been used.

must be of finite length and should ideally be small or at least matching the processing delay requirements of the task at hand. The previous section (Sect. 4.1) has dealt with these issues of segmentation.

Some feature extraction algorithms might require certain higher level knowledge to change their extraction behaviour based on it. For instance, features could only be extracted from vowels, when a vowel/non-vowel classifier is employed, or extracted for certain word classes, if a full blown ASR system combined with a part-of-speech tagger is present. In the case of music analysis, features such as Pitch Class Profiles (Sect. 2.2.14.2) should be extracted from beat or bar segments (cf. Schuller et al. 2008), for example, to obtain a better spectral accuracy. In any case, the limiting factor for incremental and real-time processing is the computational complexity and delay introduced by the secondary processing (e.g., ASR or beat tracker). Generally such features are suitable for being used in on-line processing, however, they are not considered in this thesis because it would be outside of the scope of this thesis to properly deal with the overhead introduced by combining such algorithms with the idea of incremental processing.

4.3 Architecture of the openSMILE Framework

For this thesis, extraction algorithms for all of the acoustic features and functionals described in Sects. 2.2 and 2.4 have been developed from scratch in C++. The main goals thereby were the development of an efficient, incremental framework and the extensive coverage of a wide range of descriptors. The code has been released by the author as an open-source toolkit called openSMILE[3] (Eyben et al. 2010a, 2013a). The toolkit has been used extensively to provide high quality baseline features for the series of international evaluation challenges at INTERSPEECH (see Chap. 3 and Schuller et al. 2009b, 2010, 2011b, 2012b, 2013). Alone Eyben et al. (2010a) has been cited over 250 times at the time of writing.[4] From these citations, more than 160 are publications from groups around the world which are not affiliated with the author's team. openSMILE has won an honourable mention award twice at the ACM Multimedia open-source software competition[5] (Eyben et al. 2010a, 2013a).

This section describes the software architecture and the incremental data-processing framework of openSMILE (Eyben et al. 2010b, 2013b) in detail as published in the *openSMILE book*[6] (Eyben et al. 2010c, 2013c) by the author of this thesis. openSMILE's core architecture is designed in a modular way: Feature extractors are composed of one or more components and each component has a

[3] Available at: http://opensmile.audeering.com/.

[4] According to Google scholar citations.

[5] http://sigmm.org/Resources/software/ossc.

[6] Available for the latest version at http://www.audeering.com/research-and-open-source/files/openSMILE-book-latest.pdf.

clearly defined interface, both for data and for parameter settings. A core framework, which connects the components, is built of three parts:

1. a *commandline parser* and a *configuration manager* parse commandline options and configuration files and provides the initial configuration parameters to the respective components,
2. a *component manager* takes the role of a "scheduler" and manages the component execution and asynchronous communication between components (messages) as well as pausing and resuming of the processing, and
3. a *data-memory* efficiently manages the incremental data-flow between the components.

The component manager is the most important component, as it controls the order of execution of the data processing modules (components). It is responsible of instantiating, configuring, and executing all components. The following gives a software architectural description of the interplay between component manager, the component classes, and the data-memory:

The components instantiated by the component manager (data processing components) are all descendants of the `cSmileComponent` class. They have two basic means of standardised communication: (a) directly and asynchronously, via smile messages, and (b) indirectly and synchronously via the data memory.

Method (a) is used to send out-of-line data, such as events and configuration changes directly from one smile component to another. Classifier components, for example, send a 'classificationResult' message, which can be received by other components (esp. custom plug-ins), to change their behaviour or send the message to external sources. More details on the smile messaging architecture are found in Sect. 4.3.2.

Method (b) is the standard method for handling data-flow (audio input and features) in openSMILE. Thereby a *data source* component produces a frame (vector) of data and writes it to the *data memory* to a named location (referred to as level or slot). A *data processor* reads this frame, applies some processing algorithm to it, and writes a modified frame back to a different location in the data memory. This step is usually repeated for multiple data processors. Finally, a data sink reads the frame and passes it to an external handler (file, network stream) or interprets (classifies) it in some way and sends the result as a message (Method (a)) to another component or displays it on the console, for example. The advantages of passing data in this way via the data memory are that multiple components can read the same data, and data from past frames can be stored efficiently in a central location for later use. This eliminates the need to compute the Fast Fourier Transformation (FFT) twice if both Mel-Frequency Cepstral Coefficients (MFCCs) and spectral roll-off points, for example, shall be extracted. This data-flow architecture is described in more detail in Sect. 4.3.1.

A set of source, processing, and sink components as well as their interconnections can be selected and instantiated via a text-based configuration file. Parameters of each component can also be set in this same file. To enable efficient on-line processing, openSMILE requires the parameters chosen for all algorithms, and especially the frame sizes and data-flow connections between components to remain constant throughout the run-time. This constraint enables all settings to be loaded and verified once, buffers of the right size to be allocated, and models files, parameter files, and input/output files to be loaded or opened at start-up. This is implemented by three distinct execution states, the first two involving setup of the system and the data-processing channels, and the third involving the actual processing loop:

Pre-config phase Command-line options and the configuration file(s) is/are parsed,

Configuration phase The component manager is created and instantiates all components referenced in the configuration file. Components do their individual initialisation (loading of files, creating and initialising parameters and models, optimisations to speed up computations, etc.) and allocate their I/O requirements in the data-memory. At the end of the configuration phase the ideal minimal memory requirements for the system can be allocated in the data-memory.

Execution phase When all components have been initialised successfully, the component manager starts the main execution loop.[7] In one iteration of the execution loop, the component manager runs every component once. For this purpose, every component has a single `tick()` method, which implements the main incremental processing functionality and reports on the status of the processing via its return value. The component checks for the availability of new input data (which is fully flexible and can be samples, frames, segments, etc.) and processes the new chunk of data, if available. The loop is continued as long as at least one component's tick method returns a status which indicates that data has been processed by this component.

If all components have indicated that they do not have any data to process, it can be safely assumed that no more data will arrive and the end of the input has been reached—which could be the end of an input file in the case of off-line processing, or stopping/pausing of the recording in on-line processing. In that case, the component manager signals the End-of-input (EOI) condition to the components by running one final iteration of the execution loop with an end-of-input flag set. After that, the component manager runs the execution loop again, until all components again report that no data have been processed. This second phase is referred to as end-of-input processing. It is only important for off-line processing, e.g., to compute features from the last (but incomplete) frames, to mean normalise a complete sequence, or to compute functionals from a complete sequence. Since version 2.0, openSMILE supports multiple iterations of this overall loop, i.e., the normal processing can be run again for an arbitrary number of times in a non EOI state after the execution loop has been run in a EOI state. This obviously is unsuitable for on-line processing, but brings openSMILE's algorithms and efficiency also to multi-pass batch and off-line processing.

[7] Also referred to as tick-loop in the code.

4.3.1 Incremental Processing

Figure 4.1 shows the overall incremental data-flow architecture of openSMILE, where the data memory is the central link between all components that process data. Three types of data processing components exist: the *data source* components which write data to the memory from external sources, such as devices, files, or streams; the *data processor* components which retrieve data from the data memory, modify them, and copy them back to the memory to a different location; and the *data sink* components which can only read data from the memory (and send it to devices, files, classifiers, etc.).

The incremental data-flow in openSMILE is handled by the `cDataMemory` component. This component manages multiple data memory 'levels' internally. These levels are independent storage locations, to which data can be written to by exactly one component and data can be read from by an arbitrary number of components. From the outside (the component side) the levels appear to be a $N \times \infty$ matrix, with N rows, whereby N is the frame size. Components can access frames (=columns) at any location in this virtual matrix (virtual index). Internally this matrix is either organised as a ring buffer (for on-line processing) with a predefined length or a buffer which grows dynamically to the needed length. If the matrix is internally represented by a ring-buffer, a write operation only succeeds if there are empty frames

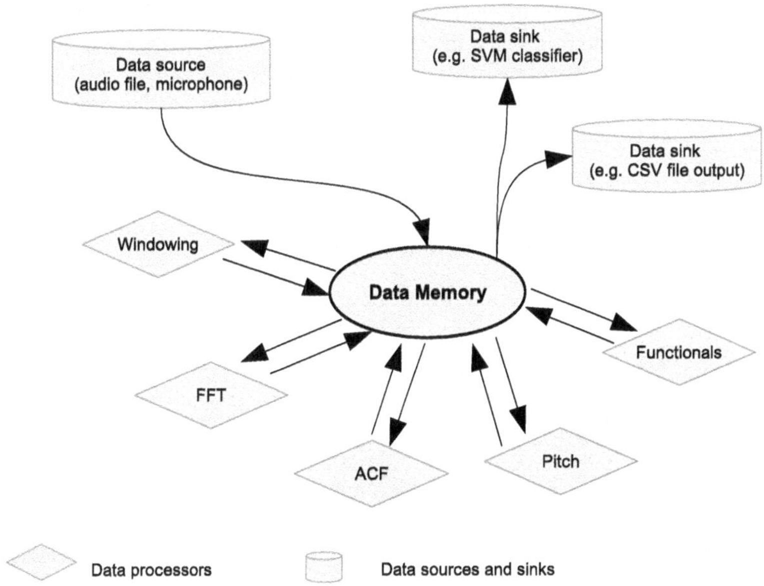

Fig. 4.1 openSMILE's basic data-flow architecture with openSMILE's generic component types. A central data memory to which data source components feed data, data processors modify the data, and data sink components pass the data on to classifiers, outputs, or external processing (e.g., a user interface); cf. Eyben et al. (2010a)

in the buffer, and a read operation only succeeds if the referred frame index lies no more than the ring buffer size in the past and has not yet been overwritten by new data.

In the case of a fixed length (non-ring) buffer, a write will always succeed as long as the referenced index is smaller than the buffer size. A read will succeed if the read position index is greater equal 0 and smaller than or equal to the highest index where data has already been written to. In case of a variable size buffer (i.e., growing dynamically) a write will succeed always, except when there is no memory left. A read will succeed under the same conditions than for the fixed buffer case.

In order to allow for a robust, synchronised, and memory efficient data-flow with ring-buffers, the data memory must track to where data have already been written to and which data have already been read. To achieve this, the data memory performs frame accounting on the internal buffer matrix: A writer pointer p_w, that stores the location to which the next sample/frame is to be written, and a set of read pointers $p_{r,i}$ (with $i = 0 \ldots N_r - 1$, and N_r representing the number of components which read from the current level) is defined for this purpose. The pointers always store a virtual index, i.e., the index in an infinite[8] buffer starting at element 0 regardless of the actual internal organisation of the buffer (cyclic/ring, fixed-length, or dynamic length).

The write pointer is updated after every successful write operation to the index n with a `max` operation: $p'_w = max(p_w, n + 1)$. The read pointers for each reader are updated similarly using the same rule: $p'_{r,i} = max(p_{r_i}, n + 1)$. For calculation of the free space in the buffer, the lowest read index $p_r^{(min)} = min(p_{r,i})$ is considered. With the buffer size N_b, the free space in a non ring buffer is given by $N_{free} = N_b - p_w$, and the number of elements available in the buffer is defined logically as $N_{avail} = p_w$. For a ring buffer, the following equations apply (ring buffer is denoted in the symbols by $\overline{N}$):

$$\overline{N}_{free} = N_b - \left(p_w - p_r^{(min)}\right), \tag{4.1}$$

$$\overline{N}_{avail} = max\left(N_b, p_w\right), \tag{4.2}$$

$$\overline{N}^{(i)}_{avail} = p_w - p_{r,i}, \tag{4.3}$$

where $\overline{N}^{(i)}_{avail}$ is the number of *unread/new* frames that are available to read for the reader i, while $\overline{N}_{avail}$ is the maximum number of frames available to read.

The ring-buffer based incremental processing is illustrated by a simplified example—which does not always use real components nor realistic frame sizes—in Fig. 4.2. Three exemplary levels are contained in this setup: wave (top), frames (middle), and energy (bottom). A `cWaveSource` component saves data samples (scalar values, frame size equal to 1) to the first level (wave). The write indices in the levels are indicated by a bold arrow. A `cFramer` produces frames of size 3 from the wave samples (non-overlapping) and saves these frames to the next level (frames). A `cEnergy` component extracts Root Mean Square (RMS) and logarithmic energy

[8] In practice limited by the range of the 'long' data-type (32-bit or 64-bit).

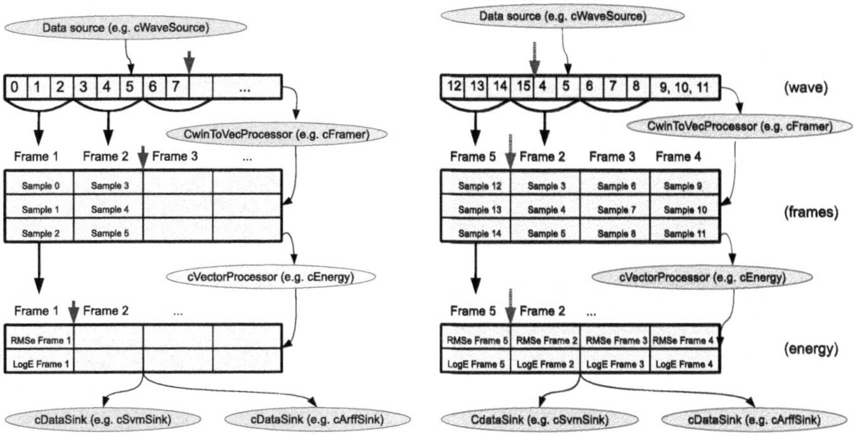

Fig. 4.2 Exemplary (simplified) incremental processing with ring-buffers as implemented in openSMILE. The plot shows partially filled ring-buffers (*left*) and filled ring-buffers with warped write pointers (*right*). Write pointers are shown by *red arrows* on the *top left corner* of the frame column; read pointers are *not* shown in order to keep the diagrams simple; cf. Eyben et al. (2010a)

features from the frames and saves them to the following level (energy). In Fig. 4.2 (right) the buffers have been filled once, and the write pointers have been warped. Data which are more than $N_b = 4$ frames in the past have been overwritten.

The *core concept* of incremental processing in openSMILE is that each component constantly checks, i.e., in each execution loop iteration, if enough new data which can be processed, is available, and—if so—immediately processes this data and makes it available for the next component(s) in the data-flow pipeline. This way, a new data sample induced into the pipeline by a data source will immediately propagate through the processing pipeline, as deep and as quickly as it is possible. In the example in Fig. 4.2 the `cFramer` component requires three input frames to compute one output frame, while the `cEnergy` component gives an output for every input frame. If it is now assumed that the `cWaveSource` component writes only a single sample to the 'wave' level during each iteration, the `cFramer` component as well as the components reading the output of the `cFramer` component are able to process data only in every third iteration.

This concept can be extended to higher levels of features without limitation. This makes the incremental processing architecture proposed for this thesis very flexible and powerful as it allows to build arbitrary high hierarchies of feature summaries. Figure 4.3 shows a simplified example for incremental computation of higher order features. Functionals (mean and variance) computed over two frames (overlapping) of the energy features (from the previous example) are extracted and saved to a functionals level. From the data in this level, for example, again functionals could be computed for the next higher level of feature summaries. In the given example, the data is read by two data sink components, where one writes the data to a Attribute Relation Feature Format (ARFF) file and the other classifies the data with a Support Vector Machine (SVM).

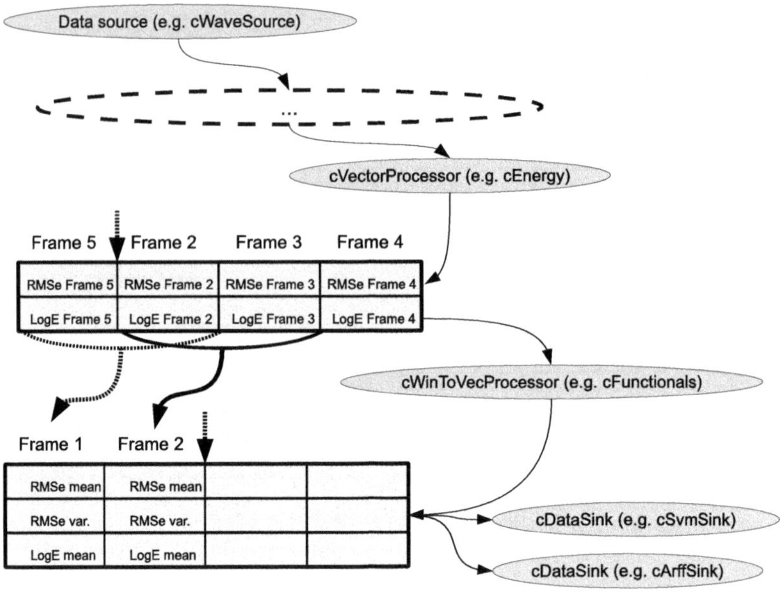

Fig. 4.3 Simplified example of incremental computation of higher-level features and feature summaries (e.g., statistical functionals). The *upper block* is left out in this figure (*dash-lined circle*), it is shown in Fig. 4.2; cf. Eyben et al. (2010a)

The size of the ring buffers of subsequent data memory levels in the processing pipeline must be set correctly according to the processing block size in order to ensure smooth processing without deadlocks. The *block size* thereby is the size of the block a reader or writer reads/writes from/to the data memory at once, i.e., in one iteration. In the above example the read block size of the functionals component is 2 because it reads two pitch frames to produce one output frame. Similarly, in the previous example the read block size of the `cFramer` component was 3. The ring buffer of the level with energy features thus must hold at least 2 frames, otherwise the functionals component will never be able to read its required input from that level and it will block the processing. This will cause the component manager to stop the processing loop because the pitch component will not be able to write new data to the pitch level, thus it will not read data from the frames level, which causes the framer component to be blocked. This process continues until all components are in a state where they do not process any data.

In order to avoid this kind of problems in complex data-flow setups, the adjustment of the buffer size is handled automatically: Reader and writer components must register with the data memory during the configuration phase and must publish their required maximum read and write block sizes, i.e., the size of the largest block the component will ever attempt to read, or to write, respectively. This might not be possible in all cases. In such cases the buffer sizes of the input and output levels have to be constrained manually in the configuration. The minimal buffer size bs_{min} is computed based on these values of minimum and maximum read/write block sizes:

$$bs_{min} = \begin{cases} 2n_w + 1 & \text{if } n_r \leq n_w \\ n_r + 2n_w & \text{if } n_r > n_w \end{cases}, \quad (4.4)$$

where n_r is the maximum read block size (from the current level) and n_w is the maximum write block size (to the current level).

4.3.2 *Smile Messages*

The data-flow framework described in the previous section enables the exchange of sequential, synchronous data between components. However, sometimes it might be required to exchange small amounts of asynchronous data between components, e.g., to change parameters at run-time, or to send events from one component to another.

To fulfil these needs, a messaging system has been implemented in openSMILE. As the messages are intended only for fast, simple exchange of a handful of parameters or small text string, a simple fixed size message structure has been defined. Compared to more flexible text-markup based messages such as XML messages this structure has very low overhead because no parsing is required. Also, no third-party dependencies are added to the source-code tree, which keeps the software simple and easy to compile and maintain.

The message structure holds eight integer values, eight 32-bit floating point values, a fixed length text field, a message name, and a message type. This is sufficient for most cases to send single values, classification results (class names or regression values), or control commands across components. In case there is need for more data to be sent, two pointers to custom extension data blocks have been reserved.

A message is sent out of the context of the sender component by calling a `sendMessage` method in the component manager. The component manager then calls the `processMessage` method in the receiving component instance, which is responsible of copying the message parameters of interest to local variables in the context of the receiving component.

4.4 Fully Continuous Speech Emotion Recognition

In this section, a novel time and value continuous modelling approach with Long Short- Term Memory Recurrent Neural Network (LSTM-RNN) is introduced (cf. Eyben et al. 2012) for a speech emotion recognition task. Thereby, modelling of dimensional affect ratings is performed based on the incremental segmentation method shown in Sect. 4.1.2 as well as directly from LLDs. Additionally, multi-task learning is applied to estimate the confidence of the automatic predictions and exploit dependencies between multiple affective dimensions.

The next section (Sect. 4.4.1) gives an overview of the state-of-the art methods in emotion recognition with continuous, dimensional labels under realistic conditions. Next, the proposed method for robust, low-latency, continuous multi-task, dimensional affect recognition is introduced and described in Sect. 4.4.2, and a suggestion for reduction of the number of acoustic parameters is given in Sect. 4.4.3. The evaluation protocol and the obtained results are presented and discussed in the next chapter in Sect. 6.4.

4.4.1 Related Work

Recent work has addressed the challenge of automatically identifying affect in natural conversations. Devillers et al. (2005) associate speech turns with multiple targets (mixtures of discrete emotion categories) based on a realistic, affective speech data set which contains non acted speech. Schuller et al. (2007c) also investigate emotion recognition on realistic and non prototypical speech. The recent INTERSPEECH challenges have attracted a lot of attention and have nourished advances in the field, but also demonstrated how challenging the topic is (Schuller et al. 2009b, 2010, 2011c, b, 2012b, 2013).

In contrast to a discrete, categorical representation of affect, such as the basic six emotion categories by Ekman and Friesen (1975), a continuous dimensional representation (Fontaine et al. 2007) could be better suited to capture a plethora of affective states and subtle differences among these (Fontaine et al. 2007; Douglas-Cowie et al. 2007; McKeown et al. 2012). Yet, automatic affect recognition in a dimensional label space is an active, but still young area of research (Grimm et al. 2007b; Wöllmer et al. 2008; Schuller et al. 2009a; Gunes and Pantic 2010b; Eyben et al. 2011; Gunes and Pantic 2010a; Schuller et al. 2012a). Currently, a common strategy is to reduce the dimensional emotion classification problem to a two-class task: e.g., positive versus negative valence or high versus low activation classification, for example (cf. e.g., Nicolaou et al. 2010; Schuller et al. 2009c; Wöllmer et al. 2010), a four-class task (classification into the quadrants of the two-dimensional valence-activation space (cf. e.g., Caridakis et al. 2006; Fragopanagos and Taylor 2005; Glowinski et al. 2008; Ioannou et al. 2005), or to a dynamic number of categories by automatic identification of clusters in the emotional space (Wöllmer et al. 2009; Lee et al. 2009; Wöllmer et al. 2010). Fixed clusters or categories, however, introduce the problem of ambiguity again: Instances with an actual (continuous) label on or near to the category/cluster boundary are likely to be misclassified. A possible solution to this problem is a regression model which directly predicts the continuous values.

Only very few publications using regression models on speech exist so far: e.g., Grimm et al. (2007a) use Support Vector Regression (SVR) to predict affect in three dimensions (activation, valence, power/dominance); Wu et al. (2010a) fuse three methods: robust regression, SVR, and locally linear re-construction; Wöllmer et al. (2008) use LSTM-RNNs and SVR; the work presented by Wöllmer et al. (2010) utilises a Bidirectional Long Short-Term Memory Recurrent Neural Network

(BLSTM-RNN) to perform regression for affect dimensions and then quantises the resulting predictions into four quadrants. The author's previous work in Eyben et al. (2010d) investigates a regression technique for continuous dimensional affect recognition which extends the work by Wöllmer et al. (2008). Alternatives to SVR include linear regression (Cohen et al. 2003), radial base function networks (Yee and Haykin 2001), and standard feed-forward neural networks. In the context of affect-sensitive virtual agents, Recurrent Neural Networks (RNNs) with Long Short-Term Memory (LSTM) (Hochreiter and Schmidhuber 1997) (see Sect. 2.5.2.2) have been applied (Peters and O'Sullivan 2002; Eyben et al. 2010d). One of the first attempts to get research teams globally to work on a comparable task of fully continuous emotion recognition was the second AVEC (Schuller et al. 2012a).

For regression analysis of emotion, the use of confidence measures has so far been—to the best of the author's knowledge—neglected. In closely related literature no experiments on confidence estimation of regression predictions have been reported. While the importance of confidence measures has risen with methods of unsupervised, semi-supervised (Deng and Schuller 2012; Zhang and Schuller 2012), and active learning (Han et al. 2013) gaining attention, this has only been explored for categorical labelling, so far (e.g., Deng et al. 2012). A method for dimensional music mood prediction which includes a confidence measure has been introduced by Schmidt and Kim (2010), who use linear regression to estimate the mood coordinates of a song excerpt in a two-dimensional valence-activation space. The uncertainty of the regressor is thereby modelled as a second regression target in a multi-target regression framework. Schmidt and Kim (2010) collected a continuously annotated database for training and evaluating the system with an on-line game application. There, participants had to label the current mood of a song by moving their mouse in a valence-activation plane while competing against another rater.

In this thesis a similar attempt to estimate the confidence for speech affect by multi-task learning is presented in Sect. 4.4.2.3: LSTM-RNNs are trained on the inter-rater standard deviation and the mean label for each dimension. Both, modelling each dimension individually and modelling all dimensions in one network is investigated. This multi-task approach is partially inspired by the work of Steidl et al. (2009), who employ a similar technique to estimate class confidences.

4.4.2 Proposed Continuous Modelling Approach

The approach towards incremental affect recognition in real-time chosen in this thesis is based on (bidirectional) LSTM-RNN regressors (see Sect. 2.5.2.2). Such networks can successfully model long-term dependencies between input data, which makes them suitable for supra-segmental-modelling of time connected segments as well as for modelling of LLDs directly. As any neural network, they are able to handle multi-dimensional target patterns, which enables multi-task learning as used here for the estimation of a confidence measure and true multi-dimensional affect prediction.

4.4.2.1 Segmentation Issues (LLD vs. Supra-Segmental)

As opposed to speech recognition, emotion recognition from isolated short-time audio frames is virtually impossible: while single phonemes are highly correlated to a specific spectral representation in short signal windows, speech emotion is a phenomenon observed over a longer time window (typically more than 1–2 s). Thus, supra-segmental modelling is typically applied (Sects. 2.4 and 2.5.1) with units of analysis that are complete sentences, sentence fragments (i.e., chunks, e.g., by grammatical rules) or words (Steidl 2009).

In order to enable output of emotion predictions at constant time intervals, independent of word or phrase level segmentations, the incremental segmentation scheme described in Sect. 4.1.2 is applied. Thereby $L_{step} = 1$ s and $L_{win} = 5$ s are chosen, and the variant (1.) of the algorithm is used. Temporal dependencies among the segments remain due to the overlap of the segments and the slow changing affect labels. These dependencies can be exploited by LSTM-RNNs, as is shown by a comparison to standard RNNs, Feed-Forward Neural Networks (FFNNs), and SVR in Sect. 6.4.2. While theoretically there is no explicit mention that long range temporal context is necessary for successful identification of emotional states, inspection of the data shows that in conversations as contained in the SEMAINE database (Sect. 6.1.9), for example, emotion does not change rapidly nor often and is dependent on the context of the discussion (cf. also Lee et al. 2009).

A major disadvantage of supra-segmental approaches is that one complete input segment is required for analysis and only a single output is produced at the end of every segment. This disadvantage is partially mitigated by the incremental segmentation approach used. However, for fully continuous affect recognition the requirement of having to define analysis segments must be abandoned. Under ideal circumstances (frame-wise) LLDs should be used as input to a classifier/regressor which can connect these inputs over time in a meaningful way and produce a prediction for every short-time frame. The regressor employed here—LSTM-RNN—is capable of this as was shown, e.g., by Eyben et al. (2010d). Thus, in this thesis it is evaluated to which extent the modelling of LLDs without any higher level feature summarisation is feasible. Thereby, each speech utterance is viewed as one sequence when training the LSTM-RNN.

4.4.2.2 Modelling with LSTM-RNNs

LSTM-RNNs with two fully connected hidden layers with 10–140 LSTM cells are used to model the five dimensional affective labels (both individually and jointly). Various uni- and bidirectional network configurations are evaluated in Sect. 6.4.2. Two cases are distinguished: modelling of frame-wise LLDs (rate of 10 ms) and supra-segmental features (rate of 1 s.). When modelling supra-segmental features, a full recording session, i.e., the series of all speech utterances in the order they occur in the original, continuous recording, represents one sequence. This means, the network has access to temporal context across the full progression of the recording session, e.g., a full conversation. In the case of LLD modelling, sequences are limited to a speech utterance or a turn (for comparison: typically $1-10$ supra segmental feature vectors), in order to avoid numerical problems when training a LSTM-RNN with Backpropagation Through Time (BPTT) on very long sequences.

The networks are trained with gradient descent BPTT (Werbos 1990) or gradient descent Resilient Propagation (rProp) (Riedmiller and Braun 1993)—for details, see Sect. 2.5.2.2. When doing so, an initial set of non-zero network weights must be chosen as starting point for the gradient descent weight updates, which is usually done with a random number generator. However, this makes the training likely to converge in local minima of the target error function, dependent of the initialisation. A possible way to reduce the influence of the initial weights is to train $N > 1$ networks with varying initialisations, and at the end take the average of all N output activations. This method is applied here, with $N = 5$ runs and with fixed random seeds $0, 1, 2, 3, 4$ for the 5 runs, respectively.

During training of the networks Gaussian white noise with $\sigma = 0.3$ is added to the features. This technique is commonly used to improve the generalisation capabilities of neural networks (cf. e.g., Fernandez et al. 2008; Graves and Schmidhuber 2005). Overall, this leads to a longer training time (more iterations). However, it helps to avoid over-fitting and therefore potentially will improve the performance on the evaluation and development sets, especially for smaller data-sets.

4.4.2.3 Multi-target Learning

A novel concept investigated by the author of this thesis for the first time is multi-target learning of dimensional affect and rater-agreement in a single model (Eyben et al. 2012). Previous work of the author on dimensional affect recognition has dealt with single target learning for the affective dimensions activation and valence only (Eyben et al. 2010d). With neural networks, multi-target learning is no different from single target learning, except for the topology of the network's output layer: when predicting one continuous dimension the output layer consists of a single linear summation unit; for multiple targets the number of linear summation units in the output layer is equal to the number of targets.

Two aspects are investigated in this thesis:

1. instead of training individual networks for every affective dimension, a single network with multiple outputs is applied to predict multiple affective dimensions simultaneously,
2. the variance of multiple individual raters' labels is learnt as a second target for each dimension.

It is assumed, that—based on this variance measure—the network might learn to predict a confidence measure for its output based on human agreement levels.

4.4.3 Acoustic Features

The INTERSPEECH 2010 Paralinguistics Challenge (IS10) baseline set (Sect. 3.2) is chosen as acoustic parameter set for the continuous modelling. It contains 1,582 features.

Because training of the LSTM-RNN with a high number of inputs (1.5 k for the supra-segmental features) suffers more from data sparsity than related work based on SVR (see Sect. 6.4.2), a Correlation-based Feature-subset Selection (CFS) is applied to the training set to determine relevant supra-segmental features for each of the five dimensions. The development and evaluation data are not used in the feature selection process to ensure fair evaluation conditions. The CFS algorithm (Hall 1998) evaluates the worth of each feature subset by considering the ability of each individual feature to predict the class or dimensional numeric label, along with the degree of redundancy between the features. Feature subsets that are highly correlated with the target while having low intra-correlation are preferred.

Out of 1,582 features, after CFS, 38 features remain for activation, 39 features for expectation, 30 features for intensity, 32 features for power, and 28 features for valence. Instead of including a full list of selected features, the most frequently occurring LLDs for each dimension sorted by frequency of occurrence (given in parentheses) are given in the following (Eyben et al. 2012):

- **Activation**: MFCCs (16), log. Mel-Frequency Bands (MFBs) (9), Line Spectral Frequencies (LSFs) (5), loudness (4), Jitter (2).
- **Expectation**: MFCCs (18), F_0 (7), LSFs (7), loudness (3), log. MFBs (2)
- **Intensity**: MFCCs (11), loudness (7), LSFs (6), log. MFBs (5)
- **Power**: MFCCs (24), log. MFBs (3), LSFs (3), F_0 (2)
- **Valence**: MFCCs (14), LSFs (7), log. MFBs (4)

It can be seen that functionals applied to MFCCs are constantly the most frequently selected features (also supported by Wu et al. 2010b, for example), which must partially be accounted to the fact that MFCCs make up a large portion of the original set. Besides MFCCs, however, there are some variations in the selected features observable across the five dimensions, which can be summarised as follows: for *activation*, spectral band energies, format frequencies (related to LSFs), and loudness seem to be most important, while for *expectation* F_0 is a key factor, which appears logical when considering that surprise is the primary related emotion category with a low value of expectation; for *intensity*, a similar result as for activation is found with a minor tendency towards loudness being more important here; for power, mostly MFCC based features are selected, which could indicate that the subject's dominance is reflected primarily by the way of articulation and least by prosody; *valence* seems to be described best by a mixture of MFCCs and LSFs, again being a hint on the possible importance of the way of phonation / vocal timbre, or even linguistic content.

For multi-target learning, the reduction of the feature set to a single set of relevant features is more complex because all targets have to be considered when evaluating the worth of the features. Instead of extending CFS to multiple targets, it was decided to use the joint feature set, i.e., the union set of the five CFS reduced feature sets (for each dimension individually). In this way, 138 features are selected. This number is below the sum of the number of features selected for each dimension individually (167), which indicates the presence of a small overlap of the reduced features sets, i.e., features which are relevant for more than one dimension.

Features which were found to be relevant for at least three dimensions are the skewness of the third MFCC (for all four dimensions), the percentage (temporal) where the MFCC 7 is above 75 % of its range (up-level time), and the Inter-Quartile Range (IQR) 2–1 of the 6th Line Spectral Pair (LSP) frequency, as well as the second quartile of the 7th LSP frequency. There are 24 features which were selected for at least two dimensions. The most frequent LLDs among these are loudness, the 0th and 6th LSP frequency, MFCC 10, and the probability of voicing.

For the direct modelling of the LLDs with the LSTM-RNN no feature reduction is performed. Due to the smaller size of the set the reduction was not necessary: The IS10 set contains 76 LLDs (including delta regression coefficients—cf. Sect. 3.2).

References

A. Batliner, D. Seppi, S. Steidl, B. Schuller, Segmenting into adequate units for automatic recognition of emotion-related episodes: a speech-based approach. Adv. Human Comput. Interact., Special Issue on Emotion-Aware Natural Interaction **2010**, 1–15 (2010). Article ID 782802 (on-line)

M. Ben-Ari, *Principles of Concurrent and Distributed Programming* (Prentice Hall, Englewood Cliffs, 1990). ISBN 0-13-711821-X

C. Busso, S. Lee, S. Narayanan, Using neutral speech models for emotional speech analysis, in *Proceedings of the INTERSPEECH 2007*, Antwerp, Belgium, August 2007. ISCA, pp. 2225–2228

G. Caridakis, L. Malatesta, L. Kessous, N. Amir, A. Raouzaiou, K. Karpouzis, Modeling naturalistic affective states via facial and vocal expressions recognition, in *Proceedings of the 8th International Conference on Multimodal Interfaces (ICMI) 2006*, Banff, Canada, 2006. ACM, pp. 146–154

J. Cohen, P. Cohen, S.G. West, L.S. Aiken, *Applied multiple regression/correlation analysis for the behavioral sciences*, 2nd edn. (Lawrence Erlbaum Associates, Hillsdale, 2003)

J. Deng, B. Schuller, Confidence measures in speech emotion recognition based on semi-supervised learning, in *Proceedings of INTERSPEECH 2012*, Portland, September 2012. ISCA

J. Deng, W. Han, B. Schuller, Confidence measures for speech emotion recognition: a start, in *Proceedings of the 10-th ITG Symposium on Speech Communication*, ed. by T. Fingscheidt, W. Kellermann (Braunschweig, Germany, September 2012). IEEE, pp. 1–4

L. Devillers, L. Vidrascu, L. Lamel, Challenges in real-life emotion annotation and machine learning based detection. Neural Netw. **18**(4), 407–422 (2005). doi:10.1016/j.neunet.2005.03.007. ISSN 0893-6080

E. Douglas-Cowie, R. Cowie, I. Sneddon, C. Cox, O. Lowry, M. McRorie, J.C. Martin, L. Devillers, S. Abrilian, A. Batliner, N. Amir, K. Karpouzis, *The HUMAINE Database*, vol. 4738, Lecture Notes in Computer Science (Springer, Berlin, 2007), pp. 488–500

P. Ekman, W.V. Friesen, *Unmasking the Face: A Guide to Recognizing Emotions from Facial Expressions* (Prentice Hall, Englewood Cliffs, 1975)

F. Eyben, B. Schuller, S. Reiter, G. Rigoll, Wearable assistance for the ballroom-dance hobbyist – holistic rhythm analysis and dance-style classification, in *Proceedings of the IEEE International Conference on Multimedia and Expo (ICME) 2007*, Bejing, China, July 2007. IEEE, pp. 92–95

F. Eyben, M. Wöllmer, B. Schuller, openSMILE – the munich versatile and fast open-source audio feature extractor, in *Proceedings of the ACM Multimedia 2010*, Florence, Italy, 2010a. ACM, pp. 1459–1462

F. Eyben, M. Woellmer, B. Schuller, openSMILE version 1.0.1 – source code, GPL. http://opensmile.sourceforge.net, 2010b

F. Eyben, M. Woellmer, B. Schuller, The openSMILE documentation v. 1.0.1. http://sourceforge.net/projects/opensmile/files/openSMILE_book_1.0.1.pdf/download, 2010c. Documentation of the the openSMILE toolkit, referred to as openSMILE book, online version 1.0.1

F. Eyben, M. Wöllmer, A. Graves, B. Schuller, E. Douglas-Cowie, R. Cowie, On-line emotion recognition in a 3-D activation-valence-time continuum using acoustic and linguistic cues. J. Multimodal User Interfaces (JMUI) **3**(1–2), 7–19 (2010d). doi:10.1007/s12193-009-0032-6

F. Eyben, M. Wöllmer, M. Valstar, H. Gunes, B. Schuller, M. Pantic, String-based audiovisual fusion of behavioural events for the assessment of dimensional affect, in *Proceedings of the International Workshop on Emotion Synthesis, Representation, and Analysis in Continuous space (EmoSPACE) 2011, held in conjunction with FG 2011*, Santa Barbara, March 2011. IEEE, pp. 322–329

F. Eyben, M. Wöllmer, B. Schuller, A multi-task approach to continuous five-dimensional affect sensing in natural speech. ACM Trans. Interact. Intell. Syst., Special Issue on Affective Interaction in Natural Environments **2**(1), 29 (2012). Article No. 6

F. Eyben, F. Weninger, F. Gross, B. Schuller, Recent developments in openSMILE, the munich open-source multimedia feature extractor, in *Proceedings of ACM Multimedia 2013*, Barcelona, Spain, 2013a. ACM, pp. 835–838

F. Eyben, F. Weninger, M. Woellmer, and B. Schuller, openSMILE version 2.0rc1 – source code, open-source research only license. http://opensmile.sourceforge.net, 2013b

F. Eyben, F. Weninger, M. Woellmer, B. Schuller, The openSMILE documentation v. 2.0 rc1. http://sourceforge.net/projects/opensmile/files/openSMILE_book_2.0-rc1.pdf/download, 2013c. Documentation of the the openSMILE toolkit, referred to as the openSMILE book, online version 2.0 rc1

S. Fernandez, A. Graves, J. Schmidhuber, *Phoneme recognition in TIMIT with BLSTM-CTC*, Technical report, IDSIA, Switzerland, 2008

J.R.J. Fontaine, K.R. Scherer, E.B. Roesch, P.C. Ellsworth, The world of emotions is not two-dimensional. Psychol. Sci. **18**(2), 1050–1057 (2007)

N. Fragopanagos, J.G. Taylor, Emotion recognition in human-computer interaction. Neural Netw., 2005 Special Issue on Emotion and Brain **18**(4), 389–405 (2005)

D. Glowinski, A. Camurri, G. Volpe, N. Dael, K. Scherer, Technique for automatic emotion recognition by body gesture analysis, in *Proceedings of the IEEE Computer Society Conference on Computer Vision and Pattern Recognition Workshops 2008 (CVPRW'08)*, Anchorage, June 2008. IEEE, pp. 1–6

A. Graves, J. Schmidhuber, Framewise phoneme classification with bidirectional LSTM and other neural network architectures. Neural Netw. **18**(5–6), 602–610 (2005)

M. Grimm, K. Kroschel, S. Narayanan, Support vector regression for automatic recognition of spontaneous emotions in speech, in *Proceedings of the ICASSP 2007*, vol. 4, Honolulu, April 2007a. IEEE, pp. 1085–1088

M. Grimm, E. Mower, K. Kroschel, S. Narayanan, Primitives based estimation and evaluation of emotions in speech. Speech Commun. **49**, 787–800 (2007b)

H. Gunes, M. Pantic, Dimensional emotion prediction from spontaneous head gestures for interaction with sensitive artificial listeners, in *Proceedings of the International Conference on Intelligent Virtual Agents (IVA)*, . (Springer, Berlin, 2010a), pp. 371–377. ISBN 978-3-642-15891-9

H. Gunes, M. Pantic, Automatic, dimensional and continuous emotion recognition. Int. J. Synth. Emot. (IJSE) **1**(1), 68–99 (2010b)

M.A. Hall, *Correlation-based Feature Subset Selection for Machine Learning*, Doctoral thesis, University of Waikato, Hamilton, New Zealand, 1998

W. Han, H. Li, H. Ruan, L. Ma, J. Sun, B. Schuller. Active learning for dimensional speech emotion recognition, in *Proceedings of INTERSPEECH 2013*, Lyon, France, August 2013. ISCA, pp. 2856–2859

S. Hochreiter, J. Schmidhuber, Long short-term memory. Neural Comput. **9**(8), 1735–1780 (1997)

S. Ioannou, A. Raouzaiou, V. Tzouvaras, T. Mailis, K. Karpouzis, S. Kollias, Emotion recognition through facial expression analysis based on a neurofuzzy method. Neural Netw., 2005 Special Issue on Emotion and Brain **18**(4), 423–435 (2005)

C.-C. Lee, C. Busso, S. Lee, S.S. Narayanan, Modeling mutual influence of interlocutor emotion states in dyadic spoken interactions, in *Proceedings of INTERSPEECH 2009*, Brighton, UK, September 2009. ISCA, pp. 1983–1986

G. McKeown, M. Valstar, R. Cowie, M. Pantic, M. Schroder, The SEMAINE database: annotated multimodal records of emotionally colored conversations between a person and a limited agent. IEEE Trans. Affect. Comput. **3**(1), 5–17 (2012). doi:10.1109/T-AFFC.2011.20. ISSN 1949-3045

E. Mower, S.S. Narayanan, A hierarchical static-dynamic framework for emotion classification, in *Proceedings of the ICASSP 2011*, Prague, Czech Republic, May 2011. IEEE, pp. 2372–2375

M. Nicolaou, H. Gunes, M. Pantic, Audio-visual classification and fusion of spontaneous affective data in likelihood space, in *Proceedings of the 2010 20th International Conference on Pattern Recognition (ICPR)*, Istanbul, Turkey, August 2010. IEEE, pp. 3695–3699

V. Parsa, D. Jamieson, Acoustic discrimination of pathological voice: sustained vowels versus continuous speech. J. Speech, Lang. Hear. Res. **44**, 327–339 (2001)

C. Peters, C. O'Sullivan, Synthetic vision and memory for autonomous virtual humans. Comput. Graph. Forum **21**(4), 743–753 (2002)

M. Riedmiller, H. Braun, A direct adaptive method for faster backpropagation learning: the RPROP algorithm, *Proceedings of the IEEE International Conference on Neural Networks*, vol. 1 (San Francisco, 1993). IEEE, pp. 586–591. doi:10.1109/icnn.1993.298623

E.M. Schmidt, Y.E. Kim, Prediction of time-varying musical mood distributions from audio, in *Proceedings of ISMIR 2010*, Utrecht, The Netherlands, 2010. ISMIR

M. Schröder, E. Bevacqua, R. Cowie, F. Eyben, H. Gunes, D. Heylen, M. ter Maat, G. McKeown, S. Pammi, M. Pantic, C. Pelachaud, B. Schuller, E. de Sevin, M. Valstar, M. Wöllmer, Building autonomous sensitive artificial listeners. IEEE Trans. Affect. Comput. **3**(2), 165–183 (2012)

B. Schuller, G. Rigoll, Timing levels in segment-based speech emotion recognition, in *Proceedings of the INTERSPEECH-ICSLP 2006*, Pittsburgh, September 2006. ISCA, pp. 1818–1821

B. Schuller, B. Vlasenko, R. Minguez, G. Rigoll, A. Wendemuth, Comparing one and two-stage acoustic modeling in the recognition of emotion in speech, in *Proceedings of the IEEE Workshop on Automatic Speech Recognition and Understanding (ASRU) 2007*, Kyoto, Japan, 2007a. IEEE, pp. 596–600

B. Schuller, F. Eyben, G. Rigoll, Fast and robust meter and tempo recognition for the automatic discrimination of ballroom dance styles, in *Proceedings of the ICASSP 2007*, vol. I, Honolulu, April 2007b. IEEE, pp. 217–220

B. Schuller, D. Seppi, A. Batliner, A. Maier, S. Steidl, Towards more reality in the recognition of emotional speech, in *Proceedings of the ICASSP 2007*, vol. IV, Honolulu, 2007c. IEEE, pp. 941–944

B. Schuller, F. Eyben, G. Rigoll, Beat-synchronous data-driven automatic chord labeling, in *Proceedings of the 34 Jahrestagung für Akustik (DAGA) 2008*, Dresden, Germany, March 2008. DEGA pp. 555–556

B. Schuller, R. Müller, F. Eyben, J. Gast, B. Hörnler, M. Wöllmer, G. Rigoll, A. Höthker, H. Konosu, Being bored? Recognising natural interest by extensive audiovisual integration for real-life application. Image Vis. Comput., Special Issue onVisual and Multimodal Analysis of Human Spontaneous Behavior **27**(12), 1760–1774 (2009a)

B. Schuller, S. Steidl, A. Batliner, F. Jurcicek, The INTERSPEECH 2009 emotion challenge, in *Proceedings of INTERSPEECH 2009*, Brighton, UK, September 2009b. pp. 312–315

B. Schuller, B. Vlasenko, F. Eyben, G. Rigoll, A. Wendemuth, Acoustic emotion recognition: a benchmark comparison of performances, in *Proceedings of the IEEE Workshop on Automatic*

Speech Recognition and Understanding (ASRU) 2009, Merano, Italy, December 2009c. IEEE,pp. 552–557

B. Schuller, S. Steidl, A. Batliner, F. Burkhardt, L. Devillers, C. Müller, S. Narayanan, The INTERSPEECH 2010 paralinguistic challenge, in *Proceedings of INTERSPEECH 2010*, Makuhari, Japan, September 2010. ISCA, pp. 2794–2797

B. Schuller, M. Valstar, F. Eyben, G. McKeown, R. Cowie, M. Pantic, AVEC 2011 - the first international audio/visual emotion challenge, in *Proceedings of the First International Audio/Visual Emotion Challenge and Workshop, AVEC 2011, held in conjunction with the International HUMAINE Association Conference on Affective Computing and Intelligent Interaction (ACII) 2011*, vol. II, ed. by B. Schuller, M. Valstar, R. Cowie, M. Pantic (Springer, Memphis, 2011a), pp. 415–424

B. Schuller, A. Batliner, S. Steidl, F. Schiel, J. Krajewski, The INTERSPEECH 2011 speaker state challenge, in *Proceedings of INTERSPEECH 2011*, Florence, Italy, August 2011b. ISCA, pp. 3201–3204

B. Schuller, A. Batliner, S. Steidl, D. Seppi, Recognising realistic emotions and affect in speech: state of the art and lessons learnt from the first challenge. Speech Commun., Special Issue on Sensing Emotion and Affect – Facing Realism in **53**(9/10), 1062–1087 (2011c)

B. Schuller, M. Valstar, R. Cowie, M. Pantic, AVEC 2012: the continuous audio/visual emotion challenge - an introduction, in *Proceedings of the 14th ACM International Conference on Multimodal Interaction (ICMI) 2012*, ed. by L.-P. Morency, D. Bohus, H.K. Aghajan, J. Cassell, A. Nijholt, J. Epps (ACM, Santa Monica, 2012a), pp. 361–362

B. Schuller, S. Steidl, A. Batliner, E. Nöth, A. Vinciarelli, F. Burkhardt, R. van Son, F. Weninger, F. Eyben, T. Bocklet, G. Mohammadi, B. Weiss, The INTERSPEECH 2012 speaker trait challenge, in *Proceedings of INTERSPEECH 2012*, Portland, OR, USA, September 2012b. ISCA

B. Schuller, S. Steidl, A. Batliner, A. Vinciarelli, K. Scherer, F. Ringeval, M. Chetouani, et al, The INTERSPEECH 2013 computational paralinguistics challenge: Social Signals, Conflict, Emotion, Autism, in *Proceedings of INTERSPEECH 2013*, Lyon, France, 2013. ISCA, pp. 148–152

S. Steidl, *Automatic Classification of Emotion-Related User States in Spontaneous Children's Speech* (Logos Verlag, Berlin, 2009)

S. Steidl, B. Schuller, A. Batliner, D. Seppi, The hinterland of emotions: facing the open-microphone challenge, in *Proceedings of the 4th International HUMAINE Association Conference on Affective Computing and Intelligent Interaction (ACII)*, vol. I, Amsterdam, The Netherlands, 2009. IEEE, pp. 690–697

P. Werbos, Backpropagation through time: what it does and how to do it. Proc. IEEE **78**, 1550–1560 (1990)

M. Wöllmer, F. Eyben, S. Reiter, B. Schuller, C. Cox, E. Douglas-Cowie, R. Cowie. Abandoning emotion classes – towards continuous emotion recognition with modelling of long-range dependencies, in *Proceedings of the INTERSPEECH 2008*, Brisbane, Australia, September 2008. ISCA, pp. 597–600

M. Wöllmer, F. Eyben, B. Schuller, E. Douglas-Cowie, R. Cowie, Data-driven clustering in emotional space for affect recognition using discriminatively trained LSTM networks, in *Proceedings of INTERSPEECH 2009*, Brighton, UK, September 2009. ISCA, pp. 1595–1598

M. Wöllmer, B. Schuller, F. Eyben, G. Rigoll, Combining long short-term memory and dynamic Bayesian networks for incremental emotion-sensitive artificial listening. IEEE J. Sel. Top. Signal Process., Special Issue on "Speech Processing for Natural Interaction with Intelligent Environments" **4**(5), 867–881 (2010)

D. Wu, T. Parsons, E. Mower, S.S. Narayanan, Speech emotion estimation in 3d space, in *Proceedings of the IEEE International Conference on Multimedia and Expo (ICME) 2010*, Singapore, July 2010a. IEEE, pp. 737–742

D. Wu, T. Parsons, S.S. Narayanan, Acoustic feature analysis in speech emotion primitives estimation, in *Proceedings of the INTERSPEECH 2010*, Makuhari, Japan, September 2010b. ISCA, pp. 785–788

P.V. Yee, S. Haykin, *Regularized Radial Basis Function Networks: Theory and Applications* (Wiley, New York, 2001), 208 p. ISBN 0-471-35349-3

Z. Zeng, M. Pantic, G.I. Rosiman, T.S. Huang, A survey of affect recognition methods: audio, visual, and spontaneous expressions. IEEE Trans. Pattern Anal. Mach. Intell. **31**(1), 39–58 (2009)

Z. Zhang, B. Schuller, Semi-supervised learning helps in sound event classification, in *Proceedings of ICASSP 2012*, Kyoto, March 2012. IEEE, pp. 333–336

Chapter 5
Real-Life Robustness

With rapidly growing interest in and market value of social signal and media analysis (Zeng et al. 2009; Eyben et al. 2012b, 2013c; Vinciarelli et al. 2009), interactive speech systems (Pittermann et al. 2010; Schröder et al. 2012), and multi-modal user profiling (Schuller et al. 2009b; Lee et al. 2010; Schuller et al. 2012b) as well as stress measurement (Lu et al. 2012), the technologies and algorithms for automatic affect recognition from speech get more and more commercial attention. While good results are reported in research papers in laboratory settings (cf. Schuller et al. 2009a) or with systems tailored towards specific databases, real-life applications still remain challenging (Schuller et al. 2011b; Mower et al. 2011; Schuller et al. 2012a) due to various factors.

These factors can be roughly summarised by three categories (Eyben et al. 2013a): The large variability of affective expression across different speakers, languages, and cultures; contextual dependencies of the meaning and significance of affective expressions; and varying and degraded acoustic conditions caused by reverberation, background noise, and acoustic properties of the recording devices used.

The variability across subjects can only be effectively addressed by analysing data from each of the target groups and deriving rules (such as by Banse and Scherer 1996) or using annotated data from the target groups (in-domain data) for training of data-driven systems. The contextual dependency has to be addressed on a higher level, when the results of a speech/music analyser are interpreted and actions of the system are planned (e.g., as in the SEMAINE system Schröder et al. 2012). In this thesis, the focus is on the robustness due to varying and unpredictable acoustic conditions—in particular additive and convolutive noise, which is an important issue for virtually all imaginable use case scenarios.

Besides robustness to changing acoustic conditions, this chapter discusses two other related robustness issues: on-line discrimination of speech and non-speech segments in a continuous audio stream (Sect. 5.1), and the normalisation of acoustic features with the goal of eliminating speaker and corpus variability (Sect. 5.2). Finally a systematic data synthesis approach to increase robustness of classification algorithms to background noise is presented in Sect. 5.3.

F. Eyben, *Real-time Speech and Music Classification by Large Audio Feature Space Extraction*, Springer Theses, DOI 10.1007/978-3-319-27299-3_5

5.1 Voice Activity Detection

Voice Activity Detection (VAD), also referred to as Speech Activity Detection (SAD), which discriminates speech and non-speech sounds, is an important first step in many speech-based systems. It is important in Automatic Speech Recognition (ASR) applications, for example, to avoid word insertions and false positives due to noise and background speech; it is also used in audio coding to improve compression and save bandwidth (Syed and Wu 2007), for example. For the speech analysis tasks investigated in this thesis, segmentation of a continuous input stream into speech segments incrementally in real-time is crucial (cf. Sect. 4.1). Moreover, it is important that this segmentation is reliable and highly accurate, even in highly noisy environments. In this section, thus, related, existing VAD methods are summarised and a novel data-driven, context aware VAD approach developed for this thesis is described in detail and evaluated on a large data set. This approach has been published and presented at the 2013 IEEE ICASSP conference by the author of this thesis (cf. Eyben et al. 2013b).

5.1.1 Related VAD Approaches

Early approaches to VAD were based on energy thresholds or pitch and Zero-Crossing Rate (ZCR) rules (e.g., Woo et al. 2000; Marzinzik and Kollmeier 2002; Hahn and Park 1992). An energy threshold VAD roughly is based on the following simple algorithm: The Root Mean Square (RMS) or logarithmic energy is computed on a frame level, and a voiced frame is found if the energy crosses a pre-defined speech energy threshold for that given frame. Only if the energy falls below the silence energy threshold (can be equal to or less than the speech energy threshold) on one of the following frames, that frame is labelled as unvoiced. All following frames below the speech energy threshold are also unvoiced until the next voiced frame is detected with an energy above the speech energy threshold.[1] Post-smoothing of the frame-wise decisions can be applied to reach a more robust final VAD decision. Commonly, for most on-line use-cases, a hysteresis based smoothing is implemented as described in Sect. 4.1.[2] Such simple approaches perform well in settings where there is little to no background noise and no non-speech sounds. They fail, however, with high levels of non-speech sounds.

A better discrimination of speech and non-speech sounds can be achieved with glottal and/or spectral features such as Linear Predictive Coding (LPC) coefficients (Rabiner and Sambur 1977) and cepstral coefficients (Haigh and Mason 1993). More recent approaches consider advanced parameters like autoregressive (AR) model parameters (Mousazadeh and Cohen 2011) and Line Spectral Frequencies (LSFs)

---

[1]In openSMILE the energy based VAD can be implemented with a `cEnergy` component and a `cTurnDetector` component.

[2]This type of smoothing is done by the `cTurnDetector` component in openSMILE.

to actively discriminate between speech-like and non-speech sounds based on static and dynamic statistics. The most promising approaches in strongly corrupted conditions seem to be data-driven methods, where a classifier is trained to predict the classes speech versus non-speech from acoustic features (cf. e.g., Misra 2012). Misra (2012) compares Gaussian Mixture Models (GMMs) with a discriminative classifier and proposes novel features instead of the standard Mel-Frequency Cepstral Coefficient (MFCC)/Perceptual Linear Prediction (PLP) frontends. Realistically noisy, manually labelled YouTube videos are used for evaluation. Many of related data-driven approaches also rely on Gaussian mixture modelling and adaptation (as is typical for ASR systems) to adapt the VAD models to new speakers (e.g., Matsuda et al. 2012) and changing background noise conditions (e.g., Deng et al. 2011; Suh and Kim 2012; Fujimoto et al. 2012). Omar (2012) adapts GMMs to both channel and noise conditions. Thambiratnam et al. (2012) propose to couple the VAD with the acoustic models in the speech recogniser, which is a first step towards inclusion of context in the VAD method. Still, the performance of such approaches goes down when background noise with spectral characteristics similar to speech is audible. Also, the non-stationarity of speech as well as the noise sounds is a big problem (Syed and Wu 2007). Very recent studies suggest that the use of long time-span features clearly improves the robustness in realistic, non-stationary noisy settings because the decision for each frame can be performed in the context of the previous frames: Ng et al. (2012) compare a standard GMM system using 14 PLP cepstral coefficients with a Multi-Layer Perceptron (MLP) based system incorporating long-span acoustic features which are computed over 0.5 s windows. MLP based speech/non-speech posteriors are then decoded over time with two ergodic Hidden Markov Models (HMMs). A similar approach has been proposed by Thomas et al. (2012) who use PLP based and similar, more advanced temporal features in combination with GMMs.

Yet, all of these systems do not use adaptive context learning as provided by Long Short-Term Memory Recurrent Neural Networks (LSTM-RNNs) (Sect. 2.5.2.2) in the approach presented below (Sect. 5.1.2). This featured approach is thoroughly benchmarked in Sect. 5.1.3, where a comparison to other state-of-the-art VAD algorithms designed for the use in noisy conditions (cf. Sohn and Kim 1999; Ramirez et al. 2005; Mousazadeh and Cohen 2011) is made (published by the author of this thesis in Eyben et al. 2013b). The reference algorithms all belong to the category of *statistical methods*, where a Likelihood Ratio (LR) test is applied to the hypotheses of speech presence (denoted as H_1) and speech absence (H_0) on each frame of the observed noisy signal $x_t = s_t + n_t$, where s_t and n_t denote the clean speech signal and the noise signal, respectively. The VAD described by Sohn and Kim (1999) (*SOHN* in the ongoing) is based on a statistical model in the time-frequency domain for the derivation of the LR test. The algorithm introduced by Ramirez et al. (2005) (*RAM*05 in the ongoing), similar to the one in (Ramirez et al. 2004), is based on the concept that more consecutive speech frames concur into the definition of the LR function—thus also taking into account a fixed amount of neighbouring frames, i.e., context information. The algorithm presented by Mousazadeh and Cohen (2011) (*ARG* in the ongoing) is based on the idea of modelling the speech signal by means of an autoregressive-generalised autoregressive conditional heteroskedasticity (AR-GARCH) model directly in the time domain. Compared to the proposed LSTM-RNN approach this method is computationally very complex because it operates on the sample level. It is thus not very suitable for on-line applications, unless implemented efficiently on a Digital Signal Processor (DSP) platform.

5.1.2 Proposed VAD Based on LSTM-RNNs

For this thesis, a novel data-driven method for voice activity detection based on (unidirectional) LSTM-RNNs (cf. Sect. 2.5.2.2 and Hochreiter and Schmidhuber 1997) has been developed by the author of this thesis and evaluated in noisy scenarios (Eyben et al. 2013b). The motivation behind the use of LSTM-RNNs is their ability to model long range dependencies between the input features and the voice activity labels and dynamically learn those dependencies from the training data. As discussed above, other common data-driven VAD approaches, such as those based on GMMs or MLPs do not consider temporal relations in the model. Delta features, modulation or long-span features (Ng et al. 2012) are used to overcome those limitations. Standard Recurrent Neural Networks (RNNs) (as applied by Gemello et al. 2005 to VAD, for example), are able to model a limited amount of temporal dependency—which, however, vanishes exponentially over time (vanishing gradient problem, cf. Hochreiter et al. 2001). LSTM-RNNs do not have this limitation due to their memory cells with constant error carousels (see Sect. 2.5.2.2 for details).

The networks implemented here have one input layer, which size-wise matches the low-level acoustic feature vectors, one or two recurrent hidden layers with Long Short-Term Memory (LSTM) blocks, and one output layer with a single linear summation unit. The networks are trained to produce a continuous voice activity score in the range $[-1; +1]$ for every input frame. Thereby $+1$ indicates voice/speech and -1 indicates silence or non-speech/noise sounds. Two network topologies have been investigated (cf. Eyben et al. 2013b), one shallow and one deep:

- *Net*1: 1 recurrent hidden layer (50 LSTM blocks)
- *Net*2: 3 recurrent hidden layers (50 LSTM blocks; 10 standard sigmoid neurons; 20 LSTM blocks)

On the input side of the networks RelAtive Spectral TrAnsform Perceptual Linear Prediction (RASTA-PLP) (Hermansky 1990) cepstral parameters 1–18 (Sect. 2.2.9.4) are applied together with their first order delta regression coefficients ($W = 2$, cf. Sect. 2.3.2). The size of overlapping audio frames is 25 ms with a frame step (period) of 10 ms. Frames are multiplied with a Hamming window. It is important to highlight that this 36 dimensional feature vector does not contain direct energy information (e.g., the 0th cepstral coefficient or RMS energy or similar). This decision was made on purpose to make the behaviour of the networks invariant to the input level. Features have been extracted with the openSMILE toolkit (Eyben et al. 2010) which was developed by the author of this thesis and z-normalisation has been applied to all features (mean 0, variance 1). The parameters for the z-normalisation are computed from the training set only in order to simulate realistic on-line evaluation conditions (for more details on feature normalisation see Sect. 5.2). The LSTM-RNNs were trained and evaluated with the *rnnlib* by Graves et al. (2007). More details on the training parameters are found in (Eyben et al. 2013b).

LSTM-RNNs are perfectly suitable for real-time on-line applications because they (a) provide frame-level decisions (low latency), and (b) the computational complexity for evaluating the networks is asymptotically linear with respect to the number of input frames, i.e., for every frame a constant number of operations must be performed. Many of these operations can be run in parallel, which is ideal for implementation on embedded hardware such as DSPs or Field Programmable Gate Arrays (FPGAs), or on multi-core processors or graphics processors.

5.1.3 Benchmarking of the Proposed Approach

For a thorough evaluation of the proposed VAD approach, a large amount of labelled data in realistic noisy conditions is required. To obtain such data for training and validating the networks, new data were synthesised by building random utterance sequences overlaid with additive and convolutive noise as first published by the author of this thesis in (Eyben et al. 2013b). Clean speech data from the Buckeye (Pitt et al. 2007) and the TIMIT corpus (Garofolo et al. 1993) are used. The Buckeye corpus consists of 26 h of spontaneous speech from 40 subjects (20 male, 20 female) recorded in informal interview settings. Only the subjects' speech is used and segments corresponding to utterances between silence parts of at least 0.5 s length are extracted according to the automatic alignment shipped with the Buckeye corpus. The corpus is split subject-independently into a training, validation, and test partition, stratified by age and gender. The segmentation and subdivision is exactly equal to the one used by Weninger et al. (2011a). TIMIT already provides a split into a training and test partition. The original TIMIT training set is further split speaker-independently into a training and validation set. Speech for the synthesised VAD test set is taken from the original TIMIT and Buckeye test sets. Four types of noises are considered: *babble*, *city*, white and pink *noise*, and instrumental *music*. The babble noise recordings are taken from the *freesound.org* website. Samples from the categories pub-noise, restaurant chatter, and crowd noise are concatenated together. The music recordings resemble instrumental and classical music pieces from the *last.fm* website. The city recordings consist of recordings conducted at the Technische Universität München (TUM) in Munich, Germany on smartphones while volunteers where cycling and walking through the city. White and pink noise samples were generated with pseudo random number generators (white noise) followed by a lowpass filter for the pink noise.

The noise samples used for synthesising the VAD training, validation, and test samples are fully disjoint (i.e., different original pieces of music, different babble samples, etc.). Noise patches for the test and validation partitions span 30 min for

Table 5.1 Length of the samples available for each noise type for corrupting the training set (Eyben et al. 2013b)

Noise-type:	Babble	City	Noise	Music
Length (h:mm)	1:34	1:56	2:00	2:56

each type, the remaining noise audio is used for the training set. The lengths of these samples vary from 94 min (babble) to 176 min (music). The lengths of these samples detailed per noise type are found in Table 5.1.

Each synthetic utterance in the VAD training set is built by concatenating $N \in \{1, \ldots, 5\}$ original clean speech utterances, which are randomly selected either from TIMIT or Buckeye. A pause (silence) before the first utterance, pauses between all utterances, and a pause after the last utterance are inserted with a randomly (uniform distribution) chosen length between 0.5 and 5 s. Each of the original utterances is level normalised to have a peak amplitude of 0 dB and then the normalised utterances are concatenated to a meta-utterance which is multiplied with a gain factor $g_{s,lin} = 10^{\frac{g_s}{20.0}}$ where $g_s \in [+3\,\text{dB}; -20\,\text{dB}]$ is randomly sampled from a uniform pseudo random number generator. For 80 % of the synthetic utterances, a random noise sample, which matches the length of the meta-utterance (N original utterances and pauses), is selected from the training noise pool and scaled to have a peak amplitude of 0 dB. In order to mix speech and noise at a controlled Signal-to-Noise Ratio (SNR), a gain factor $g_{n,lin}$ as described by Eq. (5.1) is applied multiplicatively to the noise patch:

$$g_{n,lin} = 10^{\left(log(g_{s,lin}) - \frac{SNR}{20.0}\right)}. \tag{5.1}$$

It is to note that, the SNR is based on peak signal amplitudes and no A-weighting was performed, as the goal was to measure the amount of signal distortion as it would affect linear audio signal processing and not human hearing. The SNR is randomly chosen for each mixed instance from [−6 dB; +25 dB]. The remaining 20 % of all synthetic utterances are not overlaid with noise, i.e., they remain clean speech utterances. 1,948 meta-utterances are created with speech from Buckeye. This corresponds to 15 h of total audio, where 6:43 h are non-speech and 8:17 h are speech. From TIMIT, 3,493 meta-utterances are generated, resembling 19:45 h of total audio, where 12:54 h are non-speech and 6:51 h are speech. In total there are 34:54 h of audio material in the VAD training set used for this study.

The validation set is built in a similar way, however, one single meta-utterance with a total length of 22.5 min is generated from Buckeye and TIMIT speech. The gain of each of the original utterances is varied randomly over the same range as is used for the training set, and silence segments (pauses) are added with random lengths using the same parameters. This same meta-utterance is overlaid with four continuous 22.5 min segments of babble, music, city, and white+pink noise (all normalised to 0 dB peak amplitude). A fixed gain $g_{n,lin}$ is chosen for this noise segment as $g_{n,lin} = 0.5(g^{\mu}_{s,lin} + g^{min}_{s,lin})$, where $g^{\mu}_{s,lin}$ and $g^{min}_{s,lin}$ are the mean and minimum gain factors, respectively, of all the speech utterances in the meta-utterance. This yields a total 90 min of speech audio for validation from TIMIT and Buckeye. In total, the VAD validation set has 3 h of audio, where 1:22 h are speech and 1:38 h are non-speech. The validation set is used to evaluate the neural network performance after every training epoch and stop the network training when no improvement of the performance on the validation set is obtained for more than 20 epochs. The training is also stopped when a maximum of 100 epochs has been reached.

For the VAD test set, 15 min long meta-utterances are created each from TIMIT and Buckeye speech. Thus, the total length of each test meta-utterance is 30 min. The clean version of the 30 min test audio contains 12 min of speech and 18 min of silence. A single fixed gain of −6 dB for the clean speech is applied, and noise with a peak SNR (noise gain relative to speech gain) of 0 dB is added to the speech.

In order to test the VAD in challenging real-life conditions, a second test set consisting of the full-length English audio tracks of four Hollywood film DVDs is used (Eyben et al. 2013b). The films are chosen from the official development set of the 2012 MediaEval campaign's violence detection task (Demarty et al. 2012). Speech and non-speech segments in the films were manually annotated by a single rater. The list of films and statistics on the lengths of speech/non-speech segments are given in Table 5.2.

Results for the synthetic test and validation sets are given in Table 5.3. Two evaluation metrics are used: the area under Receiver Operating Characteristic (ROC) curves (AUC) and the combined error rate (False Positive Rate (FPR) + False Negative Rate (FNR)). Fixed thresholds which correspond to the thresholds at the Equal Error Rate (EER) on the validation set are used for all test set evaluations in order to ensure fair evaluation conditions. For nets *Net*1 and *Net*2 the selected thresholds are −0.268 and −0.071, respectively. The same thresholds are applied to the DVD film test set. For computation of FPR and FNR, the predictions are binarised by applying the threshold (both for the reference and the LSTM output) and the binarised predictions are smoothed with a silence hysteresis of five frames (i.e., non-speech segments shorter than five frames are joined with the adjacent speech segments).

It can be observed that both the *Net*1 and *Net*2 network topologies outperform all baseline algorithms in terms of Area Under (ROC) Curve (AUC) and combined error rate (FNR + FPR). Notably, this is also the case for clean speech. The largest margin of improvement can be reported for music, babble, white, and pink noise. For city noise, the baselines seem relatively robust, which can be attributed to the fact that the average energy of these noise samples is much lower than the peak amplitude (e.g.,

Table 5.2 DVD film test set. Film length in [hh:mm] and percentage of parts with speech (sp); minimum, average, and maximum duration of continuous speech segments (Eyben et al. 2013b)

Film title	Duration (h:mm)	% speech	Min/avg/max (s)
I am legend	1:36	39.2	0.5/21.4/174.9
Kill bill volume 1	1:46	33.9	0.4/39.3/321.2
Saving Private Ryan	2:42	48.6	0.5/25.2/230.4
The bourne identity	1:53	40.7	0.6/32.6/185.6

Table 5.3 Area under (ROC) curve (AUC) frame-level results on the synthetic validation and test sets of LSTM-RNN approaches *Net*1 and *Net*2 and the *RAM*05, *ARG*, and *SOHN* reference algorithms as reported in (Eyben et al. 2013b)

set	AUC				
	*Net*1	*Net*2	*RAM*05	*ARG*	*SOHN*
Validation	0.814	**0.838**	0.713	0.685	0.709
Test clean	0.980	**0.985**	0.955	0.962	0.959
Test babble	0.909	**0.932**	0.877	0.875	0.826
Test city	0.968	**0.972**	0.928	0.935	0.931
Test music	0.921	**0.940**	0.725	0.675	0.677
Test noise	0.941	**0.949**	0.878	0.773	0.878
Test ALL	0.951	**0.961**	0.821	0.794	0.805

Equal error rate (EER) and combined error rate (false negative rate (FNR) + false positive rate (FPR)) are not given here; the reader is referred to (Eyben et al. 2013b). Test set: −6 dB gain applied to the original speech signal, average SNR is 0 dB

loud cars passing by). ROC curves for the proposed and the baseline algorithms are plotted in Fig. 5.1. The apparent 'smoothness' of the curves for the proposed LSTM-RNN approach compared to the curves of the baseline systems is due to the modelling as a regression task, which delivers a 'continuum' of scores in testing (by varying the binarisation threshold). As far as ROCs are concerned, the behaviour of the two network topologies is almost identical. The EER in the validation and test

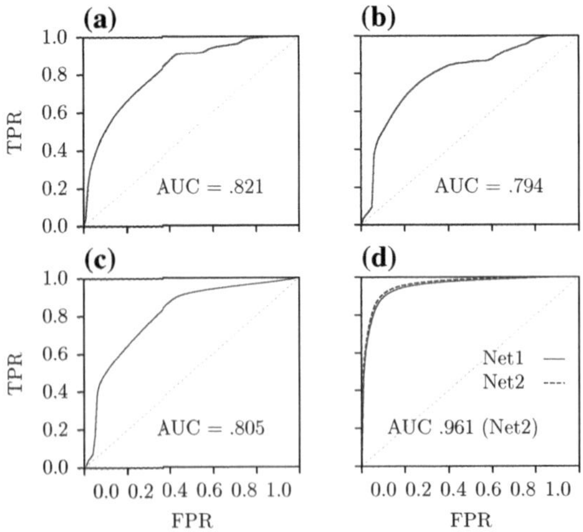

Fig. 5.1 Receiver operating characteristic (ROC) curves for all VAD algorithms on the synthetic test set: true positive rate (TPR) versus false positive rate (FPR) and area under (ROC) curve (AUC) values for Ramirez's algorithm (Ramirez et al. 2005) (**a**), AR-GARCH (Mousazadeh and Cohen 2011) (**b**), Sohn's approach (Sohn and Kim 1999) (**c**) and the novel LSTM-RNN method (**d**) using network topologies *Net*1 and *Net*2; plots drawn after (Eyben et al. 2013b)

Table 5.4 Frame-level results for the DVD film test set of nets *Net*1 and *Net*2 and the *SOHN* algorithm

Film	AUC		
	*Net*1	*Net*2	*SOHN*
I am legend	0.704	0.676	0.567
Kill bill volume 1	0.627	0.601	0.554
Saving Private Ryan	0.743	0.680	0.577
Bourne identity	0.685	0.647	0.603
ALL	0.722	0.676	0.556

Area under (ROC) curve (AUC) as reported in (Eyben et al. 2013b); more results given in (Eyben et al. 2013b). Results for *RAM*05 and *ARG* are not included due to their heavy computational load on the large DVD test set, which makes them unsuitable for real-time processing

partitions is around 10 % for both network topologies as opposed to 25 % and above for the baseline algorithms (Eyben et al. 2013b).

The results for the DVD film test set are given in Table 5.4. Compared to the synthetic test set, the performance on the film test set is clearly lower. One main reason for the lower performance on the DVD film set might be that many noise types occur that have not been seen in training, such as gunshots, fighting, or synthetic background ambience, for example. Further, noises which are likely to be confused with speech, such as animal sounds or human like sounds are found. Another possible reason might be the coarse annotation of speech segments; for the sake of efficiency, longer conversations were labelled as continuous speech segments, even though they included small pauses. In the evaluations this results in a higher miss rate than actual. In comparison to the work of Misra (2012) (25.3 % EER on YouTube videos) the EERs of the VAD approach presented in this thesis are very competitive, considering that the system of Misra (2012) was trained on in-domain data, while the approach presented here was trained on synthesised data only.

The LSTM-RNN VAD developed for this thesis outputs frame level voicing predictions. Due to the context modelling of the LSTM-RNN the predictions are already fairly smooth. However, to reliably find continuous segments of speech—which inherently include small pauses between (e.g., at a comma) and within words (e.g., plosives), an additional stage of smoothing suitable for incremental, on-line processing is implemented based on the hysteresis method as described in Sect. 4.1.

5.2 Feature Normalisation

Most models, such as Support Vector Machines (SVMs) and Neural Networks (NNs) benefit from a normalised feature space, i.e., when all features are in the same range of values, best between -1 and $+1$ or 0 and 1. For NNs the fact is obvious in two ways: (a) the input of each neuron is a weighted sum of the inputs—thus, it is best when all inputs are in a similar range of values, and (b) the gradient of the sigmoid function is highest around the origin (0)—thus, the backpropagated error will have the most effect on the weights in this case, resulting in a fast converging training and numerically optimal weights in the trained network (Sola and Sevilla 1997). Similarly, all distance based classifiers benefit from well scaled feature dimensions where all values in the same range, in order to obtain numerically well behaved and meaningful distances. For SVMs it was shown by Herbrich and Graepel (2001) that feature space normalisation is an essential pre-processing step needed prior to model training. Further, Graf and Borer (2001) investigate and discuss the effect of various normalisations on SVMs from a theoretical point of view.

Next to being a pre-requisite for training of classifier/regressor models, normalisation also serves the purpose of adaptation to diverse conditions, such as speakers, acoustic conditions, noise settings, etc. Adaptation, however, for general audio analysis is a very wide spread field, which has not yet received enough attention. For on-line processing this issue becomes even more challenging and deserves more attention in future work. In this thesis only a short summary of the challenges of adaptation and normalisation will be presented and the standard approach for (on-line) feature normalisation, that has been employed here, is presented.

Generally, feature normalisation can be applied at two different feature levels: for Low-level Descriptors (LLDs) (Sect. 5.2.1), or for supra-segmental features (Sect. 5.2.2). The following two sections discuss these two possible approaches to feature normalisation, present related work in the area, and finally briefly describe the normalisation method which is applied on the supra-segmental feature level for on-line speech and music analysis in this thesis (Sect. 5.2.3).

5.2.1 Normalisation of Low-Level Descriptors

Most work on acoustic feature normalisation has been performed in the specific field of ASR, where cepstral Mean Variance Normalisation (MVN) applied to MFCC features is a common method (Junqua and Haton 1996; Young et al. 2006): Given a set of values $x(n)$ for $n = 0 \ldots N$, μ_x and σ_x^2 are the mean and the variance of these values, respectively. The MVN for every sample x is then expressed as:

$$x' = \frac{1}{\sigma_x^2} (x - \mu_x) . \tag{5.2}$$

Thereby the values are normalised to have zero mean and a variance of one. In ASR, typically, MVN is applied to each utterance, i.e., the means and variances are computed individually for each utterance and each feature. This effectively eliminates channel noise and speaker variability in each recording separately. However, especially for short utterances, also phonetic information is discarded. Recently, more advanced feature normalisation approaches have been presented for normalising intra speaker differences in GMM/HMM speech recognition frameworks, e.g., by Sethu et al. (2007).

For paralinguistic tasks most static classification approaches apply MVN to the supra-segmental features (which is discussed in Sect. 5.2.2). While for some functionals, such as mean and variance, there is mathematically no difference in normalising the LLD before applying the functional to compute the supra-segmental feature, or normalising the supra-segmental feature directly, for others, such as linear and quadratic regression coefficients or peak amplitude based functionals, there is an inherent difference. This calls for more work in the future to investigate the difference between the two approaches, and take a deeper look at the role of individual LLDs and functionals. This thesis provides a basis for such analysis by introducing large, comprehensive feature sets. Up to now, only few publications study the effects of LLD normalisation:

For instance, Suzuki et al. (2012) normalise prosodic features to a neutral reference obtained from a speech synthesiser on a frame-level. They show the effectiveness of the approach for two affective dimensions: pleasantness and arousal. However, the method is limited by the fact that the textual content must be known in order to synthesise the neutral reference. This involves costly ASR and text-to-speech algorithms to run in parallel to the speech analyser.

Busso et al. (2009) describe and evaluate a pitch-only based emotion classification model, where emotional speech is compared to a neutral reference. The article itself does not directly describe a normalisation method, but the presented concept of the neutral pitch reference could be applied to normalisation. In that case, for instance, the mean pitch and the variance of the pitch would be computed from all neutral instances and then all instances would be scaled to have zero mean and unit variance. Alternatively this scaling could be done for each speaker separately to effectively eliminate intra-speaker pitch variability.

5.2.2 Normalisation of Supra-Segmental Features

In most work up to now, including this thesis, a quasi standard for the static classification approaches of normalising the supra-segmental feature vectors has been followed (cf. Schuller et al. 2011b, 2014). Thereby either MVN (also referred to as standardisation or z-normalisation, cf. previous section) is applied to the supra-segmental features, or Mean Range Normalisation (MRN), which is defined as follows[3]:

$$x' = \frac{2}{x_{max} - x_{min}}(x - x_{min}) - 1.0, \qquad (5.3)$$

where x_{max} and x_{min} are the maximum and minimum values in the set of all $x(n)$ values. The resulting values x' are normalised in a way that $x' \in [-1; +1]$. Compared to MVN this approach is more vulnerable to single outliers, which would significantly impact the range of $x(n)$. Thus, MRN might work well for laboratory

[3]In openSMILE, both MVN and MRN are implemented in the `cVectorMVN` component.

conditions where the data are outlier free, but for realistic conditions, MVN should be preferred.

5.2.3 *Incremental Normalisation*

A crucial question for feature normalisation is on what data set the normalisation parameters (max/min or mean/variance) should be computed. In traditional off-line experiments on a single corpus, typically the corpus is divided into a training and test partition (cf. Schuller and Batliner 2013b). The normalisation parameters are then estimated either from the training partition only and applied to the test partition without adaptation to simulate realistic test conditions where only a single unknown instance from the test partition is presented to the system at each time—e.g., for interactive systems (*on-line* normalisation). This normalisation approach was followed for the baselines of all the INTERSPEECH Challenges. Alternatively, if a batch of, or the whole test set is available at evaluation time, e.g., when doing large-scale, off-line data-mining in voice recordings, then the normalisation parameters for the test data can be computed on and only on the test data directly (*off-line* normalisation). If the test partition is large enough, this theoretically ensures the best possible adaptation (numerically, at least) to the test set conditions.

In order to eliminate inter-speaker differences, the normalisation parameters can be computed individually for data from each speaker. This is referred to as *speaker normalisation*. It, however, assumes that enough test data are available for each speaker to compute meaningful normalisation parameters.

To allow for adaptation to test conditions in an incremental set-up, i.e., where the test data are presented to the system instance by instance, a gradual adaptation to the test set or test speaker (for speaker normalisation) must be performed. This can be implemented, for instance, (a) by exponential adaptation of the parameters, (b) by implementing a fixed buffer of previous instances over which normalisation parameters are computed, or (c) by recomputing the normalisation parameters for every new instance. For method (a), the parameters for MVN, for example, will be updated with the following exponential update rule:

$$\mu_n = \alpha\mu_{n-1} + (1.0 - \alpha)x(n), \tag{5.4}$$

$$\sigma_n^2 = \alpha\sigma_{n-1}^2 + (1.0 - \alpha)(x(n) - \mu_n)^2. \tag{5.5}$$

The influence of the initial parameters decays exponentially over time with this update rule. The amount of decay is controlled by the parameter α. Typical values are $\alpha = 0.95$ to $\alpha = 0.999$. For method (c), all data recorded so far have equal influence. The update rules for the nth test sample, are (b is an initial weight bias, discussed below):

$$\mu_n = \frac{n-1+b}{n+b}\mu_{n-1} + \frac{1}{n+b}x(n), \tag{5.6}$$

$$\sigma_n^2 = \frac{n-1+b}{n+b}\sigma_{n-1}^2 + \frac{n-1+b}{(n+b)^2}\left(x(n) - \mu_{n-1}\right)^2 . \tag{5.7}$$

The above is equivalent to re-computing the mean and variance from all samples collected so far for every new test sample which is arriving, but is computationally much more efficient. In both cases (a) and (c) the initial mean μ_{n-1} and the initial variance σ_{n-1}^2 are computed from the training set. For case (c) an initial weight should be chosen for those values by adding a bias b to n in the above update equations. A choice of $b = 100$ results in the initial parameters having a weight equivalent to 100 new test instances.

A pragmatic constraint to consider when doing incremental adaptation of normalisation parameters, is the balance of classes. I.e., if a class-wise balanced training set was used to estimate the initial parameters and train the model, and then in the test case a high number of neutral samples are provided to the system before emotional samples are presented, the normalisation parameters adapt to the neutral samples. That is, the neutral test samples are stretched in the feature space to cover the full range of the normalised feature space. This will result in neutral samples being more and more misclassified as emotional samples, the further the adaptation goes. Thus, during an adaptation phase, the test samples should be balanced regarding the classes. In practical situations this, however, is rarely the case as often many more neutral cases occur. This highlights the importance for future research work to investigate these issues of incremental updates to the normalisation parameters in more detail.

5.3 Noise Robustness

As mentioned previously, a core aspect to robustness in speech and music analysis is the robustness of the algorithms against background noise. Background noise is always an issue when a system deals with live recordings in a non-studio environment. Music query services such as Shazam[4] have implemented very robust algorithms (Wang 2003) which ensure robust identification of song titles even from highly noise corrupted snippets, such as recordings of background music in a pub. For emotion analysis, for example, only very little work in this direction has been done, although it has been shown that the classification performance is highly affected by noise (cf. e.g., Eyben et al. 2013a, 2012a; Schuller et al. 2007; You et al. 2006). Although in most cases the methods developed for emotion recognition are transferable to other, related domains, such as speaker state and trait analysis, or more general, Computational Paralinguistics (cf. Schuller and Batliner 2013b), no well known specific work on noise robustness exists in these fields.

[4]http://www.shazam.com/.

In a limited sense the normalisation of acoustic parameters provides noise robustness to some extent. However, the amount of control is low, and normalisation also affects other parameters besides noise, as only a statistical scaling of the feature space is performed (see previous section). Therefore, the author of this thesis distinguishes between the following three explicit approaches to implement noise robustness, which do not include feature normalisation:

1. the *pre-processing approach*, where it is attempted to filter or transform the input audio signal in a way which reduces or removes noise. Many methods and standards exist in the areas of telecommunications and conferencing, as well as ASR. The topic is known as speech enhancement there (Benesty et al. 2005). Popular examples are spectral subtraction, Non-negative Matrix Factorisation (NMF), Wiener Filtering, or simply bandpass filtering (cf. Schuller 2013a). If a multi-channel signal exists, beam forming methods can be used to isolate speech sources spatially (cf. e.g., Parra and Alvino 2001).
2. the *feature approach* involves the design of acoustic features which are tailored very specifically to precisely reflect the target information and which are as little as possible affected by noise (anything but the target information of interest). An example would be the use of spectral peaks as for music identification as suggested by Wang (2003), or implementing an F_0 detection algorithm which is noise robust in order to extract noise robust prosodic features (e.g., Talkin 1995 or the Subharmonic Summation (SHS) algorithm used in this thesis—see Sect. 2.2.11 for details).
3. the *modelling approach* is about adapting the classification/regression models to be able to deal with noise. Typically this involves training the model in a way that it can capture the increased variance of noisy data. The best examples are matched conditions training, i.e., where the training data are corrupted with the same type (or a similar type) of noise as the training data, or multi-condition training, where a variety of corruptions are applied to the training set in order to cover a large variety of test conditions. Other methods theoretically would involve splitting the model in one part which models noise and in another part which models the actual speech or music.

Combinations of the three approaches are possible. However, noise removal in the pre-processing steps also discards target information. Especially for affective and paralinguistic speech analysis the common speech enhancement algorithms might remove too much information for highly aroused voices, or pathological voices, due to the assumptions they make on the properties of a normal speech signal. The feature approach, if well designed, does not have these problems. However, a significant amount of engineering work is required to find robust features. Then again, these features are very specific to the setting (databases, tasks, etc.) in which they were evaluated during design and might not work well in other acoustic environments or for other paralinguistic analysis tasks. Both the pre-processing and the feature extraction approaches potentially increase the computational load by a non neglectable amount, which is especially important for real-time, on-line processing. Speech enhancement methods, for example, try to re-construct the clean speech signal, and thus operate on a sample level which leads to a high number of computing operations to perform. Thus, in this thesis the modelling approach is favoured. No overhead in terms of complex pre-processing and filtering nor complex feature extraction algorithms are required at run-time. Adaptations to the model in order to make it noise robust can all be performed at design/training time. Further, the approach published by the author of this thesis in (Eyben et al. 2013a) is fully data-driven and fully unsupervised. Thus, it contains no task and data specific manual optimisations and can automatically be extended to new domains. In the light of the big data era this seems the most feasible approach for obtaining highly flexible and robust systems which work with real world data across different acoustic environments, speakers, countries, languages, cultures, and tasks.

Previous work in this direction was focussed either on additive noise (e.g., Eyben et al. 2012a) or reverberation in isolation (Schuller 2011a). Techniques from ASR for acoustic pre-processing and signal enhancement, or multi-condition training have typically been applied to boost performances in such noisy conditions (cf. Eyben et al. 2012a; Weninger et al. 2011b). The approach taken here (cf. Eyben et al. 2013a) focuses on finding acoustic features which are least degraded by noise and most correlated with the targets. This is combined with multi-condition training. In order to obtain large amounts of degraded training data for feature analysis and model training, such data are automatically synthesised on a large scale by applying realistic convolutive and additive noise to clean speech recordings.

In the following, the noise robust training method first published by the author of this thesis in (Eyben et al. 2013a) is described. Section 5.3.1 describes how large amounts of new data were synthesised in an automated and unsupervised way to produce degraded training data in four different acoustic conditions. The algorithm to select noise robust acoustic features is described in Sect. 5.3.2. An evaluation of the approach, including a discussion of feature relevance, on two affective speech corpora is given in the next chapter in Sect. 6.2.

5.3.1 Synthesis of Noisy and Reverberated Data

Realistic noise recordings of three types serve as additive noise (cf. Eyben et al. 2013a): Babble noise (*babble*), city street noise (*city*), and instrumental music (*music*). Babble noise recordings are taken from the *freesound.org* website out of the categories pub-noise, restaurant chatter, and crowd noise. Music recordings consist of instrumental and classical music from the *last.fm* website. The city recordings were recorded by the author and his coleagues in Munich, Germany with smartphones while cycling and walking through the city similarly to the procedure described by Schuller et al. (2013c). The noise samples used for synthesising training set samples are fully disjunctive of those used for the test set samples, i.e., no original sample occurs in both sets. The total length of the noise sample pool is 30 min for each noise type in the test set and 94 min for babble, 116 min for city, and 176 min for music in the training set.

To simulate convolutive noise, Room Impulse Responses (RIRs) from the Aachen Impulse Response Database (Jeub et al. 2009) were chosen. To keep the complexity of the evaluations low, and yet simulate realistic conditions, a few meaningful combinations of additive noise types and RIRs were selected:

- babble noise and lecture room (A),
- babble noise and stairway (B),
- city noise and meeting room (C),
- and music noise and chapel (Aula Carolina) (D).

The combinations in that selection range from more favourable reverberation conditions (meeting room) to heavily reverberated (chapel) conditions and at the same time represent a wide range of common non-stationary additive noises. Three different virtual microphone distances (relative) from the virtual sound sources in an azimuth angle of 90° (facing the sound source) in 'near', 'mid', and 'far' distance categories are employed to simulate various intensities of the convolutive noise. The further the virtual microphone is away from the virtual sound source, the larger the amount of convolutive noise (indirect reflections of the room) is in relation to the direct source signal. The actual distance in meters for each category depends on the room type (cf. Table 5.5).

In order to eliminate any influence of the average utterance energy, all clean speech utterances were normalised to a -1 dB peak amplitude. From these normalised clean utterances, speech samples with degraded acoustic conditions were then created. A clean (normalised) utterance is thereby convolved with a RIR, then normalised to -6 dB peak amplitude, and finally overlaid with an additive noise sample, which is scaled in order to achieve a given SNR in the created sample. The test partitions of the clean data are convolved with the 'near', 'mid', and 'far' impulse responses and noise at SNRs from 0 to 12 dB in steps of 3 dB is added resulting in 18 different test sets (including three reverberated sets without noise) for each of the four acoustic conditions (cf. Table 5.5).

Table 5.5 Four acoustic conditions: additive noise type and room impulse response (RIR) from the Aachen Impulse Response Database along with the source-microphone distances, and signal-to-noise ratio (SNR) ranges for the additive noise (Eyben et al. 2013a)

	Noise	RIR	Speaker-microphone distance (m)			SNR range (dB)
			Near	Mid	Far	
A	Babble	Lecture room	2.25	5.56	10.2	0–12
B	Babble	Stairway	1	2	3	0–12
C	City	Meeting room	1.45	1.9	2.8	0–12
D	Music	Chapel	1	5	20	0–12

The training partition for each acoustic condition has three times the size of the original training partition because each utterance is included once for the 3 RIR distances. Additive noise at random SNRs (uniformly distributed over the range 0–15 dB and with a 10 % probability of clean utterances) was overlaid. SNRs are calculated after first order high pass filtering from the difference signal of speech and noise to approximate A-weighting and thus to better match human perception. The noise samples are picked at random locations in the training and test noise sample pools, matching the length of the speech samples. These *noise samples* are then convolved with the RIR of the current acoustic condition ('far' distance only) to ensure realistic reverberant conditions also for the noise, and then normalised to −6 dB peak amplitude before additive mixing with the (reverberated) clean speech sample.

5.3.2 Acoustic Feature Analysis and Selection

A large set of acoustic features is extracted—the ComParE 2013 set (Sect. 3.5). It contains 6,368 features. From this set, features relevant for the target task are selected by computing the Pearson correlation coefficients (CC) (Pearson 1895) of each feature $x(n)$ over discrete time n with a continuous target label $y(n)$:

$$CC = \frac{\sum_{n=1}^{N} \left(x(n) - x_\mu\right)\left(y(n) - y_\mu\right)}{\sqrt{\sum_{n=1}^{N} \left(x(n) - x_\mu\right)^2}\sqrt{\sum_{n=1}^{N} \left(y(n) - y_\mu\right)^2}}, \tag{5.8}$$

where x_μ and y_μ are the arithmetic means of the time series $x(n)$ and $y(n)$ over all timesteps. This method is termed CC-FS in the ongoing. The 400 most relevant features—ranked by CC—for each of the activation, valence, and Level of Interest (LOI) tasks are chosen based on CCs computed from a training set. This set of reduced acoustic features is referred to as the CC-FS feature set.

Feature selection on clean speech data is contrasted with feature selection performed on multi-condition degraded data, in order to identify features which are robust in both clean and the degraded conditions. Results of the feature reduction, as well as classification performances with the reduced set and multi-condition training are shown and discussed in Sect. 6.2.

References

J. Benesty, S. Makino, J. Chen (eds.), *Speech Enhancement* (Springer, Berlin, 2005). ISBN 978-3-540-24039-6

R. Banse, K.R. Scherer, Acoustic profiles in vocal emotion expression. J. Personal. Soc. Psychol. **70**(3), 614–636 (1996)

C. Busso, S. Lee, S. Narayanan, Analysis of emotionally salient aspects of fundamental frequency for emotion detection. IEEE Trans. Audio Speech Lang. Process. **17**, 582–596 (2009)

C.H. Demarty, C. Penet, G. Gravier, M. Soleymani, The MediaEval 2012 affect task: violent scenes detection in hollywood movies, in *Proceedings of the MediaEval 2012 Workshop*. CEUR-WS.org, Pisa, Italy, Oct 2012

S. Deng, J. Han, T. Zheng, G. Zheng, A modified MAP criterion based on hidden Markov model for voice activity detecion, in *Proceedings of ICASSP 2011* (IEEE, Prague, 2011), pp. 5220–5223

F. Eyben, M. Wöllmer, B. Schuller, openSMILE—the munich versatile and fast open-source audio feature extractor, in *Proceedings of ACM Multimedia 2010* (ACM, Florence, 2010), pp. 1459–1462

F. Eyben, B. Schuller, G. Rigoll, Improving generalisation and robustness of acoustic affect recognition, in *Proceedings of the 14th ACM International Conference on Multimodal Interaction (ICMI) 2012*, ed. by L.-P. Morency, D. Bohus, H.K. Aghajan, J. Cassell, A. Nijholt, J. Epps (ACM, Santa Monica, 2012a), pp. 517–522

F. Eyben, F. Weninger, N. Lehment, G. Rigoll, B. Schuller, Violent scenes detection with large, brute-forced acoustic and visual feature sets, in *Proceedings of the MediaEval 2012 Workshop*. CEUR-WS.org, Pisa, Italy, Oct 2012b

F. Eyben, F. Weninger, B. Schuller, Affect recognition in real-life acoustic conditions—a new perspective on feature selection, in *Proceedings of INTERSPEECH 2013* (ISCA, Lyon, 2013a), pp. 2044–2048

F. Eyben, F. Weninger, S. Squartini, B. Schuller, Real-life voice activity detection with LSTM recurrent neural networks and an application to hollywood movies, in *Proceedings of ICASSP 2013* (IEEE, Vancouver, 2013b), pp. 483–487

F. Eyben, F. Weninger, N. Lehment, B. Schuller, G. Rigoll, Affective video retrieval: violence detection in hollywood movies by large-scale segmental feature extraction. PLoS ONE **8**(12), e78506 (2013c). doi:10.1371/journal.pone.0078506

M. Fujimoto, S. Watanabe, T. Nakatani, Frame-wise model re-estimation method based on gaussian pruning with weight normalization for noise robust voice activity detection. Speech Commun. **54**(2), 229–244 (2012). doi:10.1016/j.specom.2011.08.005. ISSN: 0167-6393

J.S. Garofolo, L.F. Lamel, W.M. Fisher, J.G. Fiscus, D.S. Pallett, N.L. Dahlgrena, V. Zue, TIMIT acoustic-phonetic continuous speech corpus (1993)

R. Gemello, F. Mana, R.D. Mori, Non-linear esimation of voice activity to improve automatic recognition of noisy speech, in *Proceedings of INTERSPEECH 2005* (ISCA, Lisbon, 2005), pp. 2617–2620

A.B. Graf, S. Borer, Normalization in support vector machines, in *Pattern Recognition*, Lecture Notes in Computer Science (Springer, Berlin, 2001), pp. 277–282

A. Graves, S. Fernández, J. Schmidhuber, Multidimensional recurrent neural networks, in *Proceedings of the 2007 International Conference on Artificial Neural Networks (ICANN)*. Lecture Notes in Computer Science, vol. 4668 (Springer, Porto, 2007), pp. 549–558

M. Hahn, C.K. Park, An improved speech detection algorithm for isolated korean utterances, in *Proceedings of ICASSP 1992* (IEEE, San Francisco, 1992), vol. 1, pp. 525–528

J.A. Haigh, J.S. Mason, Robust voice activity detection using cepstral features, in *Proceedings of the IEEE Region 10 Conference on Computer, Communication, Control, and Power Engineering* (IEEE, 1993), vol. 3, pp. 321–324

R. Herbrich, T. Graepel, A PAC-Bayesian margin bound for linear classifiers: why SVMs work, in *Advances in Neural Information Processing Systems* (MIT press, Cambridge, 2001), pp. 224–230

H. Hermansky, Perceptual linear predictive (PLP) analysis for speech. J. Acoust. Soc. Am. (JASA) **87**, 1738–1752 (1990)

S. Hochreiter, J. Schmidhuber, Long short-term memory. Neural Comput. **9**(8), 1735–1780 (1997)

S. Hochreiter, Y. Bengio, P. Frasconi, J. Schmidhuber, Gradient flow in recurrent nets: the difficulty of learning long-term dependencies, in *A Field Guide to Dynamical Recurrent Neural Networks*, ed. by S.C. Kremer, J.F. Kolen (IEEE Press, New York, 2001)

M. Jeub, M. Schäfer, P. Vary, A binaural room impulse response database for the evaluation of dereverberation algorithms, in *Proceedings of the International Conference on Digital Signal Processing (DSP)* (IEEE, Santorini, 2009), pp. 1–4

J.-C. Junqua, J.-P. Haton, *Robustness in Automatic Speech Recognition: Fundamentals and Applications* (Kluwer Academic Publishers, Boston, 1996)

C.-C. Lee, M. Black, A. Katsamanis, A. Lammert, B. Baucom, A. Christensen, P.G. Georgiou, S.S. Narayanan, Quantification of prosodic entrainment in affective spontaneous spoken interactions of married couples, in *Proceedings of INTERSPEECH 2010* (ISCA, Makuhari, 2010), pp. 793–796

H. Lu, M. Rabbi, G. Chittaranjan, D. Frauendorfer, M. Schmid Mast, A.T. Campbell, D. Gatica-Perez, T. Choudhury, Stresssense: detecting stress in unconstrained acoustic environments using smartphones, in *Proceedings of the 2012 ACM Conference on Ubiquitous Computing (Ubicomp'12)* (ACM, Pittsburgh, 2012), pp. 351–360

M. Marzinzik, B. Kollmeier, Speech pause detection for noise spectrum estimation by tracking power envelope dynamics. IEEE Trans. Speech Audio Process. **10**, 109–118 (2002)

S. Matsuda, N. Ito, K. Tsujino, H. Kashioka, S. Sagayama, Speaker-dependent voice activity detection robust to background speech noise, in *Proceedings of INTERSPEECH 2012* (ISCA, Portland, 2012)

A. Misra, Speech/nonspeech segmentation in web videos, in *Proceedings of INTERSPEECH 2012* (ISCA, Portland, 2012)

S. Mousazadeh, I. Cohen, AR-GARCH in presence of noise: parameter estimation and its application to voice activity detection. IEEE Trans. Audio Speech Lang. Process. **19**(4), 916–926 (2011)

E. Mower, M.J. Mataric, S.S. Narayanan, A framework for automatic human emotion classification using emotional profiles. IEEE Trans. Audio Speech Lang. Process. **19**(5), 1057–1070 (2011). doi:10.1109/TASL.2010.2076804

T. Ng, B. Zhang, L. Nguyen, S. Matsoukas, X. Zhou, N. Mesgarani, K. Vesel, P. Matjka, Developing a speech activity detection system for the darpa rats program, in *Proceedings of INTERSPEECH 2012* (ISCA, Portland, 2012)

M.K. Omar, Speech activity detection for noisy data using adaptation techniques, in *Proceedings of INTERSPEECH 2012* (ISCA, Portland, 2012)

L. Parra, C. Alvino, Geometric source separation: merging convolutive source separation with geometric beamforming, in *Proceedings of the 2001 IEEE Signal Processing Society Workshop on Neural Networks for Signal Processing XI* (IEEE, 2001), pp. 273–282. doi:10.1109/NNSP.2001.943132

K. Pearson, Notes on regression and inheritance in the case of two parents. Proc. R. Soc. Lond. **58**, 240–242 (1895)

M.A. Pitt, L. Dilley, K. Johnson, S. Kiesling, W. Raymond, E. Hume, E. Fosler-Lussier, in *Buckeye Corpus of Conversational Speech (2nd release)*. Department of Psychology, Ohio State University (Distributor), Columbus, OH, USA (2007). http://www.buckeyecorpus.osu.edu/

J. Pittermann, A. Pittermann, W. Minker, Emotion recognition and adaptation in spoken dialogue systems. Int. J. Speech Technol. **13**, 49–60 (2010)

L.R. Rabiner, M.R. Sambur, Voice-unvoiced-silence detection using the itakura LPC distance measure, in *Proceedings of ICASSP 1977* (IEEE, Hartford, 1977), vol. 2, pp. 323–326

J. Ramirez, J. Segura, M. Benitez, A. De La Torre, A. Rubio, Efficient voice activity detection algorithms using long-term speech information. Speech Commun. **42**(3), 271–287 (2004)

J. Ramirez, J. Segura, C. Benitez, L. Garcia, A. Rubio, Statistical voice activity detection using a multiple observation likelihood ratio test. IEEE Signal Process. Lett. **12**(10), 689–692 (2005)

M. Schröder, E. Bevacqua, R. Cowie, F. Eyben, H. Gunes, D. Heylen, M. ter Maat, G. McKeown, S. Pammi, M. Pantic, C. Pelachaud, B. Schuller, E. de Sevin, M. Valstar, M. Wöllmer, Building autonomous sensitive artificial listeners. IEEE Trans. Affect. Comput. **3**(2), 165–183 (2012)

B. Schuller, G. Rigoll, M. Grimm, K. Kroschel, T. Moosmayr, G. Ruske, Effects of in-car noise-conditions on the recognition of emotion within speech, in *Proceedings of the 33. Jahrestagung für Akustik (DAGA) 2007* (DEGA, Stuttgart, 2007), pp. 305–306

B. Schuller, B. Vlasenko, F. Eyben, G. Rigoll, A. Wendemuth, Acoustic emotion recognition: a benchmark comparison of performances, in *Proceedings of the IEEE Workshop on Automatic Speech Recognition and Understanding (ASRU) 2009* (IEEE, Merano, 2009a), pp. 552–557

B. Schuller, R. Müller, F. Eyben, J. Gast, B. Hörnler, M. Wöllmer, G. Rigoll, A. Höthker, H. Konosu, Being bored? recognising natural interest by extensive audiovisual integration for real-life application. Image Vis. Comput. **27**(12), 1760–1774, Special issue on visual and multimodal analysis of human spontaneous behavior (2009b)

B. Schuller, Affective speaker state analysis in the presence of reverberation. Int. J. Speech Technol. **14**(2), 77–87 (2011a)

B. Schuller, A. Batliner, S. Steidl, D. Seppi, Recognising realistic emotions and affect in speech: state of the art and lessons learnt from the first challenge. Speech Commun. **53**(9/10), 1062–1087, Special issue on sensing emotion and affect—facing realism in speech processing (2011b)

B. Schuller, M. Valstar, R. Cowie, M. Pantic, AVEC 2012: the continuous audio/visual emotion challenge—an introduction, in *Proceedings of the 14th ACM International Conference on Multimodal Interaction (ICMI) 2012*, ed. by L.-P. Morency, D. Bohus, H.K. Aghajan, J. Cassell, A. Nijholt, J. Epps (ACM, Santa Monica, 2012a), pp. 361–362

B. Schuller, S. Steidl, A. Batliner, E. Nöth, A. Vinciarelli, F. Burkhardt, R. van Son, F. Weninger, F. Eyben, T. Bocklet, G. Mohammadi, B. Weiss, The INTERSPEECH 2012 speaker trait challenge, in *Proceedings of INTERSPEECH 2012* (ISCA, Portland, 2012b)

B. Schuller, *Intelligent Audio Analysis*, Signals and Communication Technology (Springer, Berlin, 2013a). ISBN 978-3642368059

B. Schuller, A. Batliner, *Computational Paralinguistics: Emotion, Affect and Personality in Speech and Language Processing* (Wiley, Hoboken, 2013b). p. 344, ISBN 978-1119971368

B. Schuller, F. Pokorny, S. Ladstätter, M. Fellner, F. Graf, L. Paletta, Acoustic geo-sensing: recognising cyclists' route, route direction, and route progress from cell-phone audio, in *Proceedings of ICASSP 2013* (IEEE, Vancouver, 2013c), pp. 453–457

B. Schuller, S. Steidl, A. Batliner, F. Schiel, J. Krajewski, F. Weninger, F. Eyben, Medium-term speaker states—a review on intoxication, sleepiness and the first challenge. Comput. Speech Lang. **28**(2), 346–374, Special issue on broadening the view on speaker analysis (2014)

V. Sethu, E. Ambikairajah, J. Epps, Speaker normalisation for speech-based emotion detection, in *Proceedings of the 15th International Conference on Digital Signal Processing (DSP 2007)*, pp. 611–614, Cardiff, UK, July 2007

J. Sohn, N. Kim, A statistical model-based voice activity detection. IEEE Signal Process. Lett. **6**(1), 1–3 (1999)

J. Sola, J. Sevilla, Importance of input data normalization for the application of neural networks to complex industrial problems. IEEE Trans. Nucl. Sci. **44**(3), 1464–1468 (1997). doi:10.1109/23.589532. ISSN: 0018-9499

Y. Suh, H. Kim, Multiple acoustic model-based discriminative likelihood ratio weighting for voice activity detection. IEEE Signal Process. Lett. **19**(8), 507–510 (2012). doi:10.1109/LSP.2012.2204978. ISSN: 1070-9908

M. Suzuki, S. Nakagawa, K. Kita, Prosodic feature normalization for emotion recognition by using synthesized speech, in *Advances in Knowledge-Based and Intelligent Information and Engineering Systems—16th Annual KES Conference*, vol. 243, Frontiers in Artificial Intelligence and Applications, ed. by M. Graña, C. Toro, J. Posada, R.J. Howlett, L.C. Jain (IOS Press, San Sebastian, 2012), pp. 306–313

W.Q. Syed, H.-C. Wu, Speech waveform compression using robust adaptive voice activity detection for nonstationary noise in multimedia communications, in *Proceedings of Global Telecommunications Conference, 2007 (GLOBECOM'07)* (IEEE, Washington DC, 2007), pp. 3096–3101

D. Talkin, A robust algorithm for pitch tracking (RAPT), in *Speech Coding and Synthesis*, ed. by W.B. Kleijn, K.K. Paliwal (Elsevier, New York, 1995), pp. 495–518. ISBN 0444821694

K. Thambiratnam, W. Zhu, F. Seide, Voice activity detection using speech recognizer feedback, in *Proceedings of INTERSPEECH 2012* (ISCA, Portland, 2012)

S. Thomas, S.H. Mallidi, T. Janu, H. Hermansky, N. Mesgarani, X. Zhou, S. Shamma, T. Ng, B. Zhang, L. Nguyen, S. Matsoukas, Acoustic and data-driven features for robust speech activity detection, in *Proceedings of INTERSPEECH 2012* (ISCA, Portland, 2012)

A. Vinciarelli, M. Pantic, H. Bourlard, Social signal processing: survey of an emerging domain. Image Vis. Comput. **27**(12), 1743–1759 (2009). doi:10.1016/j.imavis.2008.11.007

A.L. Wang, An industrial-strength audio search algorithm, in *Proceedings of ISMIR* (Baltimore, 2003)

F. Weninger, B. Schuller, M. Wöllmer, G. Rigoll, Localization of non-linguistic events in spontaneous speech by non-negative matrix factorization and long short-term memory, in *Proceedings of ICASSP 2011* (IEEE, Prague, 2011a), pp. 5840–5843

F. Weninger, B. Schuller, A. Batliner, S. Steidl, D. Seppi, Recognition of non-prototypical emotions in reverberated and noisy speech by non-negative matrix factorization. EURASIP J. Adv. Signal Process. (Article ID 838790), Special issue on emotion and mental state recognition from speech (2011b)

K. Woo, T. Yang, K. Park, C. Lee, Robust voice activity detection algorithm for estimating noise spectrum. Electron. Lett. **36**(2), 180–181 (2000)

M. You, C. Chen, J. Bu, J. Liu, J. Tao, Emotion recognition from noisy speech, in *Proceedings of the IEEE International Conference on Multimedia and Expo (ICME 2006)* (IEEE, Toronto, 2006), pp. 1653–1656. doi:10.1109/ICME.2006.262865

S. Young, G. Evermann, M. Gales, T. Hain, D. Kershaw, X. Liu, G. Moore, J. Odell, D. Ollason, D. Povey, V. Valtchev, P. Woodland, *The HTK Book*, Cambridge University Engineering Department, for HTK version 3.4 edition (2006)

Z. Zeng, M. Pantic, G.I. Rosiman, T.S. Huang, A survey of affect recognition methods: audio, visual, and spontaneous expressions. IEEE Trans. Pattern Anal. Mach. Intell. **31**(1), 39–58 (2009)

Chapter 6
Evaluation

The baseline acoustic feature sets and the methods for robust and incremental audio analysis have been evaluated extensively by the author of this thesis. In this chapter, first, a set of 12 affective speech databases and two music style data-sets is introduced, which are used for a systematic evaluation of the proposed methods and baseline acoustic feature sets. Next, the effectiveness of the proposed noise robust affective speech classification approach is evaluated on two of the affective speech databases in Sect. 6.2. Then, recognition results obtained with all the baseline acoustic feature sets on a large set of 10 speech and two music databases are presented and discussed in Sect. 6.3. Finally, recognition results for continuous, dimensional affect recognition with an incremental recognition method are shown in Sect. 6.4.

6.1 Speech and Music Databases

In order to evaluate the proposed methods and the baseline acoustic features sets on a comprehensive set of audio databases, a large set of 14 well known speech and music corpora has been chosen to cover basic acted emotional speech to fully natural, subtle, spontaneous affective speech, and music style identification tasks. The sets cover acted emotional speech in basic emotional categories (DES (Sect. 6.1.4) and EMO-DB (Sect. 6.1.5)), acted emotions in more, and more subtle categories (GEMEP (Sect. 6.1.7) and GeSiE (Sect. 6.1.10)), induced emotions (ABC (Sect. 6.1.1) and eNTERFACE (Sect. 6.1.6)), spontaneous emotions in highly emotional or extreme situations (VAM (Sect. 6.1.12) and SUSAS (Sect. 6.1.11)), and subtle, spontaneous emotions (induced) in natural, every-day like situations (SAL (Sect. 6.1.8) and TUM-AVIC (Sect. 6.1.3)), as well as music style identification tasks (GEDIDB (Sect. 6.1.14) and BRD (Sect. 6.1.13)). In addition to these 12 sets, which are used for a systematic evaluation of the baseline acoustic feature sets, the FAU-AIBO set (Sect. 6.1.2) is used for additional validation, and the SEMAINE database

F. Eyben, *Real-time Speech and Music Classification by Large Audio Feature Space Extraction*, Springer Theses, DOI 10.1007/978-3-319-27299-3_6

Table 6.1 Overview of the 11 chosen speech/singing emotion databases (including the FAU-AIBO database): corpora 1–4 shown here (ABC, FAU-AIBO, TUM-AVIC, DES)

	ABC		AIBO		AVIC		DES	
Content	German, fix.		German, var.		English, var.		Danish, fix.	
Emotion type	Induced		Natural		Natural		Acted	
Emotion categories (and # of instances)	Aggressive	95	A	1,492	LOI1	553	Angry	85
	Cheerful	105	E	3,601	LOI2	2279	Happy	86
	Intoxicated	33	N	10,967	LOI3	170	Neutral	85
	Nervous	93	P	889			Sad	84
	Neutral	79	R	1,267			Surprise	79
	Tired	25						
Arousal	High	326		–		2449		250
	Low	104		–		553		169
Valence	Pos.	217		12,393		2449		250
	Neg.	213		5,823		553		169
# Inst. total	430		18,216		3002		419	
Duration [hh:mm]	01:15		09:12		01:47		00:28	
# Subjects (# f)	8 (4)		51 (30)		21 (10)		4 (2)	
Recording env.	Studio		School		Office		Office	
Sample rate (kHz)	16		16		44.1		20	

Content refers to the linguistic content, i.e., language and fixed (fix.) or variable (var.) content

(Sect. 6.1.9) is used for extensive evaluation of the proposed incremental segmentation and continuous affect regression methods.

Statistics for the 11 emotional speech and singing voice corpora (excluding the SEMAINE database) are given in Tables 6.1, 6.2, and 6.8.

6.1.1 Airplane Behaviour Corpus (ABC)

The Airplane Behaviour Corpus (ABC) introduced by Schuller et al. (2007b) was crafted for the special application of automatic public transport surveillance regarding passenger mood. During the recording, a certain mood was induced in the subjects by a script, which guided the subjects through a storyline: pre-recorded cabin announcements controlled by a hidden test-conductor were automatically played back by five speakers at different positions. Subjects had to imagine being on a vacation (and return) flight, consisting of these scenes: take-off, serving of wrong food, turbulences, sleeping, talking to the person in the next seat, and landing. The setup consisted of an airplane seat for the subject, which was located in front of a blue screen. Eight subjects in gender balance between 25–48 years (mean 32 years) old participated in the recording. The language used throughout the recordings is German. 11.5 h

Table 6.2 Overview of the 11 chosen speech/singing emotion databases; corpora 5–8 shown here (EMO-DB, eNTERFACE, GEMEP, SAL)

	EMO-DB		eNTERF.		GEMEP		SAL	
Content	German, fix.		English, fix.		No lang., fix.		English, var.	
Emotion type	Acted		Induced		Acted		Induced	
Emotion categories (and # of instances)	Anger	127	Anger	215	Amusem.	86	q1	459
	Disgust	38	Disgust	215	Pride	90	q2	320
	Fear	55	Fear	215	Joy	90	q3	564
	Happin.	64	Happin.	207	Interest	90	q4	349
	Sadness	53	Sadness	210	Relief	90		
	Boredom	79	Surprise	215	Pleasure	90		
	Neutral	78			Hot anger	89		
					Despair	90		
					Panic fear	90		
					Irritation	90		
					Sadness	90		
					Anxiety	90		
Arousal	High	246		852		535		808
	Low	248		425		540		884
Valence	Pos.	142		422		536		779
	Neg.	352		855		539		917
# Inst. total	494		1277		1075		1692	
Duration [hh:mm]	00:22		01:00		00:43		01:41	
# Subjects (# f)	10 (5)		42 (8)		10 (5)		4 (2)	
Recording env.	Studio		Office		Studio		Studio	
Sample rate (kHz)	16		48		44.1		16	

Abbreviations: happiness (happin.), amusement (amusem.); eNTERFACE (eNTERF.). *Content* refers to the linguistic content, i.e., language (lang.) and fixed (fix.) or variable (var.) content

of audiovisual material was recorded and—after pre-segmentation—was annotated independently by three experienced male labellers using a pre-defined, closed set of behavioural categories: neutral, tired, aggressive, cheerful, intoxicated, and nervous. A total of 431 clips remains with an average per-clip length of 8.4 s.

6.1.2 FAU-AIBO Database (AIBO)

The FAU-AIBO database (Steidl 2009, Chap. 5) is a corpus of recordings of children who interact with Sony's toy pet robot Aibo. It is well know to the community of speech emotion recognition researchers, since it was used for the first international Emotion Challenge in 2009 (Schuller et al. 2009b). The corpus contains emotionally coloured spontaneous, German. Emotional situations were created by making the children believe that the Aibo robot was reacting to their voice commands, whereas it was in fact controlled by a human operator, who caused the robot to perform a pre-scripted sequence of actions. Thus, by chance, in some situations Aibo seemed to obey the childrens' commands while in others it disobeyed the commands and thereby provoked emotional reactions. Recordings were conducted at two different schools, known as *MONT* and *OHM*. In total, 51 children (age 10–13, 21 male, 30 female) participated in the recordings and around 9.2 h of speech—without pauses—are contained.

Speech was recorded from a high quality, wireless head set microphone with a DAT-recorder at 48 kHz and 16-bit depth. It was then down-sampled to 16 kHz, a rate commonly used for speech databases. The continuous recordings were then segmented into turns automatically with an energy based method using a pause threshold of one second. The data was transcribed on the word level with respect to 10 emotional categories and a neutral state (default, if no emotion category was assigned to a word) independently by five advanced students of linguistics, who listened to the turns in sequential order and annotated the recording word by word. The word level labelling was chosen because many utterances are only short commands and medium pauses can occur between words due to Aibo's reaction time. A single label per word was found by majority voting of the 5 individual ratings: if three or more labellers agreed, the label was assigned to the word.

The ten emotion categories which were originally labelled (and their number of occurrence after majority voting), are: joyful (101), surprised (0), emphatic (2,528), helpless (3), touchy/irritated (225), angry (84), motherese (1,260), bored (11), reprimanding (310), rest/other (3). The majority of words, however, was found to be neutral (39,169). 4,707 words have ambiguous labellings where no majority vote could be assigned. In total, there are 48,401 words. Due to sparsely occurring emotion categories in the above, the 10 categories were combined to both a five and a two category set by Schuller et al. (2009b). The five categories are: **A**nger (containing angry, touchy, and reprimanding), **E**mphatic, **N**eutral, **P**ositive (containing motherese and joyful), and **R**est (helpless, bored, rest). The two categories represent **NEG**ative emotions/valence (angry, touchy, reprimanding, and emphatic) and **IDL**e (i.e., all non-negative emotions).

6.1.3 TUM Audiovisual Interest Corpus (AVIC)

The TUM Audiovisual Interest Corpus (AVIC) contains audiovisual recordings of spontaneous affect with non-restricted spoken content (Schuller et al. 2009a). It was used as data-set for the INTERSPEECH 2010 Paralinguistics Challenge (Schuller et al. 2010a). In the recording set-up, a product presenter agent leads a subject through a commercial presentation. The language used is English, although most of the presenters were German native speakers, and the subjects were from various European and Asian nationalities. Recordings of 21 subjects (10 female) are contained in the corpus.

The Level of Interest (LOI) is annotated for every sub-turn (pause based subdivision of speaker turns) with three labels ranging from *boredom* (the subject is bored with the conversation and/or the topic, is very passive, and does not follow the discourse; also referred to as *LOI1*), over *neutral* (the subject follows and participates in the conversation and it can not be judged, whether she/he is interested in or indifferent towards the topic; also referred to as *LOI2*) to *joyful* interaction (showing a strong wish of the subject to talk and learn more about the topic, i.e., the subject shows a high interest in the discussion; also referred to as *LOI3*). These three discrete levels were obtained from majority voting over four individual raters' opinions and after combining the original 5 level annotation to only 3 levels to avoid sparsity in some of the 5 levels.

For the evaluations of the baseline acoustic feature sets in this thesis (Sect. 6.3) all 3,002 phrases (sub-turns) as in Schuller et al. (2009c) are used—in contrast to the only 996 phrases with high inter-labeller agreement as, e.g., employed by Schuller et al. (2009a). For the analysis of features for noise robust affective speech classification (Sect. 6.2), another version of this data-set, which has continuous labels, was used. This version was provided for the INTERSPEECH 2010 Paralinguistic Challenge (Schuller et al. 2010a). The (quasi) continuous labels for the LOI were estimated by averaging the four original discrete 5-level ratings. For details, it is referred to Schuller et al. Schuller et al. (2009a, 2010a).

6.1.4 Danish Emotional Speech Database (DES)

The Danish Emotional Speech (DES) database (Engbert and Hansen 2007) contains nine Danish sentences, two words and chunks which are located between two silent segments of two fluent text passages, e.g., *"Nej" (No), "Ja" (Yes), "Hvor skal du hen?" (Where are you going?)*. The set used in this thesis has 419 speech utterances (i.e., speech segments bounded by silence pauses). These utterances were acted by four professional actors (two males and females). All utterances are equally balanced for each gender and actor. The utterances are enacted in five discrete emotional categories: *anger, happiness, neutrality, sadness, and surprise*. Twenty raters (native Danish speakers, 18–58 of age) verified the enacted emotions in a listening test with a success rate of 67 %.

6.1.5 Berlin Emotional Speech Database (EMO-DB)

A very well known and widely used set to test the effectiveness of automatic emotion classification is the Berlin Emotional Speech Database (EMO-DB) introduced by Burkhardt et al. (2005). It contains sentences spoken in the following emotion categories: *anger, boredom, disgust, fear, joy, neutrality*, and *sadness*. The spoken content is pre-defined as ten German sentences, which are emotionally neutral, such as *"Der Lappen liegt auf dem Eisschrank" (The cloth is lying on the fridge.)*. As with the DES corpus, the EMO-DB therefore provides a few common words repeated several times in diverse emotions and a neutral state. Ten (five female) professional actors act ten sentences in each of the 7 emotional states. While the whole set contains over 700 utterances, in a listening test only 494 phrases are labelled as a minimum 60 % naturally sounding and a minimum 80 % re-identifiable (in terms of emotion labelling) by 20 people. A mean accuracy of 84.3 % is achieved for identification of the emotions by the subjects in the listening experiment on this set of 494 utterances. This set is used in most other studies related to this database (cf. Schuller et al. (2009c)), thus, it is also adopted for this thesis.

6.1.6 eNTERFACE'05 Database

The eNTERFACE database (Martin et al. 2006) is a publicly available, audiovisual emotion database. The contained material consists of induced emotions of the categories: *anger*, *disgust*, *fear*, *joy*, *sadness*, and *surprise*. 42 subjects (eight of them female) from 14 nations participated in the recording at the 2005 eNTERFACE summer school. The recordings were conducted in an office environment, and have pre-defined English linguistic content. In order to induce the emotions, each subject was instructed to listen to six short stories successively, with each story designed to elicit a particular emotion. Then, they had to react to situations created by the stories by speaking previously read phrases that fit the short story in the mood that the story elicited in them. They could choose from five phrases for each emotion, such as *"I have nothing to give you! Please don't hurt me!"* in the case of fear. Two raters judged whether the reaction expressed the induced emotion unambiguously. Only in this case, that sample was added to database. In total, the database contains 1,277 verified samples.

6.1.7 Geneva Multimodal Emotion Portrayals (GEMEP)

The Geneva Multimodal Emotion Portrayals (GEMEP) corpus is a compilation of 1,260 multimodal emotion expressions acted by ten French-speaking actors (Bänziger et al. 2012). It was used in several studies of emotional speech analysis (e.g., Scherer (2013); Bänziger et al. (2014)) and, most notably, for the INTERSPEECH 2013 Computational Paralinguistics ChallengE (ComParE—cf. Schuller et al. (2013)). The list of emotions includes the ones most frequently used in the literature (e.g., anger, fear, joy, and sadness) as well as more subtle variations of those categories (e.g., anger versus irritation, and fear versus anxiety).

The actors portrayed each emotion by means of three different verbal contents (one sustained vowel and two language-free sentences) and various expression regulation strategies. During this process they were recorded by three cameras and a microphone. All devices were synchronised. In order to increase the utterances' realism and spontaneity, during the recording sessions a professional director helped the respective actor to choose one personal scenario for each emotion—e.g., by recall or mental imagery—which was relevant for the actor. The actors did not receive specific instructions how the emotions were to be expressed and they were free to use any movement and speech prosody they felt was appropriate.

In this thesis, for the evaluation of the baseline acoustic feature sets (Sect. 6.3), the 12 repeated emotions are considered, which are distributed across all four quadrants of the activation-valence space: *amusement*, *pride*, *joy*, *relief*, *interest*, *pleasure*, *hot anger*, *panic fear*, *despair*, *irritation* (cold anger), *anxiety* (worry), and *sadness* (depression). 1,075 instances ($\approx$90 per emotion) belong to this set of 12 emotions. The remaining instances contain recordings for the following five additional emotional categories *admiration*, *tenderness*, *disgust*, *contempt*, and *surprise*. However, due to the sparseness of these classes, they are excluded for the work presented in this thesis.

For the evaluations of noise robustness (Sect. 6.2), a subset of 154 instances is used for which a continuous valued dimensional observer annotation exists. Continuous dimensional annotations are required there, in order to compute Pearson correlation coefficients for the proposed feature selection approach. The continuous annotations were obtained from twenty participants (10 male) who rated each of the 154 expressions in terms of arousal and valence by using a continuous slider (range -1 to $+1$). The evaluator weighted estimator (Grimm and Kroschel 2005) of all the twenty raters is used as a single 'gold standard' for each instance. In the continuously rated part, only the pseudo-sentences are included.

6.1.8 Belfast Sensitive Artificial Listener Database (SAL)

The Belfast Sensitive Artificial Listener (SAL) database is publicly available as part of the HUMAINE database (Douglas-Cowie et al. 2007). The subset adopted by Wöllmer et al. (2008) is considered in this thesis. It comprises 25 recordings from four speakers (two female) having an average length of 20 min per speaker. The data contain audio-visual recordings from natural human-computer interactions where humans interacted with a (Wizard-of-Oz) SAL interface which was designed to direct users through a range of emotional states. The material is labelled time-continuously by four labellers in terms of valence and activation using FEELtrace (Cowie et al. 2000): thereby the labellers used a slider to annotate both emotional dimensions separately in real time while watching the videos of the interactions. The values from the sliders for valence and activation were sampled every 10 ms in order to obtain a quasi-continuum in time. To compensate for constant offsets which are different for each labeller, the labels of each annotator were shifted to zero mean globally.

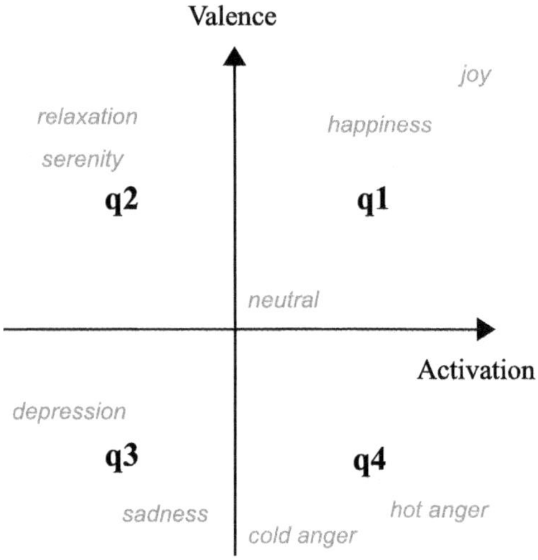

Fig. 6.1 Activation-valence space with the four quadrants (q1–4) labelled and exemplary emotion categories placed within the respective quadrants

Further, to ensure a common scale across all four annotators, each annotator's labels were re-scaled so that 98 % of all values fell into the range $[-1; 1]$ (according to Wöllmer et al. (2008)), i.e., the ratings were scaled linearly so that the 1st and the 98th percentile values were mapped to -1 and $+1$, accordingly. This ensured robustness of the scaling against single outliers. The 25 recording sessions have been segmented into user speech turns with an energy based Speech Activity Detector. A total of 1,692 user speech turns is obtained with this method. These turns are the same as were used by Wöllmer et al. (2008); Schuller et al. (2009c); Eyben et al. (2010). For each turn one label for activation and one for valence is computed by averaging the 10 ms sampled valence and activation values over the length of the turn. A great challenge of the SAL database is the fact that there is no selection of emotionally prototypical instances—one must deal with all data, i.e., natural emotions that do not necessarily match well-defined prototypical categories. As for the Vera-am-Mittag (VAM) database (cf. below), for this thesis the valence and activation labels were discretised into four categories representing the four quadrants of the activation-valence space (q1, q2, q3, and q4—see Fig. 6.1). Thereby q1 corresponds to happy, joyful emotions, q2 corresponds to relaxed, positive emotions, q3 corresponds to sadness and depressed states, and q4 to anger, rage, and aggressiveness.

6.1.9 SEMAINE Database

The SEMAINE database, as introduced by McKeown et al. (2012) and used by the author of this thesis in Eyben et al. (2012a), was recorded in order to study social interaction that occurs in natural conversations between humans and the future generation of socially intelligent and affective computing systems.[1] The scenario used for creating emotionally coloured, naturalistic interactions is the SAL scenario, which was also used for the Belfast SAL database described in the previous section. This scenario involves a user who is interacting with emotionally stereotyped 'characters'. The responses of those characters are pre-defined phrases provoked by the user's emotional state rather than by the content of the user's utterances. This model is a style of interaction which was observed in chat shows and social parties. It suggests that it seems possible that a machine with some basic emotional and conversational understanding could sustain a basic social conversation, without much linguistic knowledge. In the data used for the experiments here, which are part of the Solid-SAL part of the database, a human as a Wizard-of-Oz took the role of the virtual characters (operator). The characters did not have to adhere to the SAL protocol strictly, i.e., the main goal was to have a natural conversation going, sticking to predefined sentences was not mandatory and seldom followed. Asking questions to the virtual characters was not allowed, and the operators were instructed to not answer questions. More details can be found in McKeown et al. (2012).

The database has been constantly enlarged and extended as new data had been annotated and added to the on-line database. The version used in this thesis, is that first used by the author of this thesis in Eyben et al. (2012a). The Solid-SAL part of this database version consists of recordings of 20 trials, which are split into over 100 per-character conversations of an approximate length of 5 min each. All recorded conversations have been annotated in five core affective dimensions, namely *Activation*, *Expectation* (Anticipation), overall emotional *Intensity*, *Power*, and *Valence*. The SEMAINE database is thus among the first databases which have annotations for more than three dimensions. To the best knowledge of the author of this thesis, the only other such databases are the CINEMO corpus (Schuller et al. 2010b) and the RECOLA database (Ringeval et al. 2013).

The choice of these dimensions is based on psychological findings reported by Fontaine et al. (2007): in his work, the four dimensions activation, valence, expectation, and power, were obtained by a Principal Component Analysis (PCA) which was applied to 144 hand assigned "emotion features" which in turn were derived from adjectives people use to verbally describe emotional events. Activation (or arousal) indicates the level of arousal, i.e., the level of active engagement or readiness for action, versus passiveness as found in boredom, for example. The valence dimension contrasts pleasant against unpleasant emotions, i.e., it indicates how positive (i.e., pleasant) or negative (i.e., unpleasant) the emotion is ranked. The 'Power' dimension characterises whether the subject feels power and in control of a situation or not. Expectation is a quantification of unpredictability versus expectedness or familiarity of an event. Surprise, fear, and disgust, for example, are associated with a low expectation value, while emotions, such as stress or neutrality, are associated with a higher value because they occur in situations the subject is more familiar with. The intensity dimension, was added by the creators of the SEMAINE database as

[1] The database is publicly available for scientific research for free from http://semaine-db.eu.

a measure of the perceived emotional intensity, i.e., the distance of an instance from the point of general neutrality, regardless on which dimension.

The annotations were performed by using trace like continuous ratings (similar to FeelTrace (Cowie et al. 2000)). Thereby, the raters could move a slider continuously in a given range when listening to the recording to rate their subjective opinion for a single affective dimension at a time. They did this by watching the video and listening to the audio of a single, full recording session and thereby adjusting the trace slider—as the video went along—for a single selected dimension as they personally thought it would best match the emotional state of the user in the video at each time instant. Corrections were not allowed and not possible. It is important to highlight that the ratings are based on the video and audio at the same time and no individual ratings exist. Therefore, it can be expected that certain regions of the ratings are more correlated to the audio and others are more correlated to the video. Details and annotation guidelines can be found in Douglas-Cowie et al. (2007). The ratings from the slider are sampled at a rate of 50 samples per second (matching the video frame rate) and with a numerical accuracy of 0.001. Full trace style ratings for the five core affective dimensions exist from eight raters in the database at present. However, at the time when the experiments were conducted not all raters had rated all sessions yet. Thus, only those sessions were considered in the experiments where the number of raters available was at least three (Eyben et al. 2012a).

A single target for each dimension was obtained by averaging the ratings of the individual raters as in Eyben et al. (2012a). For all experiments in the ongoing this mean label is used as the target. It is referred to as *mean label* or *mean dimension label* in the following text. In addition to this mean label, the standard deviation of all individual raters is computed as an inter-rater confidence measure. Wöllmer et al. (2008) used a normalisation of the labels for each rater before averaging in order to compensate for inter-rater scale differences and biases. However, in contrast to the SAL database used by Wöllmer et al. (2008), for the SEMAINE database no major scale differences or biases were observed among the raters. Thus, it was decided by the author of this thesis in Eyben et al. (2012a) not to apply any normalisation to the ratings.

Here, 20 recordings with 3–4 sessions on average are considered. From these, 57 sessions remain with 3 or more raters for each dimension. From these, 36 sessions are used for training, 14 for evaluation, and 7 as a development set which was used for tuning model parameters. The sequence IDs of the training sessions as they appear in the publicly available SEMAINE database are 34–37, 40–43, 46–49, 58–61, 70–73, 76–79, 82–85, 88–91, and 94–97. The IDs of the development set are 19–22, and 29–31, and the evaluation set IDs are 13–16, 25–27, 52, 53, 55, and 64–67 (cf. Eyben et al. 2012a). In total, 2,189 user speech turns are contained. Table 6.3 gives a summary of the database size and the contents. Gender balance of the subjects in the evaluation set was ensured by selecting sessions from two male and two female subjects. The training, development, and evaluation sets are disjunctive in terms of subjects, i.e., speech data from no single subject occurs in more than one set, ensuring speaker independence for the experiments.

Table 6.3 The SEMAINE database: size and statistics of the data used for experiments in this thesis (Eyben et al. 2012a)

No. of characters	4 (Obadiah, Poppy, Prudence, and Spike)
No. of subjects	21
No. of recording sessions	57
No. of user speech turns	2,189
Total time of user speech	≈5 h

Table 6.4 Dimensional affect ratings' statistics for the union set of training and development set (SEMAINE database) as used in this thesis

Dimension	Minimum	Maximum	μ	σ
A_μ	−0.798	0.656	−0.043	0.222
A_σ	0.000	0.594	0.259	0.102
E_μ	−0.769	0.604	−0.358	0.190
E_σ	0.000	0.720	0.223	0.127
I_μ	−0.899	0.697	−0.144	0.188
I_σ	0.000	0.597	0.254	0.105
P_μ	−0.747	0.749	0.417	0.189
P_σ	0.000	0.664	0.165	0.116
V_μ	−0.965	0.887	0.040	0.320
V_σ	0.000	0.499	0.124	0.075

Dimensions: A(ctivation), E(xpectation), I(ntensity), P(ower), and V(alence). Mean of all raters of the mean turn label (subscript μ) and inter-rater standard deviation for the average turn label (subscript σ); minimum and maximum value, mean (μ), and standard deviation (σ) of μ and σ (subscripts) given for each dimension (Eyben et al. 2012a)

The training set contains 1,584 user speech turns. A turn is defined as a continuous segment of user speech preceded or followed either by the initial or final silence of the recording session or a segment of agent speech. The turn start and end points have been manually transcribed. The development set has 169 user speech turns. Table 6.4 shows the distribution of the "gold standard" dimensional affect labels (average human ratings) for all user turns in the training and development set as published by the author of this thesis in Eyben et al. (2012a). Table 6.5 shows the distribution of the "gold standard" dimensional affect labels (mean human ratings) for the *evaluation* set (436 user speech turns). The figures approximately match those of the whole corpus, which proofs that the data in the evaluation set (with respect to emotionalness) reflects the conditions in the training and development sets sufficiently.

In Tables 6.4, and 6.5, the minimum values of the inter-rater standard deviation for each dimension are close to zero. This, in theory, indicates ideal rater agreement in certain (possibly small) regions of the sessions. Although this can be a random effect, it is more likely due to the simple fact that during the rating process the sliders were

Table 6.5 Dimensional affect ratings' statistics for the *evaluation* (test) set of the SEMAINE database as used in this thesis

Dimension	Minimum	Maximum	μ	σ
A_μ	−0.582	0.480	−0.027	0.250
A_σ	0.000	0.547	0.247	0.087
E_μ	−0.730	0.441	−0.351	0.229
E_σ	0.000	0.773	0.254	0.140
I_μ	−0.548	0.648	−0.118	0.270
I_σ	0.000	0.425	0.189	0.080
P_μ	−0.350	0.718	0.339	0.237
P_σ	0.000	0.610	0.220	0.141
V_μ	−0.715	0.659	0.008	0.321
V_σ	0.000	0.401	0.119	0.075

Dimensions: A(ctivation), E(xpectation), I(ntensity), P(ower), and V(alence). Mean of all raters of the mean turn label (subscript μ) and inter-rater standard deviation for the average turn label (subscript σ); minimum and maximum value, mean (μ), and standard deviation (σ) of μ and σ (subscripts) given for each dimension (Eyben et al. 2012a)

initialised to the same value at the beginning of each session. These initial values remained for a few seconds, until the labellers moved the sliders. Therefore, these low standard deviation values can not be regarded as a good indicator of the actual minimum values of the inter-rater standard deviation (Eyben et al. 2012a).

In addition it is to note that in Eyben et al. (2012a) the dimension *Expectation* (E) was scaled from its original range [0; 100] by Eq. (6.1) in order to match the range $[-1; +1]$ of the other four dimensions:

$$E^* = \frac{E}{50} - 1 \tag{6.1}$$

From Table 6.4 it can be seen that for the training and development sets the mean inter-rater standard deviation (subscript σ) for each turn is about 0.2, with a maximum up to about 0.7. This emphasises the subjectivity and individual differences of the ratings and suggests that the disagreement is far more pronounced for some sentences than for others.

A more fine grained method to systematically analyse pairwise rater agreement is to compute correlation coefficients between all rater pairs. Table 6.6 displays pairwise correlations between four raters (R1, R3, R5, and R6).[2] The inter-rater correlations are given for the evaluation set, for reasons of comparability with the correlations of automatically obtained predictions discussed in Sect. 6.4.2. Interestingly, the human agreement is best for the valence dimension and worst for the power dimension. For automatic recognition of affect it is commonly observed (and also confirmed

[2]The names of the raters are as used in the SEMAINE database, therefore the rater IDs are not continuous.

Table 6.6 (Pearson) Correlation Coefficient (CC) between all labellers for each of the five dimensions, computed from the evaluation set sessions

Labeller: Dimension	R1-R3	R1-R5	R1-R6	R3-R5	R3-R6	R5-R6	Mean
Activation	0.655	0.552	0.541	0.667	0.494	0.495	**0.567**
Expectation	0.331	0.116	0.216	0.394	0.233	0.212	**0.250**
Intensity	0.694	0.635	0.401	0.696	0.576	0.535	**0.590**
Power	0.306	0.172	0.217	0.452	0.064*	0.156	**0.228**
Valence	0.756	0.720	0.750	0.790	0.829	0.779	**0.771**

CCs marked with * are *not* statistically significant (i.e., can be seen as uncorrelated) on a level of $p = 0.05$ using a 2-tailed test (Eyben et al. 2012a)

Table 6.7 Pairwise (Pearson) Correlation Coefficient (CC) between all five dimensions, computed on the evaluation set sessions

Dimension	**E**	**I**	**P**	Valence
Arousal	−0.136	0.673	0.126	−0.125
Expectation		0.132	−0.659	0.004
Intensity			−0.287	−0.496
Power				0.220

All correlations, except for the V-E combination) are significant (i.e., can be seen as correlated more than two random sequences) on a level of $p = 0.05$ based on a 2-tailed test (Eyben et al. 2012a)

in this thesis) that—from acoustic parameters alone—valence is the most difficult dimension to predict. Presumably the human raters rely on the linguistic content, i.e., the meaning of the words, for their judgement of valence. Acoustic parameters, on the other hand, are very reliable indicators of activation and intensity. Given the low inter-rater correlations for Power and Expectation (roughly 0.25), it can be questioned whether there is a reliable gold standard to train models on for these two dimensions. It is assumed by the author of this thesis in Eyben et al. (2012a) that there is still some validity as some raters agree far better than other raters. This could show that each rater judged these dimensions based on different cues but each single rater could be consistent with itself though. Also, this could indicate problems of the raters with understanding the meaning or definition of these dimensions.

Table 6.7 gives the inter-dimension correlation coefficients on the evaluation set, to better understand the relations between the five dimensions. The highest correlation appears for activation and intensity (0.673). This suggests that a high overall emotional intensity occurs together with high activation. It can be further seen that expectation and power are negatively correlated, i.e., a high value for expectation is associated with a low value of power/control. A negative correlation is also observed for intensity and valence. This indicates that—at least in the SEMAINE database—negative emotions are expressed with a higher intensity than positive emotions. The remaining pairs are considered uncorrelated due to their low correlation coefficients.

6.1.10 Geneva Singing Voice Emotion (GeSiE) Database

A database of singing emotional speech was first introduced by Scherer et al. (2015). In this thesis, an extended set of the database is used (abbreviated as GeSiE in the ongoing) as introduced by the author of this thesis in Eyben et al. (2015): Compared to the original set which contains three singers, additional recordings of five professional opera singers have been added here (eight recordings in total) following the same protocol. There now are five male and three female singers. The singers had to sing three different phrases and tone scales in ten emotion categories: panic/fear, passionate love, tense arousal, animated joy, triumphant pride, anger, sadness, tenderness, calm/serenity, condescension, and a neutral state (no expression). Each recording session was recorded as a whole without pause. Then, the recordings were manually split into the phrase and scale parts. In this way, a set of 300 instances was obtained. The distribution of the instances across the emotion classes is almost balanced (cf. Table 6.8).

Table 6.8 Overview of the 11 chosen speech/singing emotion databases; corpora 9–11 shown here (GeSiE, SUSAS, VAM)

	GeSiE		SUSAS		VAM	
Content	Singing, fix.		English, fix.		German, var.	
Emotion type	Acted		Induced		Spontaneous	
Emotion categories (and # of instances)	Neutral	24	High stress	1202	q1	21
	Fear	30	Medium stress	1276	q2	50
	Pass. love	24	Neutral	701	q3	451
	Tense aro.	24	Scream	414	q4	424
	Anim. joy	31				
	Pride	30				
	Anger	29				
	Sadness	30				
	Tenderness	30				
	Calm	24				
	Condescension	24				
Arousal	High	144		2892		445
	Low	156		701		501
Valence	Pos.	163		1977		71
	Neg.	137		1616		875
# Inst. total	300		3593		946	
Duration [hh:mm]	01:39		01:01		00:47	
# Subjects (# f)	8 (3)		7 (3)		47 (32)	
Recording env.	Studio		Field, noisy		Tv studio	
Sample rate (kHz)	96		8		16	

Content refers to the linguistic content, i.e., language and fixed (fix.) or variable (var.) content

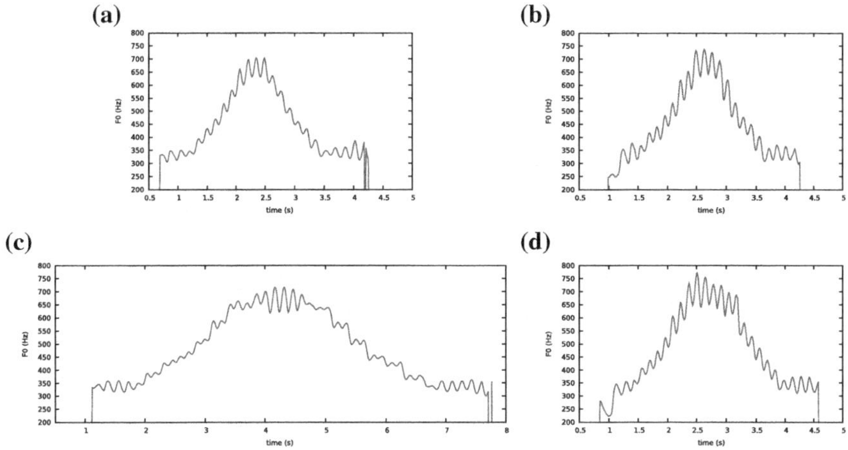

Fig. 6.2 Pitch (F_0) contour plots (extracted with Subharmonic Summation (SHS) and Viterbi post-smoothing) for the same tonal scale sung by one female opera singer in three different emotional styles: **a** Neutral, **b** angry, **c** sad, and **d** proud. Plots after Eyben et al. (2015)

Pitch contour plots for one of the female singers, singing the same tonal scale forwards and backwards emotionally neutral, sad, angry, and proud is shown in Fig. 6.2. In these plots clear differences of the style and type of vibrato can be seen across the emotions. For sadness a large variation in the strength of vibrato over time is observed and very little vibrato is seen in the ascending scale; also, the overall tempo is slower. Most vibrato is seen for anger, closely followed by pride. This supports the findings, that vibrato is likely an indication of arousal and enthusiasm (Nakano et al. 2006; Daido et al. 2014).

6.1.11 Speech Under Simulated and Actual Stress (SUSAS)

The Speech Under Simulated and Actual Stress (SUSAS) database (Hansen and Bou-Ghazale 1997) was one of the first databases to contain recordings of spontaneous emotions. To increase the challenge, speech is partially masked by field noise. In this thesis the 3,593 actual stressed speech samples are used, which were recorded in motion fear and stress tasks. Recordings from seven subjects (three female) in actual stress situations (roller coaster and free fall) are contained in this set. Next to *neutral* speech and speech under *fear*, two different stress conditions were collected: *medium stress*, and *high stress*. Additionally, a small set of *screaming* utterances (presumably very high stress) was collected—but is excluded in the evaluations here. The texts in SUSAS are also restricted to pre-defined content of 35 English air-traffic commands, e.g., *"brake"*, *"help"* or *"no"*. Thus, only single words are contained.

6.1.12 Vera-Am-Mittag (VAM)

The VAM corpus was prepared at the University of Karlsruhe, Germany by Grimm et al. (2008). It consists of audio-visual recordings extracted from the German TV show "Vera am Mittag". In this show, the host (Vera) moderates discussions between the guests, e.g., by using questions to guide the discussion. The database contains 947 spontaneous speech and emotionally rich utterances sampled from 47 talk show guests. The discussions were authentic and not scripted and due to the nature of the show and the selection of guests the discussions between the talk show guests are quite often rather affective and contain a large amount of emotional states. The topics discussed in the show were mostly personal issues, like friendship crises, fatherhood questions, or love affairs. At the time of the recording of the TV show, the subjects were not aware that the recordings were ever going to be analysed in scientific studies. The emotions within the VAM corpus are described in terms of three affective dimensions: activation, valence, and dominance/power. More details on these dimensions can be found in Sect. 6.1.9. The dominance dimension is not used in this thesis as is was found to be highly correlated with activation.

For annotation, the audio recordings were manually segmented into utterances. 17 annotators rated one half of the data, and 6 raters the other half. The raters used an icon-based method where they could choose an image from an array of five images (Self-Assessment Manikins) for each emotion dimension (Grimm and Kroschel 2005). Each rater had to listen to each utterance and then choose an icon for each emotion dimension which best described the emotion in that utterance. The choice of these icons was afterwards mapped onto a discrete five category scale for each dimension evenly distributed across the range $[-1; 1]$ and averaged over annotators with some weighting function to consider annotator certainty as described by Grimm et al. (2007b). The average standard deviations of the annotators are 0.29, 0.34, and 0.31 for valence, activation, and dominance, respectively (Schuller et al. 2009c). The average correlation coefficients between the annotators are 0.49, 0.72, and 0.61, respectively (Schuller et al. 2009c). As for the SAL database, for this thesis, the valence and activation labels were discretised into four categories representing the four quadrants of the activation-valence space (q1, q2, q3, and q4—see Fig. 6.1).

6.1.13 Ballroom Dance-Style Database (BRD)

So far, the spoken and singing voice emotion databases have been introduced. This section and the following one cover the databases which were used for the evaluation of automatic music style recognition.

The BallRoom Dance-style (BRD) database was first introduced by Schuller et al. (2007a) for evaluation of a tempo detection algorithm and automatic ballroom dance-style classification. It was subsequently used by Eyben et al. (2007) and Schuller et al. (2008) for evaluation of beat-tracking and a refined version of the tempo detec-

Table 6.9 Number of instances for each dance-style in the BallRoom Dance-style database

Cha Cha	Foxt.	Jive	Quickst.	Rhumba	Samba	Tango	V. Waltz	Waltz
211	245	138	242	217	188	185	136	293

Abbreviations: V. Waltz (Viennese Waltz), Foxt. (Foxtrott), Quickst. (Quickstep). (Schuller et al. 2007a)

Table 6.10 Number of instances for each music style in the GeDiDB

Chill	Classical	Live	Meditative	Pop/Rock	Electronic	Unplugged
107	94	99	86	102	126	79

tion algorithm presented in (Schuller et al. 2007a). The database (cf. Schuller et al. (2008)) contains 1,855 pieces of ballroom and Latin dance music sampled from the ballroomdancers.com web-shop.[3]

The set consists of 9 ballroom dance-style classes being the standard styles Waltz, Viennese Waltz, Tango, Quick Step, and Foxtrot, as well as the Latin styles Rhumba, Cha Cha Cha, Samba, and Jive. See Table 6.9 for the distribution of dance-styles. The set has a wide tempo range from 68 Beats per Minute (BPM) to 208 BPM. 30 s of each song are available in a low-quality compressed audio format (real media). The total playing time of this set, is 15 h. This set is referred to as BRD in the ongoing text. Ground truth for tempo and the ballroom dance-style are included with the music tracks, also taken from *ballroomdancers.com*. An additional label, indicating the metrical grouping (duple or triple) is also contained. It is implicitly deduced from the dance-style: all waltzes are assigned triple metrical grouping, while all other styles are assigned duple metrical grouping.

6.1.14 Genre Discrimination Database (GeDiDB)

The Genre Discrimination Database (GeDiDB) is a collection of 694 audio tracks of the following 7 music genres/styles: Chill-out, classical, live performances, meditative music, rock and pop music, electronic music, and unplugged recordings. The recordings are stored as MP3 files with an average bitrate of 128 kBit/s.

The total length of all songs together is 52 h and 14 min. The number of songs per category is shown in Table 6.10. For the purpose of music genre recognition a whole song is seen as one instance, i.e., one single static feature vector is extracted for each song.

[3] 30 s preview audio tracks taken from https://secure.ballroomdancers.com/music/style.asp, Nov. 2006; a detailed list of these songs can be downloaded from http://www.audeering.com/research-and-open-source/files/brd.txt.

6.2 Noise Robust Affective Speech Classification

The approach to noise robust classification of affective speech proposed in Sect. 5.3 is evaluated on two corpora, in order to cover both natural, spontaneous affect, and professionally acted emotion categories: The TUM AVIC (Sect. 6.1.3) and the GEMEP (Sect. 6.1.7) database, respectively.

For TUM-AVIC, the partitioning of the INTERSPEECH 2010 Challenge was adopted and the training and development partitions were jointly used for training. The corpus in this version contains data form 21 speakers (3,880 utterances in total, 10 female speakers). The pseudo continuous LOI labels are used for CC-FS and feature analysis (Sect. 6.2.1); the ternary LOI labels (obtained from majority voting and binning of LOI levels 1–3 from the 5 level labels) are used for the classification experiments.

For GEMEP, a pre-defined training and evaluation partitioning is used in which actors 1, 2, 3, 4, 6, and 7 are assigned to the training set and actors 5, 8, 9, and 10 are assigned to the test set. Acoustic feature selection is performed on the continuous activation and valence labels. Classification experiments, however, are performed on discretised binary activation and valence labels (high versus low; positive versus negative). These were derived from the continuous labels (*not* from the categories) with a discretisation threshold of -0.1 applied for both dimensions.

6.2.1 *Analysis of Acoustic Features*

The features obtained by CC-FS (Sect. 5.3.2) on the clean training set are compared to those obtained on the multi-condition training sets (training sets of all four noisy conditions combined) in Fig. 6.3. For both activation and valence, it can be seen that the relative importance of prosodic features decreases when noise and reverberation is present. For instance, features that describe changes in the signal energy seem to be much less reliable in the reverberated and noisy cases, such as the root quadratic mean of loudness delta coefficients ($CC = 0.651$ with activation on clean speech, and $CC = -0.052$ on multi-condition corrupted speech), which is expected due to the strong corruption of loudness by additive noise.

Regarding the types of functionals, it can be seen that the relative importance of the temporal functionals (e.g., up-level and rise times) increases in the reverberant and noisy cases. This might indicate that in the presence of noise the overall distribution of Lowlevel Descriptors (LLDs) over time remains more intact than other parameters like means and moments. Overall, however, there is more change in feature group relevance between the individual tasks (activation, valence, and LOI) than between selecting features on the clean or noisy multi-condition sets.

6.2.2 Classification Performance

In classification experiments it is shown how the reduction of the feature set size to 400 features affects the automatic classification performance and how the data set (clean or corrupted) on which the CC based feature selection is performed on influences the performance of the classifier. In this respect, clean and Multi-condition Training (MCT) were compared. In MCT the training sets of three degraded acoustic conditions (e.g., A, B, and C—see Sect. 5.3.1) were joint and evaluated for all test sets of the remaining condition (e.g., D) in the case of *mismatched* evaluations and on the test sets for each of the three training conditions (e.g., A, B, and C) for *matched* evaluations. The evaluations were repeated four times, in order to have each condition once as a mismatched test set (leave one condition out). It is worth noting, that the training set in MCT is nine times the size of the original (clean) training set (three acoustic conditions by three Room Impulse Response (RIR) distances).

As classifier Support Vector Machines (SVMs) with a linear kernel function were used. They were trained with the Sequential Minimal Optimization (SMO) algorithm implemented in the WEKA toolkit (Hall et al. 2009). The SVM complexity parameter was set to $C = 0.1$ for the GEMEP database (activation and valence) and to $C = 0.01$

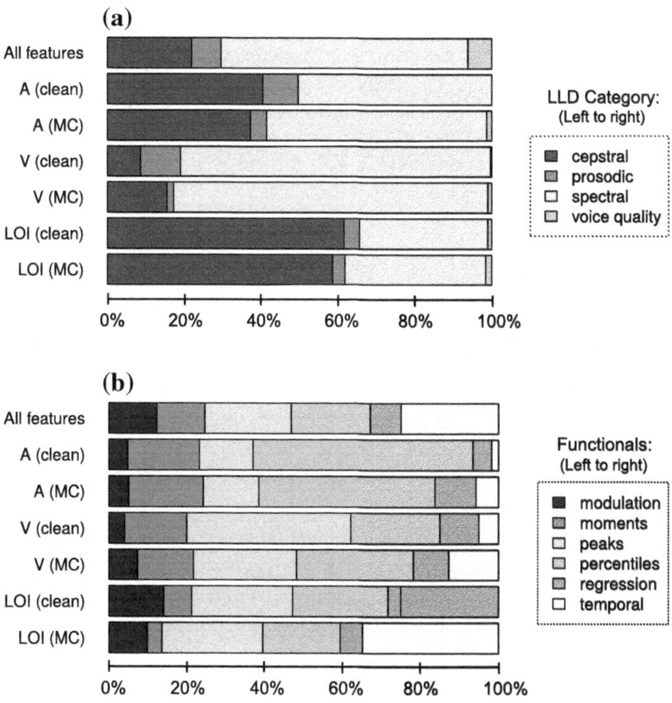

Fig. 6.3 Full ComParE feature set (All) versus 400 best features selected by CC-FS on clean and multi-condition corrupted (MC) data-sets for activation (A), valence (GEMP) (V), and Level of Interest (LOI) classification (AVIC). Shown is the percentage of selected Low-level Descriptors (LLDs) and types of functionals. Figure drawn after (Eyben et al. 2013). **a** Low-Level Descriptors. **b** Functionals

Table 6.11 Average classification results (Unweighted Average Recall (UAR)) for each acoustic condition averaged over Room Impulse Response (RIR) distances and Signal-to-Noise Ratios (SNRs) (*top*), as well as over all acoustic conditions (*bottom*); for binary activation (Act.) and valence (Val.) as well as ternary Level of Interest (LOI)

	Noise/reverberation conditions (Sect. 5.3.1)					
UAR (%)	A		B	C		D
Activation/Valence (binary)						
MCT mi	74.0/59.2		78.8/65.0	75.3/63.2		77.7/58.9
MCT ma	76.4/63.5		79.1/61.8	77.4/66.1		77.3/61.1
Clean T	73.4/52.0		73.7/54.9	73.4/56.6		70.2/52.4
Level of Interest (ternary)						
MCT mi	50.2		50.6	53.0		46.1
MCT ma	50.6		51.6	52.2		49.3
Clean T	42.9		44.0	48.7		40.8
UAR (%)	Act.	Val.	LOI	Act.	Val.	LOI
	All features			CC-FS		
MCT mi.	74.4	56.9	49.5	**76.5**	**61.6**	**50.0**
MCT ma.	75.7	56.1	49.9	**77.6**	**63.1**	**50.9**
Clean	74.7	53.1	44.2	72.6	54.0	44.1
Clean MC-FS	–	–	–	74.8	59.9	45.1

Training (T) on clean speech, and Multi-condition Training (MCT) on all other conditions than the test condition, i.e., leave-one-condition-out (mismatched, *mi.*) and all conditions including the test condition (matched, *ma.*). *Top* CC-FS feature selection only. *Bottom* Comparison of all features and best 400 features from CC-FS. Feature selection for clean training on clean speech set (clean) and on joint training set of all four conditions (MC-FS); the two best results per target are in bold face font. (Eyben et al. 2013)

for the (larger) TUM-AVIC database. Before training the classifier the features were standardised to zero mean and unit variance on the training set. The means and variances computed from the training set are then used to scale the features in the test sets, simulating realistic incremental processing conditions wrt. feature normalisation (cf. Sect. 5.2.2). To avoid over-fitting towards the majority class for the TUM-AVIC database (1.88 k instances of *LOI2* versus 349 instances *LOI1*), the training set was balanced by random instance sub-sampling of the two majority classes to the number of instances of the minority class.

The upper benchmark of theoretical performance for each database is given by training on the clean training set and evaluating on the clean test set. These results in terms of Unweighted Average Recall (UAR) are as follows (Eyben et al. (2013)): 77.2 % for binary activation, and 61.7 % for binary valence (GEMEP corpus), and 56.0 % for the three levels of interest in TUM-AVIC. The obtained result for TUM-AVIC is below results reported in other publications (e.g., Schuller et al. (2009a, c)). This is due to two factors: a) here, the full set of instances as in the 2010 paralinguistic Challenge is used (the other studies used a set where instances with low

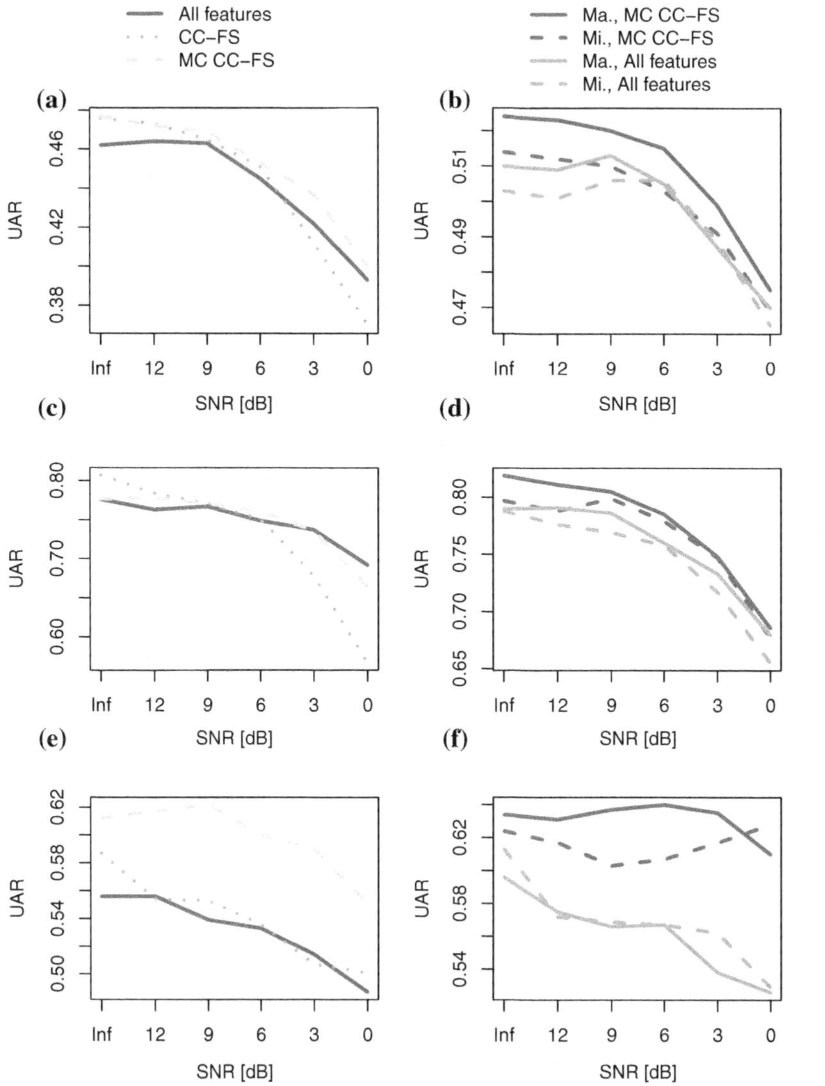

Fig. 6.4 Plots of Unweighted Average Recall (UAR) for binary activation, valence and ternary Level of Interest (LOI) classification: Influence of the Signal-to-Noise Ratio (SNR) across all acoustic conditions. Clean or Multi-condition Training (MCT) including (matched) or not including (mismatched) the test noise condition. Full ComParE 2013 feature set (All) versus CC-FS on clean set or multi-condition sets. Figure drawn after (Eyben et al. 2013). **a** LOI clean training, **b** LOI, MCT, **c** Activation, clean training, **d** Activation, MCT, **e** Valence, clean training, **f** Valence, MCT

labeller agreement were removed) and b) in the experiments here, each utterance is normalised to a peak amplitude of -1 dB, which eliminates the mean utterance energy as discriminative cue.

Table 6.11 (top) shows the results for the four test set acoustic conditions (A–D), each averaged over all the 18 sub-conditions (6 Signal-to-Noise Ratios (SNRs) levels and 3 RIR distances). Chapel reverberation and music noise is clearly the most challenging one of the four conditions, both for clean training and MCT cases. Overall, MCT boosts the performance compared to clean training. This finding is in line with earlier findings in Eyben et al. (2012b). In MCT, matched conditions training is only marginally better than mismatched conditions training, which shows the strong potential of MCT for unknown test conditions.

Table 6.11 (bottom) shows the results averaged over all acoustic conditions and sub-conditions in order to contrast CC based feature selection (CC-FS) with using the full set of features. For clean training there is only little difference between CC-FS and the full feature set. However, the best result is achieved with combining MCT and CC-FS for selection estimated on the multi-condition training set. Just by selecting features on the multi-condition training set, the performance of clean training can be boosted without actually using the multi-condition data to train the classifier. This strongly suggests that the multi-condition feature selection will actually select features that are more noise robust, than those selected by a feature selection on the clean data only, for example.

In Fig. 6.4, the UARs on the binary activation and valence as well as the ternary LOI tasks obtained with clean training or MCT, and with optional feature selection, are shown. For activation it appears that there is only a slight performance difference between clean training and MCT if all features are used; however, a drop in performance at lower SNRs is observed (below 60 % UAR) for clean training and feature selection on the clean data. If features are selected from the multi-condition set, results are similar to training on the multi-condition set. This shows again that the feature selected on the clean data do not generalize well to unfavourable acoustic conditions, yet multi-condition feature selection can remedy this. For valence, unfortunately, no robust results even in the noise free case (only up to 63 % UAR) are obtained. This can be assigned to the general difficulty of estimating valence from pure acoustic parameters. It is encouraging, though, that through MCT a result mostly independent of the SNR can be obtained. Finally, for the task of LOI determination, a huge influence of MCT can be seen, yet a minor influence of feature selection. Interestingly, the worst result (at 0 dB SNR) with MCT is very close to the best result (at SNR $= \infty$) without MCT. This fact indicates a deterioration due to reverberation. Overall, the LOI results indicate that there are only few features which stand out across multiple acoustic conditions for that task—apparently there one has to learn how the feature behaviour changes in unfavourable acoustic conditions, which multi-condition training provides.

Regarding feature selection, it can be concluded that in most cases similar or better results with respect to the full feature set are obtained with only a fraction (6.3 %) of the features. The combination of MCT and feature selection appears to be especially powerful in coping with very challenging acoustic conditions, even under mismatched training and test settings.

6.3 Evaluation of the Baseline Feature Sets

Extending the work previously published by the author of this thesis in Eyben et al. (2012c), the 10 standard, baseline acoustic feature sets described in Chap. 3 have been systematically evaluated for this thesis on a set of 12 databases which contain speech emotion recognition and music genre and dance-style identification tasks. These are: 8 speech emotion databases as used by Eyben et al. (2012c) and Schuller et al. (2009c) (excluding the SmartKom database), leaving ABC (Sect. 6.1.1), TUM-AVIC (Sect. 6.1.3), DES (Sect. 6.1.4), EMO-DB (Sect. 6.1.5), eNTERFACE (Sect. 6.1.6), SAL (Sect. 6.1.8), SUSAS (Sect. 6.1.11), and VAM (Sect. 6.1.12). Additionally the GEMEP database for speech emotion recognition (Sect. 6.1.7) and the Geneva Singing Voice Emotion (GeSiE) database for recognition of emotions in the singing voice (Sect. 6.1.10) are included. These 10 databases are referred to as the set of 10 databases in the ongoing. In total they contain 13,228 instances (=segments) of emotional speech and cover a well balanced mixture of diverse emotion portrayals ranging from basic acted to fully natural, spontaneous, and subtle emotions (cf. Sect. 6.1).

In order to evaluate the baseline feature sets and the parametrisations tuned on the 10 databases on an eleventh, large, independent database, the FAU-AIBO database (Sect. 6.1.2) was included. Further, the suitability of the proposed feature sets for music style classification was evaluated on two data-sets: the BRD (Sect. 6.1.13) and the GeDiDB (Sect. 6.1.14).

6.3.1 Mapping of Emotions

In order to better compare the results across databases (cf. Schuller et al. 2010c), the emotion categories of the 10 databases were mapped onto two binary labels: negative and positive valence, and low and high activation. These mappings are given in Table 6.12. Note, that these mappings are by no means straight forward and standardised. They are based on the original mappings for most databases, as suggested by the author of this thesis in Schuller et al. (2009c) and Eyben et al. (2012c) and adopted by Schuller et al. (2010c), the mapping for the GEMEP database as suggested by Bänziger et al. (2012), and a mapping for the GeSiE database proposed by the author of this thesis. Likely, the most controversial mapping is the mapping of *neutral*, which should be mapped to a third state. However, because not all databases provide a neutral state (e.g., the GEMEP database) it was decided for a binary mapping here in order to be able to evaluate performances across all databases using the same labels and have two more balanced binary labels for each database. Neutral is commonly mapped to low activation and positive valence. Thus, as in the FAU-AIBO database (Sect. 6.1.2), the positive valence label is rather a non-negative valence, or idle label.

6.3.2 *Evaluation Method*

An extensive series of experiments was conducted on the 10 emotion databases and 2 music style databases in a systematic way to evaluate the influence of several parameters, including the choice of the acoustic feature set, the SVM model complexity, the method with which the class distribution in the training set is balanced, and the way the features are scaled. Following Eyben et al. (2012c), SVMs are chosen as classifier, but the described parameters are systematically varied.

For all experiments on the 10 emotion databases, Leave-One-Subject-Out (LOSO) cross-validation evaluations as in Eyben et al. (2012c) were performed to have fair, speaker independent evaluation conditions. For databases with more than 8 speakers (TUM-AVIC, EMO-DB, eNTERFACE'05, GEMEP, and VAM), the speakers or

Table 6.12 Mapping of emotion categories to binary activation labels (low/high) and binary valence labels (negative/positive) for the 10 databases

Corpus	Activation		Valence	
	Low	High	Negative	Positive
ABC	Neutral, tiered	Aggressive, cheerful, intoxicated, nervous	Aggressive, nervous, tired	Cheerful, intoxicated, neutral
AVIC	LOI1	LOI2, LOI3	LOI1	LOI2, LOI3
DES	Neutral, sad	Angry, happy, surprise	Angry, sad	Happy, neutral, surprise
EMO-DB	Boredom, disgust, neutral, sadness	Anger, fear, happiness	Angry, sad	Happy, neutral, surprise
eNTER- FACE	Disgust, sadness	Anger, fear, happiness, surprise	Anger, disgust, fear, sadness	Happiness, surprise
GEMEP	Pleasure, relief, interest, irritation, anxiety, sadness	Joy, amusement, pride, hot anger, panic fear, despair	Hot anger, panic fear, despair, irritation, anxiety, sadness	Joy, amusement, pride, pleasure, relief, interest
SAL	q2, q3	q1, q4	q3, q4	q1, q2
SING	Neutral, tense arousal, sadness, tenderness, calm/serenity, condescension	Fear, passionate love, animated joy, triumphant pride, anger	Fear, tense arousal, anger, sadness, condescension	Neutral, passionate love, animated joy, triumphant pride, tenderness, calm/serenity
SUSAS	Neutral	High stress, medium stress, screaming	High stress, screaming	Medium stress, neutral
VAM	q2, q3	q1, q4	q3, q4	q1, q2

some of the speakers were combined to yield 8 speaker groups. The data was thereby split into the 8 (or less) speaker (group) partitions, each partition containing data only from one (or more, in case of groups) speakers. Data from no speaker was present in more than one partition, i.e., the partitions are speaker disjunctive. For the FAU-AIBO database, the training and test partitions as defined by Schuller et al. (2009b) were used.

For the 2 music style databases for obvious reasons there is no speaker ID information available, therefore a pseudo-random, stratified 8-fold cross-validation was used: The instances (songs) were randomly split into 8 equal sized partitions, while enforcing that the distribution of the instances across the classes (music styles) remained the same for each fold as it is for the whole data set (stratification).

The classifier of choice for all experiments is SVM (Sect. 2.5.1) with a linear kernel function. Due to the high dimensionality of the static feature vectors, polynomial or other higher-dimensional kernel functions would produce a task which is computationally too complex. The WEKA data mining toolkit was chosen as SVM implementation—specifically the implementation of SVM with the SMO training algorithm. The following 18 complexity constants C were evaluated: 0.00001, 0.000025, 0.00005, 0.000075, 0.0001, . . ., 0.075, 0.1, 0.25. Higher complexities were avoided because it was assumed they would only lead to more over-fitting.

Further, it was ensured that the class distribution in the training partition (*not* in the test partition) was always balanced. Two different methods were investigated for achieving this: random *up-sampling* of each class to the size of the majority class, and random *down-sampling* of each class to the size of the minority class.

The following acoustic feature sets were evaluated: The INTERSPEECH 2009–2013 Challenge baseline feature sets (see Sect. 3.1 through Sect. 3.5), the Audio-visual Emotion Challenge (AVEC) 2013 baseline feature set (Sect. 3.7), the Geneva Minimalistic Acoustic Parameter Set (GeMAPS) and the extended version thereof (Sect. 3.8), as well as the music genre feature set (*Genre* in the ongoing—Sect. 3.9), and the music genre feature set without the rhythmic descriptors.

Further, five different strategies for the normalisation (cf. Sect. 5.2.2) of the feature values are evaluated in this thesis on all corpora and all acoustic feature sets. These are based on the two approaches *standardisation* (Mean Variance Normalisation (MVN)—Eq. (5.2)) and *normalisation* (Mean Range Normalisation (MRN)—Eq. (5.3)). The parameters μ_x and σ_x, as well as x_{min} and x_{max}, respectively, are computed either from the training set instances only (referred to as MVN-ON and MRN-ON, where ON stands for 'on-line'), or from only the training set instances during training, and from all test set instances during testing (referred to as MVN-OF and MRN-OF, where OF stands for 'off-line'). The fifth method is per speaker standardisation (MVN-SPK), where μ_x and σ_x are computed individually from the data of each speaker and are used to normalise only exactly this data. As the training and test partitions are speaker disjunctive due to the LOSO evaluation scheme, this way of normalisation is valid. It is important to note, that the normalisation parameters from the training set are computed *after* class-wise instance balancing by up- or down-sampling. Normalisation parameters computed from the test set are always computed

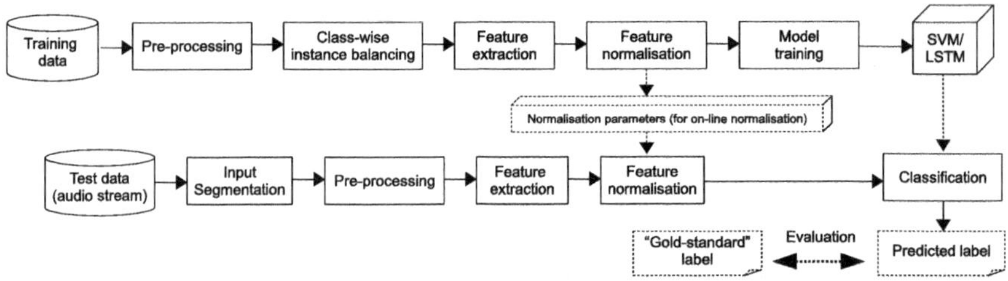

Fig. 6.5 Detailed steps of processing for systematic speech and music classification evaluations

from an unbalanced class distribution (unless the test set is balanced by default). The whole chain of processing is summarised in Fig. 6.5.

The evaluation measure is UAR (Schuller 2013), which in contrast to Weighted Average Recall (WAR) is a better indicator of the true performance of the system for evaluation sets where the occurrence of class labels is very unbalanced across the classes. WAR is also known as the standard *Accuracy* and is defined as:

$$WAR = \frac{\sum_{i=1}^{I} N_c^{(i)}}{N}, \tag{6.2}$$

with N being the total number of instances in the test set and the number of correct instances being $N_c^{(i)}$ for the ith class ($i \in [1; I]$). In contrast, UAR is defined as:

$$UAR = \frac{1}{k} \sum_{i=1}^{I} \frac{N_c^{(i)}}{N^{(i)}}, \tag{6.3}$$

with k being the number of classes and $N^{(i)}$ the total number of instances in the test set belonging to the ith class (according to the ground truth or gold standard labels). The term $N_c^{(i)}/N^{(i)}$ represents the recall-rate for the ith class, hence, UAR according to the above equation is the unweighted average of the class-wise recall-rates.

In Eyben et al. (2012c) it was argued that UAR cannot be used when results are averaged across databases. This is not the case in the set-up here, as the averaging across databases is done in an unweighted way, i.e., UAR is computed for each database individually, then the unweighted average over all databases is computed. This weights down the influence of larger databases, which otherwise would be too dominant. As the databases are well balanced across various tasks, emotion types, and recording conditions, this is a valid measure.

6.3.3 *Results*

By systematic, brute force, evaluation of all parameter combinations, a large number of results was produced on the 10 speech/singing voice corpora. 10 feature sets, 10 databases, 2 balancing methods, five feature normalisation methods, and 18 SVM complexity constants were systematically evaluated, resulting in a total of 18,000 individual combinations. These were evaluated for the original categories of each corpus, the binary activation labels, and the binary valence labels. Thus, 54,000 individual results were obtained, each of these in 8-fold cross validation. A total of roughly 430 k train/test runs was thus executed. All 10 corpora combined have a total of 13,228 instances.

It would be overwhelming and not practical to show and discuss all 54 k individual results. The results presented here are thus averaged over several runs in order to highlight parameters of interest. E.g., to meaningfully evaluate the best feature normalisation method, for example, for activation, all 3,000 UARs obtained with each of the five normalisation methods are averaged (unweighted, i.e., not weighted by the amount of instances per corpus). The same method is applied for the balancing method, the SVM complexity constant, and the feature set. The averaging is thereby done both globally over all the 10 databases (see Sect. 6.3.3.1), and individually for each database (see Sect. 6.3.3.2). Results averaged over all 10 databases are only discussed for the binary activation and valence labels. Emotion category results are only discussed per database, due to the varying types and numbers of categories across databases.

6.3.3.1 Results Averaged Across Databases

In this section the average results for each parameter setting over all 10 databases are shown and discussed for the binary activation and binary valence tasks.

To estimate the statistical significance of the results, approximate upper bounds for the minimally required UAR differences are computed for each case with a one-tailed significance test based on the number of differing true instances between two results (Smucker et al. 2007), assuming 50 % UAR for system A and 50 % + x accuracy for system B. x is then chosen in a way that the two systems are significantly different at a given significance level. The minimally required difference x in UAR at which two results are significantly different at a level $\alpha = 0.05$ is denoted by Δ_{p5} in the ongoing. The total number of trials, which is required for this significance estimation, is computed as the total number of instances in the evaluation set multiplied by the number of parameter combinations the results were averaged over. When using cross-validation and when averaging over all ten databases the total number of instances is 13,228. Thus, for a result with a single parameter setting averaged over all ten corpora a difference of $\Delta_{p5} = 1.02\,\%$ is required for statistical significance at $\alpha = 0.05$. It is to note that, this estimate is strictly only valid when weighted averaging of the results over the 10 databases would be used. In the case of unweighted averaging the

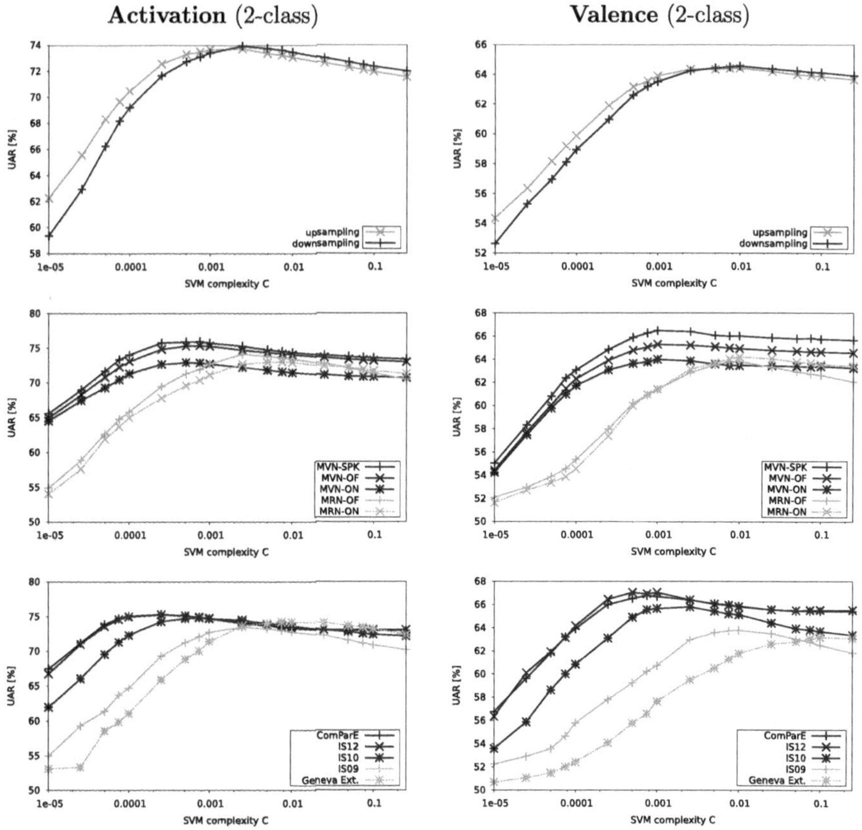

Fig. 6.6 Unweighted Average Recall (UAR) performance over a range of 18 SVM complexities, for binary activation and binary valence classification, average over all databases; comparison of down- and up-sampling for balancing of the training set, 5 feature normalisation methods, and 5 selected feature sets (INTERSPEECH 2013 ComParE, INTERSPEECH 2012 Speaker Trait, INTERSPEECH 2010 Paralinguistic, INTERSPEECH 2009 Emotion, Geneva Ext. = eGeMAPS)

randomness of the smaller databases is underestimated by the significance test, and the larger databases are overestimated. However, the goal at this point is to have an approximate significance estimation, for which it is believed that this consideration is sufficient.

In Fig. 6.6, the influence of the model complexity on UAR is shown for each balancing method (thereby averaging over all corpora, feature normalisation methods and feature sets), feature normalisation method, and feature set (in both cases averaging over the remaining parameters, respectively).

Regarding down- and up-sampling, it can be seen that for low complexities up-sampling gives better performance, while for high complexities down-sampling is marginally better. Both reach a maximum around $C = 0.0025$ for activation, and $C = 0.01$ for valence, where both balancing methods perform almost equally well.

Table 6.13 Results (Unweighted Average Recall (UAR)) averaged (unweighted) over all ten emotion databases and all configuration parameters (including 13 SVM complexities) other than the one differentiated above in reach row-set (balancing, normalisation, or feature set)

UAR (%)	Activation	Valence
Balancing	Instances: 8,598,200, $\Delta_{p5} = 0.04\,\%$	
Up-sampling	72.83	63.76
Down-sampling	**72.93**	**63.82**
Normalisation	Instances: 3,439,280, $\Delta_{p5} = 0.07\,\%$	
MVN-SPK	**74.68**	**65.91**
MVN-ON	71.75	63.52
MVN-OF	74.22	64.81
MRN-ON	71.55	62.56
MRN-OF	72.22	62.16
Feature set	Instances: 1,719,640, $\Delta_{p5} = 0.09\,\%$	
GeMAPS	71.44	59.42
eGeMAPS	72.24	60.14
IS09	71.85	61.97
IS10	73.66	64.61
IS11	73.61	65.70
IS12	**73.89**	**66.09**
ComParE	**73.81**	65.97
AVEC 2013	73.76	64.98
Genre set	72.26	64.59
Genre set (w/o rh.)	72.31	64.44

Best results for each setting and task (binary activation/valence) are highlighted in bold face. w/o rh. = Genre set without rhythmic descriptors

This marginal difference is confirmed by the global average (over 13 complexity values) in Table 6.13. It can thus be concluded, that in general the method to balance the training set does not notably influence the results. However, as the individual results show (Sect. 6.3.3.2), this conclusion depends on the database size and is only true for larger sets.

Regarding feature normalisation, a great difference can be seen between the min/max (MRN) scaling and the z-normalisation (MVN) for $C < 0.0025$ for activation and $C < 0.005$ for valence: z-normalisation is significantly superior. The best overall normalisation is per speaker MVN, followed by *off-line* MVN. Generally, off-line normalisation, i.e., where the parameters for the evaluation set/partition are computed independently from the training set is—as expected—superior to *on-line* normalisation, where the scaling parameters computed from the training set are applied to evaluation set data and no statistics from the evaluation set are considered.

Large differences can also be seen for the feature sets: The smaller sets (IS09 and eGeMAPS) give the lowest performance. This is most pronounced for valence, where they are roughly 4–5 % behind the best performing sets (which are also the largest sets)—ComParE and IS12. This is also an indication that the ComParE and IS12 sets might contain features which are more relevant for valence, and which are missing in the smaller sets. The performance of the IS10 set lies in between the other four sets, and is slightly below ComParE and IS12 for valence. For activation, the smaller sets are almost as good as the larger sets, however, at a SVM model complexity which is about a 100 times higher. This makes sense and is expected, since the larger sets lead to a more "complex" model due to the increased number of parameters, thus requiring a less complex model regarding the number of support vectors. The IS10 set shows a very similar performance to ComParE and IS12, especially for higher complexities. All three are outperformed at higher complexities by the IS09 and the eGeMAPS sets. The eGeMAPS set notably outperforms the IS09 set, although eGeMAPS is much smaller than IS09 (87 feature versus 384 features, respectively), which shows that the features in eGeMAPS are very well chosen.

For valence classification, on average, approximately a 2–3 times higher complexity is required as for activation (for the best UAR). This shows that activation is an "easier" task, and valence from acoustic features alone, is a much more complex task which requires more support-vectors.

It was found that the SVM complexity in general has a great influence on the results and a notable drop in UAR averaged over all settings can be observed for complexities below 0.00025. Thus, for the following investigations of the balancing and feature normalisation methods, as well as the acoustic parameter sets, it was seen as acceptable to discard all results obtained with complexity values below 0.00025, i.e., the averaging is only performed over the remaining 13 complexity values which are greater than or equal to 0.00025. With these 13 complexity values,

Table 6.14 Unweighted Average Recall (UAR) results per feature set and balancing method, average (unweighted) over all 10 databases, 13 SVM complexities, and all 5 feature normalisation methods

	Activation		Valence	
	Up-sampling	Down-sampling	Up-sampling	Down-sampling
GeMAPS	**71.69**	71.19	**59.70**	59.15
eGeMAPS	**72.50**	71.97	**60.40**	59.88
IS09	**71.93**	71.77	**62.09**	61.86
IS10	73.56	**73.76**	**64.67**	64.55
IS11	73.50	**73.72**	65.57	**65.83**
IS12	73.77	**74.01**	65.95	**66.24**
ComParE	73.60	**74.02**	65.84	**66.09**
AVEC 2013	73.59	**73.92**	64.91	**65.06**
Genre set	72.08	**72.45**	64.64	64.54
Genre set (w/o rh.)	72.09	**72.53**	64.47	64.41

$\Delta_{p5} = 0.11\,\%$ (859,820 instances). Best result per feature set and for each task in bold-face (if significant). w/o rh. = Genre set without rhythmic descriptors

Table 6.15 Unweighted Average Recall (UAR) results per feature set and feature normalisation method, average (unweighted) over all 10 corpora, 13 SVM complexities, and both up-sampling and down-sampling

UAR (%)	Activation				
	MVN-SPK	MVN-ON	MVN-OF	MRN-ON	MRN-OF
GeMAPS	**75.14**	71.84	73.64	67.92	68.67
eGeMAPS	**75.65**	72.47	74.17	69.20	69.71
IS09	**73.22**	71.25	**73.14**	70.34	71.32
IS10	**74.77**	72.17	74.53	72.98	73.87
IS11	**74.73**	72.16	**74.85**	73.06	73.27
IS12	**75.44**	72.32	75.03	73.09	73.55
ComParE	**75.59**	72.04	75.10	72.80	73.52
AVEC 2013	**75.02**	71.94	**75.20**	72.74	73.88
Genre set	**73.56**	70.66	73.20	71.69	72.19
Genre set (w/o rh.)	**73.65**	70.66	73.30	71.71	72.25
UAR (%)	Valence				
	MVN-SPK	MVN-ON	MVN-OF	MRN-ON	MRN-OF
GeMAPS	**62.26**	60.29	60.63	56.95	56.99
eGeMAPS	**63.39**	60.67	61.42	57.56	57.64
IS09	**64.07**	62.37	63.35	59.94	60.15
IS10	**66.02**	64.23	65.74	63.67	63.38
IS11	**67.79**	64.54	66.93	65.12	64.12
IS12	**68.04**	65.22	67.03	65.62	64.56
ComParE	**67.95**	65.01	67.04	65.43	64.40
AVEC 2013	**67.09**	64.81	65.50	64.02	63.50
Genre set	**66.28**	64.15	65.35	63.76	63.41
Genre set (w/o rh.)	**66.19**	63.91	65.13	63.49	63.48

$\Delta_{p5} = 0.2\,\%$ (343,928 instances). Best result per feature set and task in bold-face, if significant; if two results are bold-face, they are both best, because one is not significantly better than the other ($\alpha = 0.05$). w/o rh. = Genre set without rhythmic descriptors

detailed averaged results are shown in Table 6.13 in order to find the (on average) best configurations for acoustic feature set, feature normalisation, and balancing strategy. The trends seen from Fig. 6.6, are confirmed by Table 6.13: Down-sampling is marginally (but significantly) better than up-sampling, MVN per speaker is superior to the normalisation of the whole training/test sets, and IS12 is the on average best feature set, closely followed by ComParE (explicable by the large overlap between the two sets). For valence, IS12 is marginally and significantly better than ComParE, but for activation, there is no significant difference between IS12 and ComParE.

The choice of the best balancing method and feature normalisation method might vary with the acoustic feature set. In Table 6.13 this cannot be seen because the results are averaged over all feature sets. Therefore, more detailed results for balancing and feature normalisation are shown per feature set in Tables 6.14 and 6.15, respectively.

Here, it can be seen that up-sampling is significantly superior for all the small feature sets, up to IS09 for activation, and IS10 for valence, but down-sampling is superior for the larger feature sets. This result, however, has to be treated with caution, as the influence of model complexity is not accounted for here. The averaging for all settings in Table 6.14 is done over the same range of complexities. However, the complexity has a great influence on the choice of up-/down-sampling (see Fig. 6.6 (top)).

Regarding feature normalisation, Table 6.15 confirms, that for all feature sets the speaker based MVN performs best. Alone for the IS09, IS11, and AVEC 2013 sets, there is no significant difference between speaker normalisation (MVN-SPK) and off-line MVN for activation. While on average the IS12 and ComParE sets were the best, now, when only looking at speaker normalisation, the best feature set for activation is in fact the eGeMAPS set, although it is not significantly better than the ComParE set (0.06 % difference versus 0.2 % required for significance at $\alpha = 0.05$). For valence, IS12 and ComParE are clearly the winners. There is no significant difference between these two sets, but both are ≈4.5 % absolute better than eGeMAPS, 2 % better than IS10, and only 0.2 % better than IS11.

The Genre set performs surprisingly well, although it was not primarily designed for speech emotion recognition. There is no significant difference between the Genre set with the 87 rhythmic descriptors and the set without these. A slight tendency, however, can be seen that the set without the rhythmic descriptors performs better, which is expected, as the rhythmic descriptors are designed for longer segments of music with a regular rhythm, while the set of general statistical descriptors, besides the CHROMA Energy-distribution Normalised Statistics (CENS) descriptors, also contains general descriptors such as Mel-Frequency Cepstral Coefficients (MFCCs) and spectral band energies.

6.3.3.2 Results for Each Database

Results per database are evaluated as average results per database (over all configurations, except the one of interest—as for the averaging over all corpora, cf. previous section), which are shown and discussed in detail in Sect. 9.2. Here, the overall best result for each database is shown in Table 6.16 and a summary of the per database average results is presented in Table 6.17. In order to reduce the number of displayed results for each database, feature sets with similar performance over all databases are not shown in the tables in Sect. 9.2. However, they were not excluded from the averaging for the other parameters. The feature sets removed include: GeMAPS (similar but inferior to extended Geneva Minimalistic Acoustic Parameter Set (eGeMAPS)), the INTERSPEECH 2011 Speaker State Challenge (IS11) set (similar to the INTERSPEECH 2012 Speaker Trait Challenge (IS12) set), the AVEC 2013 set (similar to ComParE and IS11, but inferior to ComParE), and the Genre set without rhythmic descriptors (similar to the full Genre set).

Table 6.16 Best single per category result for each speech/singing emotion database and the settings with which this result was obtained: *up* and *down* for balancing with up-/down-sampling; feature set (Ft. Set) and feature normalisation method (Ft. Norm.); SVM complexity constant C

Database	Best		Settings			
(# classes)	UAR (%)	UAR_n (%)	Ft. Set	Bal.	Ft. Norm.	C
ABC (5)	59.1	48.9	IS10	Up	SPK	0.0005
AVIC (3)	70.7	56.1	AVEC 2013	Up	SPK	0.00001
DES (5)	62.7	53.4	IS12	Down	SPK	0.0005
EMODB (7)	86.1	83.8	AVEC 2013	Up	SPK	0.00025
eNTERF. (6)	80.5	76.6	ComParE	Down	SPK	0.0005
GEMEP (12)	43.7	38.6	IS11	Down	SPK	0.001
SAL (4q)	37.3	16.4	Genre (no rh.)	Up	SPK	0.0005
GeSiE (11)	39.8	33.8	AVEC 2013	Up	SPK	0.05
SUSAS (4)	63.7	51.6	IS10	Up	SPK	0.0025
VAM (4q)	44.2	25.6	IS10	Up	MVN	0.00075

eNTERF. = eNTERFACE'05 database; no rh. = without rhythmic features; Bal. = Balancing method. UAR_n: class number normalised Unweighted Average Recall (UAR), see text. SPK = MVN-SPK

In order to be able to better compare results across databases with a varying number of categories/classes n, a class number normalised UAR is proposed here, which is defined as (for UAR values $0 \ldots 1$):

$$UAR_n = \frac{(n-1)\,(UAR - 1/n)}{n}. \tag{6.4}$$

UAR_n scales the range of UAR in a way that 0 is chance level and 1 is 100 % UAR. A UAR of 75 % for 2 classes will result in 50 % UAR_n, and a UAR of 55 % for 10 classes will also result in a UAR_n of 50 %.

The best results per database (Table 6.16) vary highly across the databases. Affect classification on the prototypical and acted databases (EMO-DB, eNTERFACE) gives high accuracy ($\approx$80 % UAR_n), while the naturalistic affect databases (SAL, VAM, and TUM-AVIC) range between 16 % (SAL) and 56 % (AVIC) UAR_n. Acted databases such as GEMEP and Geneva Singing Voice Emotion (GeSiE) seem to be more challenging, presumably due to their high number of (sometimes only subtly differing) categories. All results reported here, outperform earlier results in terms of UAR as reported by Schuller et al. (2009c) (both Hidden Markov Model (HMM) and SVM classification results).

In a similar fashion, the setting with which the best results were obtained vary, and no clear recommendation for a winning configuration can be given from the individual results. Regarding feature sets, the best result is achieved with IS10 in three cases, AVEC 2013 in another three cases, and then once with each IS11, IS12, ComParE, and the Genre set. Complexities range from 0.00001 up to 0.05, and 0.0005 appears most frequently. The highest complexities are required for GEMEP and GeSiE, presumably again due to the high number of sometimes only subtly differing classes. No clear trend can be seen regarding the balancing method: up-sampling appears in the best configurations for small (ABC) and large data-sets (TUM-AVIC), contradicting the assumption that up-sampling is better for smaller data-sets, and down-sampling for larger data-sets. Speaker normalisation is again the winning method for feature normalisation. All best results, except for the VAM database, were obtained with speaker normalisation.

Looking at the average results in Sect. 9.2 as summarised in Table 6.17, it can be seen that the configurations which give the average best results for the classification of categories are not always identical to the single best configurations. This suggests—at least on a per database level—that there is a high interdependency between the parameters, and an analysis of single parameters based on an average over all others might not give the optimal result for a very specific task.

Overall, the summarised best settings per corpus confirm the observations made from the average results in the previous section: Up-sampling is slightly better than down-sampling, and per speaker MVN (MVN-SPK) clearly outperforms all other feature normalisation methods. In terms of best feature sets, it appears that the Computational Paralinguistics ChallengE (ComParE) set is the best in most cases (for categories), closely followed by the IS10 set, which is the best in many cases for binary valence. For the SAL database, which is the most challenging database, the GeMAPS gives the best average results. This could possibly be an indication that this set is suitable for natural affect, in real-world conditions, due to its small size and possibly higher generalisation abilities (less over-fitting). The larger sets are more prone to over-fitting and adapting to specific conditions of the database due to their vast number of acoustic descriptors. This indication has to be evaluated in future work in cross-database experiments. If it is true, the cross-database experiments should show superior performance for the GeMAPS compared to the larger sets.

6.3.3.3 Results on the FAU-AIBO Database

In order to verify whether the on average best configurations found on the 10 databases also apply to a held out database, additional evaluations on the FAU-AIBO database (Sect. 6.1.2) are conducted and the results are compared to the findings on the 10 databases.

Due to the large size of the FAU-AIBO database, only down-sampling is considered over all feature sets, as up-sampling would be computationally not feasible with the high dimensional features sets, especially for the 5-class task, due to the pairwise binary classification in multi-class SVM as implemented in WEKA (Hall et al. 2009). The down-sampled version of the 2-class FAU-AIBO database contains $2\cdot 5{,}823 = 11{,}646$ instances, for which a single binary SVM pair has to be trained. The down-sampled version of the 5-class FAU-AIBO database contains $5\cdot 889 =$

Table 6.17 Best configuration for each database from average results per database

Database	Setting											
	Bal.		Ft. Norm					Ft. Set				
	Up-sampling	Down-sampling	MVN-SPK	MVN-ON	MVN-OF	MRN-ON	MRN-OF	eGeMAPS	IS09	IS10	ComParE	Genre
ABC	✓	V	V		✓					✓	V	
AVIC		✓	✓			A V			A V	✓		
DES	V		✓		A						✓	
EMO-DB	✓		✓								✓	
eNTERFACE	✓		✓								✓	
GEMEP			✓					A		V	✓	
SAL	✓	A V	✓					✓		V		
GeSiE	✓	A	✓							V	✓	
SUSAS	✓		✓							✓		
VAM	✓	A	V	✓						A	✓	
Sum #	**8**	5	**10**	1	2	1		2	1	**7**	**7**	0

✓denotes the best configuration for categories. If the best configuration for binary activation and valence differs, it is denoted by A and V, respectively. See Sect. 9.2 for detailed results

4,445 instances, for which $4! = 24$ SVM pairs have to be trained. For up-sampling, the 5-class version would have $5\cdot 10{,}967 = 54{,}835$ instances (on which 24 SVM pairs have to be trained!) and the 2-class version would have $2\cdot 12{,}393 = 24{,}786$ instances (for 1 binary SVM).

Figure 6.7 shows results over the full range of 18 complexities, contrasting feature normalisation and feature sets. Regarding feature normalisation, the results are mostly in line with the average results on the 10 databases (see Sect. 6.3.3.1): MVN slightly outperforms MRN, and for MRN normalised features a higher model complexity ($\approx$factor 50) is required. Regarding on-line versus off-line normalisation, an interesting observation is made: for MVN off-line normalisation is best, while for MRN the situation is significantly reversed, off-line MRN falls far behind on-line MRN. This might indicate the presence of a few "corrupt" instances in either the evaluation or the training partition, which cause outliers in the features which break the normalisation parameters in one of these partitions. MVN is not affected by a few outliers, and thus, is more robust. For the average results over the 10 databases, the performance of normalisation is mostly coherent in a sense that off-line is always better than on-line and MVN is always better than MRN. The only case where a reversal effect as for FAU-AIBO can be seen, is for valence at higher complexities (see Fig. 6.6 (middle), (right)). Again, this is likely caused by outliers, which "break" the normalisation parameters, i.e., cause a mismatch between training and evaluation conditions in terms of feature value scaling. At higher complexities, such outliers (when present in the training partition) might additionally have a larger effect on the model, due to the higher number of support-vectors in the model.

More detailed results for FAU-AIBO are given in Sect. 9.2.11. From these results the best performing configuration (on average over the 13 lowest SVM complexities of the total of 18 complexities) for FAU-AIBO is: feature normalisation with off-line MVN and the IS09 feature set for the 2-class task and the ComParE feature set for the 5-class task.

The overall, single best result (only considering down-sampling) for the 5-class task is 43.57 % UAR. It is obtained with the IS12 feature set, on-line MRN, and a complexity $C = 0.001$. For the 2-class task 69.55 % UAR is obtained with the IS11 set and off-line MVN at a complexity of $C = 0.00005$. By exploring parameters on the FAU-AIBO evaluation set, these results, however, cannot compete with other results (such as in the challenge), which were produced with parameters not optimised on the evaluation set. Thus, these results can only be seen as an upper performance mark, and should by no means be treated as a reference.

Rather, the best result for FAU-AIBO should be estimated with the best parameter settings estimated from the consensus of the other 10 databases. For valence, this is: IS12 feature set, feature normalisation with MVN-OF, and down-sampling. The best complexity for MVN-OF is read from Fig. 6.6 as $C = 0.00075$. The best complexity estimated with the overall average over all feature normalisation methods, balancing methods, and feature sets, is found to be $C = 0.01$ (avg. UAR 64.49 %). For 2-class valence classification 66.26 % UAR are obtained on FAU-AIBO with $C = 0.00075$ and 62.1 % UAR with $C = 0.01$. This indicates that there is a strong dependency between the parameters, and the on average, individual selection does not give the

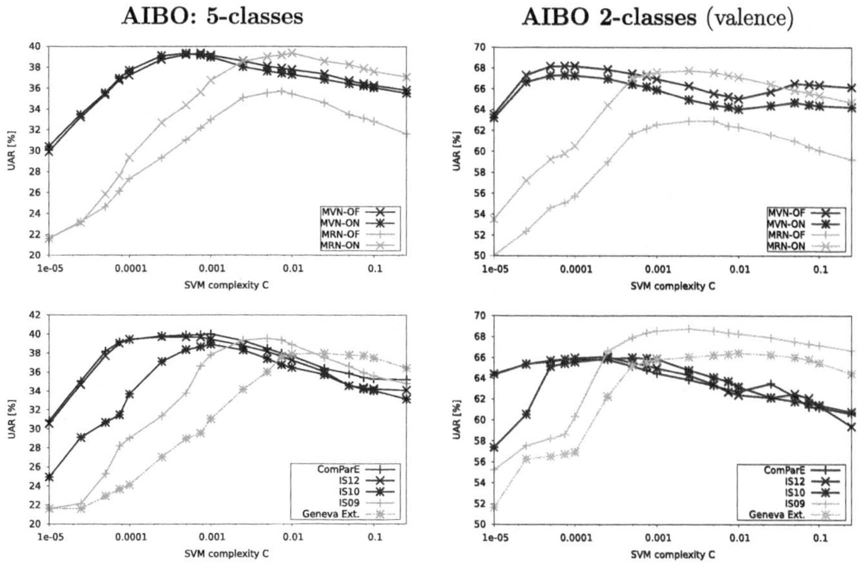

Fig. 6.7 Unweighted Average Recall (UAR) multi-class performance over a range of 18 SVM complexities; FAU-AIBO database, 5 classes (*left*) and 2-class valence (*right*); only down-sampling for balancing of the training set, 4 feature normalisation methods (no MVN-SPK), and 5 selected feature sets; Geneva Ext. = eGeMAPS

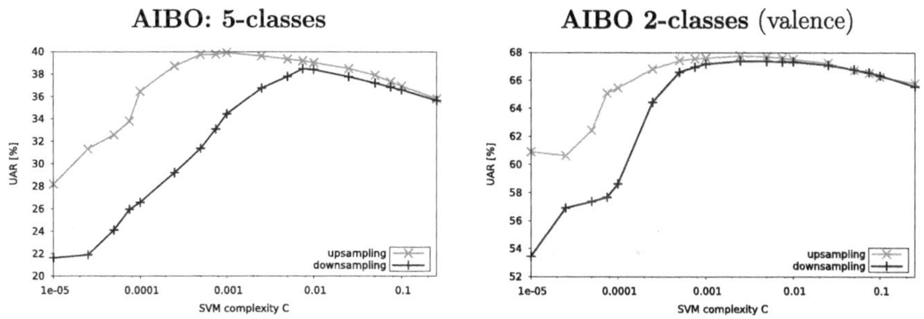

Fig. 6.8 Unweighted Average Recall (UAR) multi-class performance over a range of 18 SVM complexities; FAU-AIBO database, 5 classes (*left*) and 2-class valence (*right*); comparison of up- and down-sampling for balancing of the training set, averaged over 4 feature normalisation methods (no MVN-SPK) and 2 feature sets: eGeMAPS and IS09

overall best combination of parameters when simply combining the best average parameters. A more fine-grained analysis, such as picking the best complexity or range of complexities for each feature set, and feature normalisation method, should be considered in future work.

In order to compare up- and down-sampling, results from evaluations using only the GeMAPS extended and INTERSPEECH 2009 Emotion Challenge (IS09) feature set are analysed and plots over all 18 complexities are shown in Fig. 6.8. Up-sampling is best at low complexities, presumably because at lower model complexities more instances do not contribute to the model, thus discarding more information. At higher complexities the difference between up-sampling and down-sampling vanishes. It is still most present for the 5-class task, but negligible for the 2-class task for $C \geq 0.025$. Over all complexities, up-sampling gives the absolute best result.

Overall, it can be concluded that the trends found for the 10 databases in Sects. 6.3.3.1 and 6.3.3.2 also hold for the FAU-AIBO database: up-sampling is globally the best balancing method, and off-line MVN is the best feature normalisation method (at lower complexities). Both, up-sampling and off-line MVN perform best at complexities around $C = 0.0005$, which suggests that in combination they will also perform well around this complexity. Regarding the feature set, the INTERSPEECH 2013 ComParE set wins for the 5-class task, in line with the results for activation and valence on the other 10 databases (only slightly behind the IS12 set there). For the 2-class task, surprisingly the IS09 set wins by far. This might indicate that the IS09 set, due to its smaller size, performs better on the large training set with the parameters explored so far. A further parameter tuning, including up-sampling, might show superior performance of the ComParE set in future work.

Compared to related work on FAU-AIBO, the results obtained here, clearly can compete. Schuller et al. (2011) report UARs between 66.4 and 71.2 % for the 2-class task. The result obtained with parameters from the 10 databases is only slightly below this, at 66.3 %. The best result on the FAU-AIBO evaluation set in this thesis was found to be 69.55 %. Hassan et al. (2013) report a SVM baseline of 67.0 % UAR and improves up to over 72 % UAR by applying domain adaptation methods to adapt the evaluation set feature space to the training set feature space.

6.3.3.4 Results for Music Style Identification

The two music style tasks haven been evaluated with the same protocol as the speech and singing affect classification, described in Sect. 6.3.2. They are, for obvious reasons, not discussed together with the speech results, nor are they included in the averaging.

Figure 6.9 shows the UAR results plotted over 18 SVM complexities, for both datasets (BRD and GeDiDB), differentiating the balancing method, the feature normalisation method, and the feature set. In contrast to the speech affect tasks, up-sampling is always superior to down-sampling for the music style identification tasks. As for speech, MVN, especially at lower complexities, highly outperforms MRN. Except for MRN-ON (which is inferior), all normalisation methods perform equally well for higher model complexities close to 0.5.

Regarding the choice of feature sets, interesting observations are made for the music style classification and the rhythmic descriptors:

For ballroom dance-style identification (BRD database), the Genre set with rhythmic descriptors gives more than 20 % absolute better performance than the Genre set without rhythmic descriptors. This proves the effectiveness of the rhythmic features for the task they were originally designed for (cf. Schuller et al. 2007a), i.e., ballroom

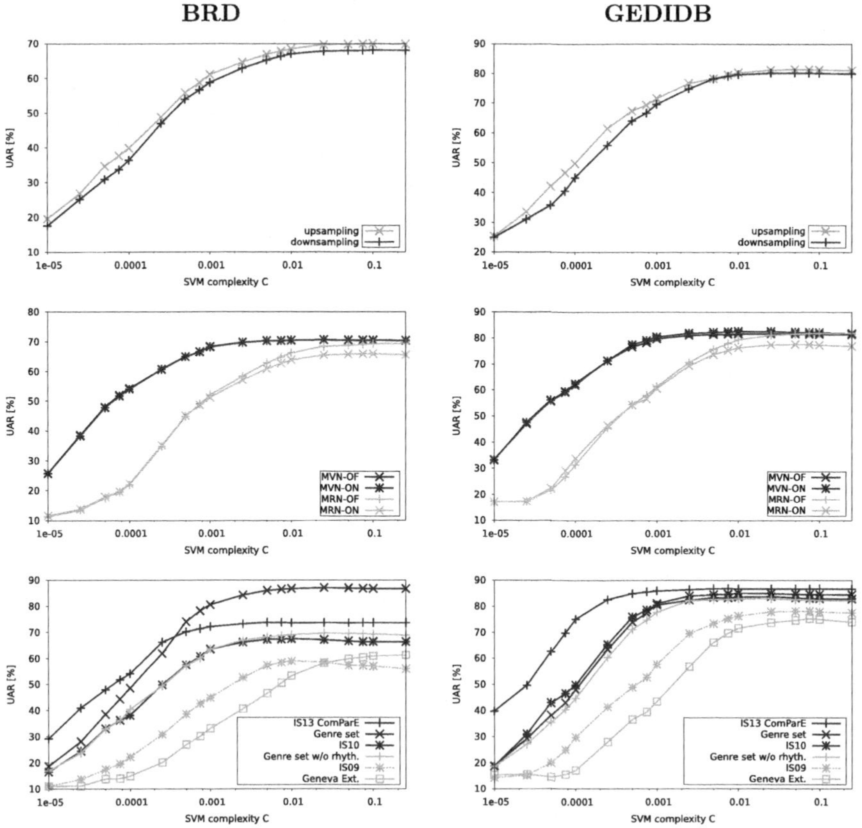

Fig. 6.9 Unweighted Average Recall (UAR) performance over a range of all 18 SVM complexities, for both music style classification databases (BRD—ballroom dance-style; and GEDIDB—music genres); comparison of down- and up-sampling for balancing of the training set, 4 feature normalisation methods, and 6 selected feature sets (INTERSPEECH 2013 ComParE, INTERSPEECH 2010 Paralinguistic, INTERSPEECH 2009 Emotion, Geneva Ext. = eGeMAPS, Genre set, Genre set without the rhythmic descriptors). For BRD, all results with a difference $\geq 1.0\,\%$ are significant at $\alpha = 0.05$, for GeDiDB a difference $\geq 1.6\,\%$ is required

dance-style identification, over other, statistical feature sets. From the statistical sets, the ComParE set gives the best performance for dance-style identification, but is still roughly 18 % below the Genre set with rhythmic descriptors. In future work, a combination of the 6k descriptors from the ComParE set with the 87 rhythmic descriptors from the Genre set would be interesting to evaluate.

For music genre discrimination (GeDiDB), there is only a marginal benefit ($\approx 2\,\%$) from including the rhythmic descriptors. The ComParE set clearly outperforms the Genre set (even with rhythmic descriptors), which only scores second place for this task.

Table 6.18 Average Unweighted Average Recall (UAR) results for the BRD and GeDiDB database

Parameter		UAR (%)	
		BRD (9)	GeDiDB (7)
Balancing	Up-sampling	**64.8**	**76.3**
	Down-sampling	62.9	74.6
Ft. Norm.	MVN-ON	**68.8**	80.0
	MVN-OF	**68.8**	**80.7**
	MRN-ON	59.9	71.7
	MRN-OF	58.0	69.3
Ft. Set	GeMAPS	43.2	56.7
	eGeMAPS	46.3	60.5
	IS09	51.5	68.1
	IS10	64.1	80.7
	IS11	71.9	85.2
	IS12	72.3	85.5
	ComParE	72.6	**86.2**
	AVEC 2013	68.6	71.0
	Genre	**82.5**	81.4
	Genre (w/o rh.)	65.4	79.1

Results averaged over 13 SVM complexities, and the respective other settings (balancing, feature normalisation (Ft. Norm.), and feature set (Ft. Set)). Number of classes for BRD and GeDiDB given in (). Results with $\Delta_{p5} = 0.3\,\%$ UAR for BRD and $\Delta_{p5} = 0.5\,\%$ UAR are significant at $\alpha = 0.05$. Best result(s) in bold-face

The IS09 and eGeMAPS sets, which do not contain any specifically music related features, perform (as expected) worst on both tasks. The eGeMAPS set, which was specifically designed for voice analysis, scores last place for genre identification. Surprisingly, it outperforms the IS09 set (at a complexity of 0.5) for the ballroom dance-style classification, for which no explanation can be given.

Table 6.18 shows the results averaged over complexities $C \geq 0.00025$ for both the BRD and the GeDiDB database. These results are below the single best results from Table 6.19, due to the averaging over lower, sub-optimal SVM complexities. It was decided, however, to stick to the same range of complexities as for the speech tasks, even though higher complexities yield better results for the music tasks, especially for the smaller feature sets.

Due to the relatively small size of the two data-sets, and the rather imbalanced original distribution of dance-styles and music genres in the sets, up-sampling gives the better results here. Despite the individual best result (cf. below) on BRD being obtained with MRN normalisation, on average, for both data-sets, MVN normalisation (off-line, closely followed by on-line) is best. As already mentioned in the discussion of Fig. 6.9, the Genre set is highly superior for the BRD database (10 % over the second best set, ComParE), while for the GeDiDB, the ComParE set gives the overall best performance, closely followed by the IS12 and IS11 sets. The range

Table 6.19 Best single result for each music style database (UAR and WAR) and the settings with which this result was obtained: *up* and *down* for balancing with up-/down-sampling; feature set (Ft. Set) and feature normalisation method (Ft. Norm.); SVM complexity constant C

Database	UAR (%)	WAR (%)	Settings			
			Ft. Set	Balancing	Ft. Norm.	C
BRD	90.0	90.2	Genre set	Up	MRN-ON	0.025
GEDIDB	89.4	93.8	AVEC 2013	Up	MVN-OF	0.00005

of obtained average UAR performance on BRD for the 10 feature sets is very large, covering a range of absolute 40 %: from 43 to 83 %. This again highlights the importance of choosing the optimal feature set. Generally—in line with the findings from the speech/singing voice tasks—the larger feature sets perform better, with the clear exception of the Genre set for BRD here. This indicates the potential of a feature set tailored specifically for the task at hand, but also shows the general potential of the proposed large-scale speech and music sets.

The best result in Table 6.19 outperforms the best results on the same data so far. Schuller et al. (2007a) report a recall (WAR) for dance-style recognition of 83.3 % and Schuller et al. (2008) reports 89.1 % recall rate. Thus, the best result so far (89.1 %) is outperformed here, by 1.1 % absolute with 90.2 % WAR. This improvement is presumably mostly due to the statistical descriptors in the set and the optimisation of feature normalisation and SVM complexity. I.e., if the original features from Schuller et al. (2008) were used, and another parameter optimisation was performed, similar results could possibly be obtained with a smaller set of features.

6.4 Continuous Dimensional Affect Recognition

In the previous section the recognition of discrete emotion categories, binary activation and valence labels, ballroom dance-styles, and music genres has been considered. Automatic emotion recognition from acted, emotionally prototypical speech (e.g., EMO-DB database, cf. Sects. 6.1.5 and 6.3.3.2) gives results comparable to human performance (cf. Burkhardt et al. (2005); Schuller et al. (2009c)) and thus appears to be solved. However, reliable affect recognition in natural, dynamic environments from spontaneous, natural speech, remains a strong challenge (Schuller et al. 2009b, 2011, 2013).

In this section, evaluation results of the proposed approach for fully continuous modelling described in Sect. 4.4 are shown. The results are obtained on the SEMAINE database (Sect. 6.1.9). Modelling of SEMAINE's five dimensional affective ratings directly from LLDs is compared to modelling with incrementally extracted suprasegmental features. Additionally, multi-task learning is applied in order to estimate a confidence measure for the automatic predictions and exploit dependencies between all five dimensions in a single model.

Table 6.20 Six (bi-directional) Long Short-Term Memory Recurrent Neural Network topologies used for prediction of dimensional affect ratings

Name	Bi-directional	# Cells in hidden layer 1, 2
$T1$	No	140, 40
$T1^b$	Yes	70, 20
$T2$	No	100, 20
$T2^b$	Yes	50, 10
$T3$	No	40, 20
$T3^b$	Yes	20, 10

Two LSTM hidden layers in each network (Eyben et al. 2012a)

The remainder of this section is structured as follows: first, the experiment setup is described (Sect. 6.4.1), and second, the results obtained are shown and discussed (Sect. 6.4.2).

6.4.1 Experimental Setup

The incremental segmentation approach (Sect. 4.1.2) adapted for continuous affect recognition (Sect. 4.4.2) was applied to the SEMAINE database (Sect. 6.1.9): 7,313 (overlapping) turn segments were created from the 1,584 turns in the training set, 981 (overlapping) turn segments from the 169 turns in the development set, and 1,330 (overlapping) turn segments from the 436 turns in the evaluation set.

For supra-segmental modelling of continuous, dimensional affect as described in Sect. 4.4.2, six differently sized Long Short-Term Memory Recurrent Neural Networks (LSTM-RNNs) with two hidden layers (fully connected), both unidirectional and bidirectional (see Table 6.20 for details), were evaluated. Further, two Recurrent Neural Network (RNN) topologies, and two feedforward Neural Network (NN) topologies were evaluated, both with two fully connected hidden layers of 140, and 40 units, respectively (for the uni-directional networks). The hidden layers in the bi-directional networks were chosen to be half the size of those in the uni-directional networks, to ensure an equal number of parameters (weights) in related bi-directional and uni-directional networks.

For direct modelling of LLD, one single network topology is selected: $T1$ (Table 6.20) with two Long Short-Term Memory (LSTM) hidden layers with 140 and 40 LSTM units, respectively. In preliminary trials larger networks were investigated, however, due to the rather small data-set, networks with more hidden units did not give an increased performance. Uni-directional networks are preferred for LLD modelling, since bidirectional networks can not achieve the goal of real-time, low-latency output because they require the complete input sequence (i.e., one user speech turn, in the case of the SEMAINE database). Nonetheless, both network types (uni- and bi-directional) are evaluated here, for the sake of comparison.

Both Resilient Propagation (rProp) and standard Backpropagation Through Time (BPTT) training algorithms are compared (see Sect. 4.4.2). Networks marked with the r subscript are trained using rProp, all other networks with standard BPTT. I.e., $T2_r$ denotes a network with topology $T2$ trained using rProp.

In order to have a fair comparison of the results of single task learning to multi-task learning, all results for supra-segmental modelling are computed on the joint set of 138 Correlation-based Feature-subset Selection (CFS) selected features across all dimensions (see Sect. 4.4.3 for details). To avoid single, sub-optimal local minima of the gradient descent based learning approach, 5 networks with different random initialisation were trained and the resulting predictions averaged before scoring against the ground-truth.

Four configurations for multi-target learning are investigated: a single target (rater mean for each dimension), two targets (rater mean and inter-rater standard deviation for each dimension), five targets (rater mean for all five dimensions), and ten targets (rater mean and inter-rater standard deviation for all five dimensions).

As performance measure the Pearson Correlation Coefficient (CC) (cf. Eq. (5.8)) is used, which is computed from the automatic predictions and the "gold standard" labels (mean label of raters) of SEMAINE, as was suggested by Schuller et al. (2010a) and was applied by the author of this thesis in Eyben et al. (2010, 2011, 2012a).

6.4.2 Results

All four multi-target learning configurations are evaluated for all six LSTM topologies and both rProp and BPTT. Due to the high number of resulting individual evaluations (48), (Pearson) Correlation Coefficients (CCs) averaged over various configurations are reported here. These are in particular: average CCs for each topology (Table 6.23), and for each multi-/single-task learning configuration (Table 6.24). The best individual results are shown in Table 6.21. A comparison with related approaches is printed in Table 6.22. These related methods are Support Vector Regression (SVR), RNNs, and feedforward NNs (the latter two both having the same hidden layer size as the $T1$ LSTM-RNN).

A clear trend is observed (Table 6.21) that the worst performance is obtained for the valence (V) dimension, and the best performance is obtained for activation (A). This is confirmed throughout the related literature (e.g., Eyben et al 2010, Grimm et al. 2007a). The performance for the new dimensions, expectation (E), intensity (I), and power (P) lies in between A and V, in terms of CC. Further, in Table 6.21 for all dimensions, except for A and V, the best result is obtained with a Bidirectional Long Short-Term Memory Recurrent Neural Network (BLSTM-RNN). However, the best unidirectional networks are only ≈0.02–0.05 absolute CC behind the BLSTM-RNN.

When comparing the results from Table 6.21 with the average correlation between the human raters on the evaluation set (Table 6.6), it can be seen that the automatic system based on LSTM outperforms the human abilities for all dimensions except valence. This is in line with the findings reported by the author of this thesis in Eyben et al. (2011). Interestingly, valence has the highest agreement among human raters but is most difficult to predict with an automatic system which relies on acoustic cues only. Much valence information might be carried by the linguistic content and the context an utterance is presented in. The acoustic-only system can exploit neither of these well. Given the low human agreement of the dimensions *Expectation* and *Power*, the results of the automatic recognition seem very good, which is surprising, but seems to show that the average of the human ratings does provide a ground truth which is correlated to some acoustic correlates and thus is more valid than one would assume from the high ambiguity of the ratings. In particular for *Expectation*, the best result was obtained with multi-target learning of all five ratings and rater standard deviations, which further supports the fact that each individual rater is consistent with itself, even though the inter rater agreement is low. Each rating therefore adds additional information to the model.

A comparison of LSTM-RNN to other related neural networks and to SVR is shown in Table 6.22. The NN and the RNN have the same number of hidden units in the two hidden layers as the LSTM-RNN. SVR has been trained with the SMO algorithm and the WEKA data-mining toolkit by Hall et al. (2009). Linear kernel SVR was used with a complexity parameter $C = 1.0$. In three out of five cases the LSTM outperforms the other methods, in the other two cases the RNN delivers the best performance, indicating the importance of short and long-term context, especially for expectation and valence.

In order to find out which topology performs best, the results were averaged over all five dimensions (see Table 6.23). No clearly best performing topology can be identified, as all topologies achieve an average CC roughly between 0.4 and 0.5. The network $T1$ is the best according to this table, with topology $T1_r^b$ being the best for single target runs and $T1^r$ the best for multi-target runs. Presumably a larger bidirectional network would make $T1_r^b$ also the winning topology for the multi-target setting. As $T1$ is the largest network, this suggests that the bigger networks are better. However, at the same time the much smaller networks ($T3$) do not perform much worse, especially when trained with resilient propagation.

Table 6.21 Best individual results ((Pearson) Correlation Coefficient (CC)) for the automatic prediction of the mean affect rating obtained with the given configurations (topology, multi-/single-target learning)

Dimension	Topology	Num. targets	CC
A	$T1_r$	2	0.805
E	$T2_r^b$ ($T3_r$)	10 (10)	0.624 (0.592)
I	$T1^b$ ($T1$)	1 (1)	0.606 (0.593)
P	$T3_r^b$ ($T1_r$)	1 (10)	0.657 (0.610)
V	$T1_r$	2	0.308

Multi-target learning: 2 targets (rater mean, and inter-rater variance), 5 targets (5 dimensions, rater means), 10 targets (rater means and inter-rater variances for all 5 dimensions). ^{b}: bi-directional network; $_r$: resilient propagation. For best results with bi-directional networks, the best uni-directional result is given in brackets

Table 6.22 Comparison of performance (CC with mean rater label) of Long Short- Term Memory Recurrent Neural Network (LSTM-RNN), Recurrent Neural Network (RNN), Feed-Forward Neural Network (FFNN), and Support Vector Regression (SVR)

Dimension	LSTM (CC)	RNN (CC)	FFNN (CC)	SVR (CC)
A	**0.785**	0.747	0.735	0.653
E	**0.470**	0.420	0.278	0.190
I	0.489	**0.559**	0.458	0.503
P	**0.527**	0.427	0.438	0.367
V	0.115	**0.152**	−0.069	−0.085

Topology $T2_r$, single target (mean of each affective dimension). LSTM-RNNs, RNNs, and NNs have the same number of hidden units in 2 hidden layers and are all trained with resilient propagation (topology $T2_r$). Best result marked in boldface font (Eyben et al. 2012a)

For most cases across all topologies it is evident that networks trained with resilient propagation perform better than those trained with BPTT. Bi-directional networks (although they have exactly the same amount of parameters as the respective uni-directional networks) perform better in most cases across all three topologies. Again, the results in Table 6.23 confirm that the RNN and especially the NN networks perform worse than the LSTM networks. This again shows that context is important for automatic dimensional affect recognition.

Overall, the differences between the topologies are not very large. From this finding, one can assume that a further, more fine grained investigation of other topologies is not necessary or will not yield any large improvement over the current results. In future studies the most interesting experiment in this respect will be to reduce the network size further in order to find the point where performance significantly drops and thus develop a minimal network size recommendation for low-resource environments.

A clear difference in performance can be seen for single target versus multi-target learning. Table 6.24 shows the averaged results for the four cases of (a) learning of each dimension separately, (b) learning of each dimension together with the inter-rater standard deviation as confidence measure, (c) multi-target learning of all five dimensions together, and (d) multi-target learning of all five dimensions and corresponding inter-rater standard deviations (10 targets). Only including the inter-rater standard deviation apparently degrades the results, while multi-target learning of all five dimensions clearly shows an improvement. Adding inter-rater deviation in the multi-target learning seems to improve the performance further, which suggests that the cross dimensional information is complemented by the inter rater agreement information. A possible explanation for the fact that the inclusion of the inter-rater standard deviation improves the results for multi-dimension learning while it does not improve it for single dimension learning could be that, the joint uncertainty information of all dimensions helps the network, while single dimension uncertainty is not useful for the prediction of the dimension's mean rating. The best performing method is found to be multi-target learning where all five dimensions and the corresponding inter-rater standard deviations are learnt by a single network.

Table 6.23 Average (Pearson) Correlation Coefficient (CC) per topology for prediction of mean ratings; CC averaged over all five affective dimensions

Dimension	CC-1	CC-10
$T1$	0.449	0.519
$T1^b$	0.513	0.523
$T1_r$	0.490	**0.566**
$T1_r^b$	**0.563**	0.542
$T2$	0.432	0.479
$T2^b$	0.490	0.518
$T2_r$	0.477	0.518
$T2_r^b$	0.478	0.572
$T3$	0.481	0.465
$T3^b$	0.499	0.506
$T3_r$	0.516	0.541
$T3_r^b$	0.495	0.541
$T2^{nn}$	0.361	0.431
$T2^{rnn}$	0.403	0.462
$T2_r^{nn}$	0.350	0.381
$T2_r^{rnn}$	0.461	0.481

CC-1: CC averaged over single target runs (dimension mean as target); **CC-10**: CC averaged over multi-target runs (mean and inter-rater variance of all five dimensions as targets); ^{b}: bi-directional network. Bottom part: feedforward and standard recurrent neural network topologies trained with backpropagation through time and resilient propagation (subscript r). Best result per column in boldface font

It further seems that for single dimension learning the net is able to better predict the standard deviation (i.e., the uncertainty)—see Table 6.25, but the performance for predicting the actual dimensional label is degraded, while in the multi-dimension learning, the performance for predicting the actual labels is improved but the prediction of uncertainty suffers. Overall, the uncertainty of the dimensions intensity and power can be predicted best. For the other dimensions the prediction of uncertainty is rather low. It can be thus concluded that, in order to use the automatically pre-

Table 6.24 Average (Pearson) Correlation Coefficient (CC) for the mean rating (over all five dimensions, and (B)LSTM topologies) for single- versus multi-target learning: **CC-1**: single target runs (dimension mean as target); **CC-2**: single target runs (dimension mean and inter-rater variance as target); **CC-5**: multi-target runs (mean of all five dimensions as targets); **CC-10**: multi-target runs (mean and inter-rater variance of all five dimensions as targets)

	CC
CC-1	0.490
CC-2	0.422
CC-5	0.513
CC-10	0.524

dicted inter-rater standard deviation as a confidence measure for all five dimensions, more work is required. Still, with the multi-dimensional learning presented here, it appears beneficial to include the inter-rater standard deviation in order to improve the prediction of the actual dimensional label (mean).

In order to justify the use of an automatic classifier/regressor for dimensional affect recognition on the SEMAINE data, correlation coefficients between the maximum, mean, and standard deviation (over a segment/user speech turn) of the LLDs loudness (approximated by Eq. (2.43) in Sect. 2.2.2) and F_0 (Autocorrelation function (ACF) and Cepstrum based, see Sect. 2.2.11.3) and the mean ratings for the five dimensions on the evaluation set are shown in Table 6.26. For some features and the dimensions activation and intensity high CCs are obtained, while for the other three dimensions the correlations are not significant (t-test). The maximum loudness has a correlation of 0.65 with the activation dimension, which is above the average human rater agreement (0.57), yet below the best result obtained with LSTM-RNNs (0.81). I.e., the use of LSTM-RNNs is justified. Further, these findings are compatible with the result of the CFS feature selection (Sect. 4.4.3), where loudness related features were shown highly relevant for activation and intensity. For F_0, the mean value per segment shows a stronger correlation with activation and intensity, while for loudness the maximum value per segment seems to be better correlated. Minor correlations between the expectation dimension label and F_0 mean and standard deviation (0.18 and 0.2, respectively) are observed.

The best way to judge the actual performance of the networks and to visualise what is actually happening to the outputs, is to plot the activations and the gold standard as is done in Fig. 6.10 for three dimensions and the respective best performing networks. In the plots all the user speech turns are concatenated and the gaps caused by the operator turns are removed in order to keep the plot easy to view.

The results for the LLD based modelling are given in Table 6.27. Instead of the IS10 feature set, as used for the other experiments, the LLDs of the feature set implemented in the SEMAINE 3.0 system have been used to obtain these results (cf. Eyben et al. (2012a), where this set is described in detail). No feature selection was performed. The correlations obtained by LLD modelling are rather low compared to those obtained with the supra-segmental approach and those obtained with the

Table 6.25 Best results obtained for the prediction of the inter-rater standard deviation for single dimension learning (1-dim, 2-targets) and multi-dimension learning (5-dim, 10 targets); all best results were obtained with two target learning (mean and standard deviation of each dimension individually); ^{b}: Bi-directional network; resilient propagation: subscript r

Dimension	Topology	CC 1-dim	CC 5-dim
A	$T3_r$	0.184	0.140
E	$T1_r^b$	0.297	0.271
I	$T1_r^b$	0.445	0.173
P	$T1_r^b$	0.522	0.287
V	$T3^b$	0.123	0.077

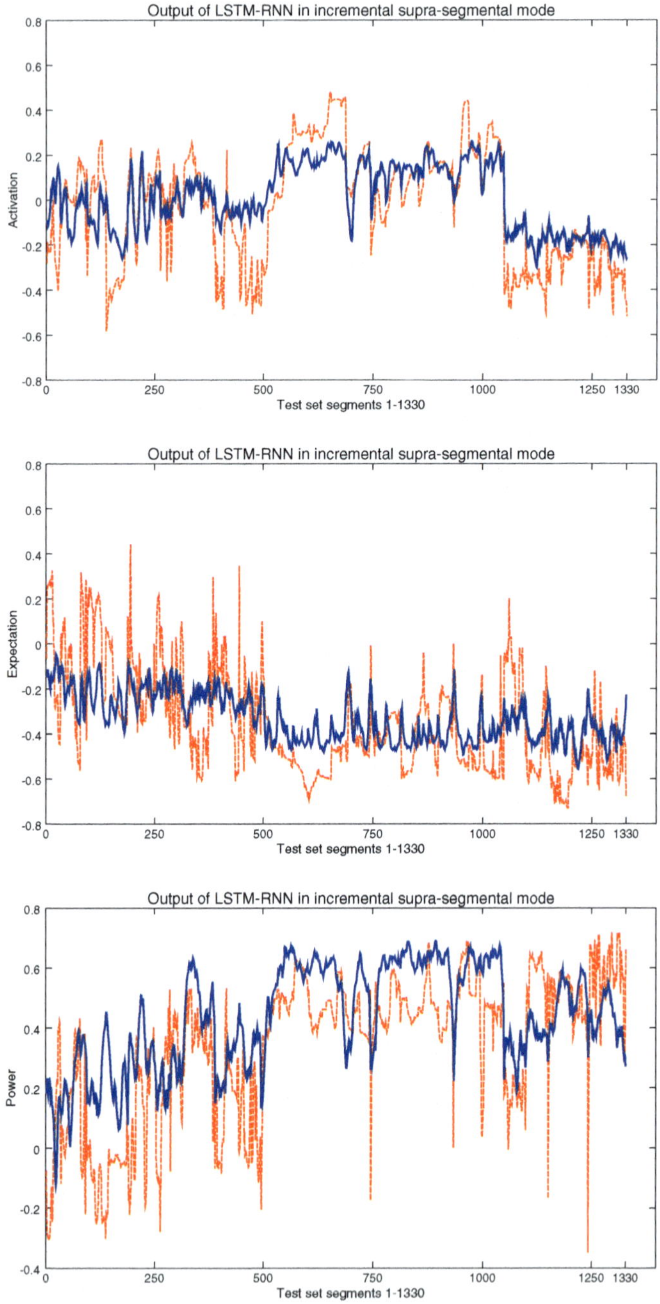

Fig. 6.10 Plots of the automatic predictions for the configurations that gave the best results for activation (*top*), expectation (*middle*) and power (*bottom*) (Table 6.21). The thick line is the automatic prediction (average over 5 runs), the thin line is the gold standard (rater mean). All segments of the evaluation set are arranged in chronological order on the x-axis. Incremental supra-segmental modelling. For more details see (Eyben et al. 2012a)

Table 6.26 Correlations of mean, maximum, and standard deviation of loudness and F_0 with the five dimensional labels (mean of raters)

Pearson [CC]	Maximum loudness	Mean loudness	Std. dev. loudness
A	0.65	0.63	0.60
E	0.06	−0.03	0.10
I	0.57	0.49	0.52
P	−0.00	0.11	−0.03
V	0.16	−0.12	−0.12
	Maximum F_0	Mean F_0	Std. dev. F_0
A	0.39	0.50	0.22
E	0.09	0.18	0.20
I	0.33	0.49	0.26
P	−0.08	−0.15	−0.18
V	−0.04	−0.09	0.01

Statistics computed over the incremental (5 s) segments. Cf. (Eyben et al. 2012a)

Table 6.27 Results of Low-level Descriptor (LLD) level modelling of mean rater label

Configuration	Activation	Expectation	Valence
1-d learning $T1_r$	0.082	0.355	-0.006
1-d learning $T1_r^b$	0.123	0.323	0.003
1-d learning $T1^b$	0.271	0.279	0.090
5-d learning $T1_r$	0.560	0.110	0.116
5-d learning $T1_r^b$	0.469	0.056	0.295
5-d learning $T1^b$	0.378	0.056	0.296

(B)LSTM (bi-directional: ^{b}), topology $T1$, Resilient Propagation (subscript r) or standard Backpropagation Through Time for training. Dimensions A(ctivation), E(xpectation), V(alence). Correlation coefficient averaged over 5 network trainings with different initial weights (same as for the supra-segmental results). Single dimension (1-d) learning compared to multi-dimension (5 dimensions) learning (5-d). (Eyben et al. 2012a)

loudness and F_0 to label correlation presented in Table 6.26. Thus, it can be concluded that the incremental supra-segmental modelling should be preferred unless the LLD modelling will be improved.

Regarding the best configuration for LLD modelling, no clear trend can be deducted from Table 6.27. However, it can be concluded that multi-dimensional learning is also beneficial for LLD modelling, at least for for activation and valence. A high variance in the results across different settings can be seen for activation, for example (CC 0.082 to 0.560). This may be an indication that the training algorithm is not fully stable for this complex task. In future work on LLD based modelling the activations should thus be smoothed over far more than five individual training runs to obtain more stable results, or alternative training algorithms, such as the Extended Kalman Filter training (Pérez-Ortiz et al. 2003), should be investigated.

References

T. Bänziger, M. Mortillaro, K.R. Scherer, Introducing the Geneva multimodal expression corpus for experimental research on emotion perception. Emotion **12**(5), 1161–1179 (2012)

T. Bänziger, S. Patel, K.R. Scherer, The role of perceived voice and speech characteristics in vocal emotion communication. J. Nonverbal Behav. **38**(1), 31–52 (2014). doi:10.1007/s10919-013-0165-x. ISSN 0191-5886

F. Burkhardt, A. Paeschke, M. Rolfes, W. Sendlmeier, B. Weiss, A database of german emotional speech, in *Proceedings of the INTERSPEECH 2005 (ISCA)* (Lisbon, Portugal, 2005), pp. 1517–1520

R. Cowie, E. Douglas-Cowie, S. Savvidou, E. McMahon, M. Sawey, M. Schröder, Feeltrace: an instrument for recording perceived emotion in real time, in *Proceedings of the ISCA Workshop on Speech and Emotion* (Newcastle, Northern Ireland, 2000), pp. 19–24

R. Daido, M. Ito, S. Makino, A. Ito, Automatic evaluation of singing enthusiasm for karaoke. Comput. Speech Lang. **28**(2), 501–517 (2014)

E. Douglas-Cowie, R. Cowie, I. Sneddon, C. Cox, O. Lowry, M. McRorie, J.C. Martin, L. Devillers, S. Abrilian, A. Batliner, N. Amir, K. Karpouzis, *The HUMAINE Database*, Lecture Notes in Computer Science (Springer, Heidelberg, 2007)

I.S. Engbert, A.V. Hansen, *Documentation of the danish emotional speech database DES. Technical report, Center for PersonKommunikation* (Aalborg University, Denmark, 2007)

F. Eyben, B. Schuller, S. Reiter, G. Rigoll, Wearable assistance for the ballroom-dance hobbyist—holistic rhythm analysis and dance-style classification, in *Proceedings of the IEEE International Conference on Multimedia and Expo (ICME) 2007 (IEEE)* (Bejing, China, July, 2007), pp. 92–95

F. Eyben, M. Wöllmer, A. Graves, B. Schuller, E. Douglas-Cowie, R. Cowie, On-line emotion recognition in a 3-D activation-valence-time continuum using acoustic and linguistic cues. J. Multimodal User Interfaces (JMUI) **3**(1–2), 7–19 (2010). doi:10.1007/s12193-009-0032-6

F. Eyben, M. Wöllmer, M. Valstar, H. Gunes, B. Schuller, M. Pantic, String-based audiovisual fusion of behavioural events for the assessment of dimensional affect, in *Proceedings of the IEEE International Workshop on Emotion Synthesis, rePresentation, and Analysis in Continuous spacE (EmoSPACE) 2011, held in conjunction with FG 2011* (Santa Barbara, CA, USA, March, 2011), pp. 322–329

F. Eyben, M. Wöllmer, B. Schuller, A multi-task approach to continuous five-dimensional affect sensing in natural speech, special issue on affective interaction in natural environments. ACM Trans. Interact. Intell. Syst. **2**(1), Article no 6, p. 29 (March 2012a)

F. Eyben, B. Schuller, G. Rigoll, Improving Generalisation and Robustness of Acoustic Affect Recognition, in *Proceedings of the 14th ACM International Conference on Multimodal Interaction (ICMI) 2012*, ed. by L.-P. Morency, D. Bohus, H.K. Aghajan, J. Cassell, A. Nijholt, J. Epps (ACM, Santa Monica, CA, USA, 2012b), pp. 517–522

F. Eyben, A. Batliner, B. Schuller, Towards a standard set of acoustic features for the processing of emotion in speech. Proc. Meet. Acoust. (POMA) **9**(1), 1–12 (2012c)

F. Eyben, F. Weninger, B. Schuller, Affect recognition in real-life acoustic conditions—a new perspective on feature selection, in *Proceedings of the INTERSPEECH 2013 (ISCA)* (Lyon, France, August 2013), pp. 2044–2048

F. Eyben, G.L. Salomão, J. Sundberg, K.R. Scherer, B. Schuller, Emotion in the singing voice—a deeper look at acoustic features in the light of automatic classification. EURASIP J. Audio, Speech, Music Process, special issue on scalable audio-content analysis 2015 (2015, in press), p. 14

J.R.J. Fontaine, K.R. Scherer, E.B. Roesch, P.C. Ellsworth, The world of emotions is not two-dimensional. Psychol. Sci. **18**(2), 1050–1057 (2007)

M. Grimm, K. Kroschel, Evaluation of natural emotions using self assessment manikins, in *Proceedings of the IEEE Workshop on Automatic Speech Recognition and Understanding (ASRU) 2005 (IEEE)* (Cancun, Maxico, November, 2005), pp. 381–385

M. Grimm, K. Kroschel, S. Narayanan, Support vector regression for automatic recognition of spontaneous emotions in speech, in*Proceedings of the ICASSP 2007 (IEEE)*, vol. 4 (Honolulu, HI, USA, April, 2007a), pp. 1085–1088

M. Grimm, E. Mower, K. Kroschel, S. Narayanan, Primitives based estimation and evaluation of emotions in speech. Speech Commun. **49**, 787–800 (2007b)

M. Grimm, K. Kroschel, S. Narayanan, The Vera am Mittag German audio-visual emotional speech database, in *Proceedings of the IEEE International Conference on Multimedia and Expo (ICME) 2008* (Hannover, Germany, 2008), pp. 865–868

M. Hall, E. Frank, G. Holmes, B. Pfahringer, P. Reutemann, I. Witten, The WEKA data mining software: an update. SIGKDD Explor. Newsl. **11**(1), 10–18 (2009). doi:10.1145/1656274.1656278

J.H.L. Hansen, S. Bou-Ghazale, Getting started with SUSAS: a speech under simulated and actual stress database, in *Proceedings of the EUROSPEECH-97 (ISCA)* (Rhodes, Greece, 1997), pp. 1743–1746

A. Hassan, R. Damper, M. Niranjan, On acoustic emotion recognition: compensating for covariate shift. IEEE Trans. Audio, Speech Lang. Process. **21**(7), 1458–1468 (2013)

O. Martin, I. Kotsia, B. Macq, I. Pitas, The eNTERFACE'05 audio-visual emotion database, in *Proceedings of the IEEE Workshop on Multimedia Database Management (IEEE)* (Atlanta, GA, USA, April 2006)

G. McKeown, M. Valstar, R. Cowie, M. Pantic, M. Schroder, The SEMAINE database: annotated multimodal records of emotionally colored conversations between a person and a limited agent. IEEE Trans. Affective Comput. **3**(1), 5–17 (2012). doi:10.1109/T-AFFC.2011.20. ISSN 1949-3045

T. Nakano, M. Goto, Y. Hiraga, An automatic singing skill evaluation method for unknown melodies using pitch interval accuracy and vibrato features, in *Proceedings of the INTERSPEECH 2006 (ISCA)* (Pittsburgh, PA, USA, 2006), pp. 1706–1709

J.A. Pérez-Ortiz, F.A. Gers, D. Eck, J. Schmidhuber, Kalman filters improve LSTM network performance in problems unsolvable by traditional recurrent nets. Neural Netw. **16**(2), 241–250 (2003). doi:10.1016/s0893-6080(02)00219-8. ISSN 0893-6080

F. Ringeval, A. Sonderegger, J. Sauer, D. Lalanne, Introducing the RECOLA multimodal corpus of remote collaborative and affective interactions, in *Proceedings of the 2nd International Workshop on Emotion Representation, Analysis and Synthesis in Continuous Time and Space (EmoSPACE), held in conjunction with FG 2013 (IEEE)* (Shanghai, China, April 2013), pp. 1–8

K.R. Scherer, Vocal markers of emotion: comparing induction and acting elicitation. Special issue on paralinguistics in naturalistic speech and language. Comput. Speech Lang. **27**(1), 40–58 (2013). doi:10.1016/j.csl.2011.11.003. ISSN 0885-2308

K.R. Scherer, J. Sundberg, L. Tamarit, G.L. Salomão, Comparing the acoustic expression of emotion in the speaking and the singing voice. Comput. Speech Lang. **29**(1) (2015). doi:10.1016/j.csl.2013.10.002

B. Schuller, *Intelligent Audio Analysis*, Signals and Communication Technology (Springer, Heidelberg, 2013)

B. Schuller, F. Eyben, G. Rigoll, Fast and robust meter and tempo recognition for the automatic discrimination of ballroom dance styles, in*Proceedings of the ICASSP 2007 (IEEE)* (Honolulu, HI, USA, April 2007a), pp. 217–220

B. Schuller, M. Wimmer, D. Arsić, G. Rigoll, B. Radig, Audiovisual behavior modeling by combined feature spaces, in *Proceedings of the International Conference on Acoustics, Speech and Signal Processing, 2007 (IEEE)*, vol. 2 (Honolulu, HI, USA, 2007b), pp. 733–736

B. Schuller, F. Eyben, G. Rigoll, Tango or Waltz?—putting ballroom dance style into tempo detection, special issue on intelligent audio, speech, and music processing applications (Article ID 846135). EURASIP J. Audio, Speech, Music Process. (April 2008). doi:10.1155/2008/846135

B. Schuller, R. Müller, F. Eyben, J. Gast, B. Hörnler, M. Wöllmer, G. Rigoll, A. Höthker, H. Konosu, Being bored? recognising natural interest by extensive audiovisual integration for real-life application. Image Vis. Comput., Spec. Issue Vis. Multimodal Anal. Hum. Spontaneous Behav., **27**(12), 1760–1774 (2009a)

B. Schuller, S. Steidl, A. Batliner, F. Jurcicek, The INTERSPEECH 2009 emotion challenge, in *Proceedings of the INTERSPEECH 2009* (Brighton, UK, Sept. 2009b), pp. 312–315

B. Schuller, B. Vlasenko, F. Eyben, G. Rigoll, A. Wendemuth, Acoustic emotion recognition:a benchmark comparison of performances, in*Proceedings of the IEEE Workshop on Automatic Speech Recognition and Understanding (ASRU) 2009, IEEE* (Merano, Italy, December, 2009c), pp. 552–557

B. Schuller, S. Steidl, A. Batliner, F. Burkhardt, L. Devillers, C. Müller, S. Narayanan, The INTERSPEECH 2010 paralinguistic challenge, in *Proceedings of the INTERSPEECH 2010 (ISCA)* (Makuhari, Japan, September, 2010a), pp. 2794–2797

B. Schuller, R. Zaccarelli, N. Rollet, L. Devillers, CINEMO—a french spoken language resource for complex emotions: facts and baselines, in *Proceedings of the 7th International Conference on Language Resources and Evaluation (LREC) 2010, European Language Resources Association (ELRA)* (Valletta, Malta, 2010b). ISBN 2-9517408-6-7

B. Schuller, B. Vlasenko, F. Eyben, M. Wöllmer, A. Stuhlsatz, A. Wendemuth, G. Rigoll, Cross-corpus acoustic emotion recognition: variances and strategies. IEEE Trans. Affective Comput. (TAC) **1**(2), 119–131 (2010c). doi:10.1109/T-AFFC.2010.8

B. Schuller, A. Batliner, S. Steidl, D. Seppi, Recognising realistic emotions and affect in speech: state of the art and lessons learnt from the first challenge, special issue on sensing emotion and affect—facing realism in speech processing. Speech Commun. **53**(9/10), 1062–1087 (2011)

B. Schuller, S. Steidl, A. Batliner, A. Vinciarelli, K. Scherer, F. Ringeval, M. Chetouani, The INTERSPEECH 2013 computational paralinguistics challenge: social signals, conflict, emotion, autism, in *Proceedings of the INTERSPEECH 2013 (ISCA)* (Lyon, France, 2013), pp. 148–152

M.D. Smucker, J. Allan, B. Carterette, A comparison of statistical significance tests for information retrieval evaluation, in *Proceedings of the sixteenth ACM conference on information and knowledge management (CIKM'07)* (ACM, 2007), pp. 623–632

S. Steidl, *Automatic Classification of Emotion-Related User States in Spontaneous Children's Speech* (Logos, Berlin, 2009)

M. Wöllmer, F. Eyben, S. Reiter, B. Schuller, C. Cox, E. Douglas-Cowie, R. Cowie, Abandoning emotion classes—towards continuous emotion recognition with modelling of long-range dependencies, in *Proceedings of the INTERSPEECH 2008 (ISCA)* (Brisbane, Australia, September, 2008), pp. 597–600

Chapter 7
Discussion and Outlook

This chapter summarises the methods presented and the results obtained, and concludes this thesis (Sect. 7.1). Further, it is discussed if and how the aims defined at the beginning (cf. Sect. 1.3) were achieved (Sect. 7.2). Open issues for future work are discussed in Sect. 7.3.

7.1 Summary

The main contribution of this thesis are several standard acoustic parameter sets, which were proposed by the author and evaluated at several, mainly INTERSPEECH, challenges from 2009 to 2014. Implementations of these feature sets along with a generic framework for incremental, real-time acoustic feature extraction were published as an open-source toolkit (first released as openEAR (Eyben et al. 2009a) including affect recognition models (Eyben et al. 2009b), and later released as openSMILE (Eyben et al. 2010, 2013a)). This toolkit in its latest version (Eyben et al. 2013b) implements all of the proposed standard acoustic feature sets, as well as Support Vector Machine (SVM) and Long Short-Term Memory Recurrent Neural Network (LSTM-RNN) classifiers. At the time of writing, the three publications which introduce and evaluate this toolkit have altogether been cited already over 700 times[1] and the toolkit is highly accepted by the community and used frequently. The numerous speech and music related acoustic descriptors contained in the standard sets, as well as the functionals applied in order to summarise descriptors over given temporal segments, are described formally in detail in Chap. 2. The two most important modelling methods, which are also implemented (as decoders) in the openSMILE toolkit, were described in this thesis (Sect. 2.5): SVMs and Support Vector Regression (SVR) for static prediction, and LSTM-RNNs for dynamic time-series prediction of context sensitive data.

[1]According to Google scholar citations.

F. Eyben, *Real-time Speech and Music Classification by Large Audio Feature Space Extraction*, Springer Theses, DOI 10.1007/978-3-319-27299-3_7

The second contribution of this thesis, is methods for speech and music analysis under real-world acoustic conditions and in real-time, continuous stream processing, which were proposed in Chaps. 4 and 5 evaluated in Sects. 6.2 and 6.4, respectively. Namely, these are: noise robust, context-aware, data-driven Voice Activity Detection (VAD) based on LSTM-RNNs and multi-condition training on large, synthesised data-sets, a real-time, incremental input segmentation approach building on the noise robust VAD output, and multi-condition classification for increased noise and reverberation robustness. Both methods were evaluated extensively for speech affect classification tasks. However, they are general methods, which are applicable to any speech and/or music recognition task.

The 10 proposed acoustic feature sets (Chap. 3) were systematically evaluated on 13 databases containing speech affect and music style classification tasks (Sect. 6.3). These databases cover a wide range of affective speech recordings, ranging from acted and prototypical to natural, spontaneous, and continuous affect, and music genre and ballroom dance-style collections (cf. Sect. 6.1). Experiments over a wide range of settings were performed. These include comparison of up- and down-sampling for training instance balancing, five methods for feature value normalisation, and 18 SVM model complexity settings.

Overall, the reported results for the 13 databases can compete with or outperform the state-of-the-art (Sects. 6.3.3.2 and 6.3.3.3). The larger feature sets gave better classification accuracy, especially for the affect dimension valence. For the dimension activation, one of the smallest sets (extended Geneva Minimalistic Acoustic Parameter Set (eGeMAPS)) also gave results competitive to the larger sets, given an optimised SVM model complexity constant C. This confirms that the features in the eGeMAPS set are highly relevant for analysis of vocal activation, while (at least) some of the features contained in the larger sets, such as the INTERSPEECH 2010 Paralinguistics Challenge (IS10) set or the Computational Paralinguistics ChallengE (ComParE) set, seem to be (reasonably) good acoustic indicators of vocal valence. It is left to future work to identify and extract the essential ones, and add them to the Geneva Minimalistic Acoustic Parameter Set (GeMAPS) sets. Normalisation of acoustic features to a per-speaker zero mean and variance of one was shown to be the best method of scaling features. Second best is the independent normalisation of the training and test set.[2] As expected, the normalisation to mean zero and variance one was shown to outperform the normalisation to the minimum and maximum value. The latter is more vulnerable to single outliers, which can drastically change the minimum and/or maximum value of each feature. A single outlier instance can thus affect the scaling of the whole data-set!

The proposed methods for real-time, noise robust, incremental input segmentation, and noise robust multi-condition classification were extensively evaluated on suitable databases (the TUM Audiovisual Interest Corpus (AVIC) and the Geneva Multimodal Emotion Portrayals (GEMEP) databases). It was shown that the proposed Multi-

[2]It is to note that, a joint normalisation of training and test set might be best, however, it is not possible, as at training time the test set is not known (or must not be used!) but would be required to normalise the training set.

condition Training (MCT) approach is highly effective and improves classification accuracy even for unknown noise types in testing. For speech/non-speech segmentation, the proposed VAD delivered accuracy superior to state-of-the-art algorithms when evaluated in realistic noisy conditions. Thus, it enables the robust, real-time and low-latency (typically ≤ 400 ms, depending on the amount of post-smoothing, cf. N_{post} variable in Sect. 4.1.1), identification of speech segments in challenging, non-stationary noisy audio streams. The speech segments are then incrementally sub-segmented with the sliding-window approach described in Sect. 4.1.2, and static feature vectors at any arbitrary quasi-continuous time scale can be obtained. With these, fully continuous (time and value), automatic recognition of affect in a five dimensional space with LSTM-RNN is evaluated on a database of natural and spontaneous affect expressions (SEMAINE). It is shown that the LSTM-RNN outperforms human ratings (measured as the average inter-rater correlation/agreement) for all dimensions, except for valence, which is a remarkable result.

7.2 Achievements

The aims of this thesis, as given in Sect. 1.3, were:

1. Development of an open-source framework and toolkit for incremental speech and music analysis with a strong focus on real-time, incremental feature extraction algorithms and a broad coverage of acoustic descriptors,
2. Definition of novel, large-scale standard acoustic feature sets for incremental speech and music analysis,
3. Design of speech and music analysis methods which can robustly handle real-life conditions,
4. Evaluation of the framework, the methods, and the standard acoustic feature sets on a wide range of music classification and speech affect recognition tasks.

Aim (1) was fully achieved through the implementation of the openSMILE toolkit by the author of this thesis. The framework described in Sect. 4.3 implements all the acoustic descriptors described in Sects. 2.2 and 2.4, the classifiers for static and dynamic modelling described in Sect. 2.5. All analysis with this framework can be performed incrementally in on-line analysis applications (Chap. 4), as well as in off-line, batch processing applications. The success of reaching aim (1) is further undermined by the fact that the publications which introduce and describe the toolkit were cited more than 400 times at the time of writing of this thesis. It was used for several research challenges in which the author and his colleagues were involved, i.e., all the INTERSPEECH challenges (Schuller et al. 2009, 2010, 2011a, 2012a, 2013, 2014a), as well as the Audio-visual Emotion Challenge (AVEC) challenges (Schuller et al. 2011b, 2012b; Valstar et al. 2013), and challenges in which the author was not involved, such as (Dhall et al. 2013). Further, it was used for or cited in the context of a multitude of speech and music analysis applications. Due to the multitude of publications and use-cases, only a few diverse examples of different use-cases are named here: Analysis of the speech wrt. behavioural cues in interactions of married couples by Black et al. (2013, 2011) or regarding entrainment in these interactions (Lee et al. 2014), mood and depression detection from spontaneous speech by Alghowinem et al. (2012, 2013), voice communication (e. g., telephony) speech quality experience evaluation based on affect by Bhattacharya et al. (2012), recognition of personality traits from speech by, e. g., Ivanov et al. (2011), estimation of speaker height by Mporas and Ganchev (2009), detection of intoxication from speech (Schuller et al. 2011a; Bone et al. 2014), analysis of stress (Muaremi et al. 2013), music playlist generation (Lukacs et al. 2013), music emotion classification (Nikolaou 2011), laughter analysis (Eyben et al. 2012; Jacykiewicz and Ringeval 2014), attempts to more general recognition of non-linguistic vocal outbursts (Eyben et al. 2011; Schuller et al. 2013), and controlling the actions and responses of a virtual agent based on the user's speech (Reidsma et al. 2011).

For achieving aim (2), 8 major large-scale (and two minimalistic) acoustic feature sets were proposed which have been already widely accepted and used by the community. This includes use by the participants of the various INTERSPEECH and AVEC challenges (e. g., Lee et al. 2009; Jeon et al. 2010; Ivanov et al. 2011; Bone et al. 2011; Kim et al. 2012; Räsänen and Pohjalainen 2013), as well as by follow-up work and work by non-challenge participants (e. g., Lukacs et al. 2013; Suzuki et al. 2013). The success of these sets and of the developed openSMILE toolkit is also owed to the free and open-source availability of openSMILE for research. Further, it was shown in Sect. 6.3, that the large feature sets, which contain a large range of speech and music acoustic descriptors, perform well on a wide range of tasks. It can thus be concluded that aim (2) is more than well achieved.

Aim (3) was achieved by successful application of the methods for multi-condition training of speech analysers (Sect. 5.3), and a novel, robust segmentation algorithm (Sect. 5.1.2) which enables fully time and value continuous analysis—cf. Sects. 6.2 and 6.4, respectively, for the evaluation results which show the benefits of the proposed approaches for near real-world recognition tasks. The evaluation of the proposed methods on fully realistic data is still lacking, as such data does not exist yet on a large scale. All databases which are available at present are recorded in some set-up where emotions are induced, acted, or foreseen in some way by the choice of participants or situations. Labelled everyday data from call-centres, personal video blogs, or person-to-person communication is very rare. Still, aim (4) was achieved well with the data and resources available.

Aim (4) was achieved mainly through the extensive evaluations presented in Sect. 6.3, and also by the large acceptance of the feature sets by the research community and the manifold results on various tasks reported on with these sets. Suitability of all proposed sets for selected speech and music classification tasks—even in degraded acoustic conditions—was proven in Sects. 6.2 and 6.3, It was shown that, the most recently proposed sets, i. e., the most complete sets of acoustic descriptors, gave the best performance on average on all the selected tasks (Sect. 6.3). Various recommendations for parametrisations for speech and music analysis systems have also been proposed based on the results. An open issue still remains, however, which is the analysis of the suitability of the individual features in the sets for various tasks.

While this sounds trivial at first, it is a very complex task indeed. Single features are seldom good for discrimination of complex speech attributes such as affect or personality, it is rather the right combination of several features that is important. Thus, more work on the analysis of feature groups and subsets for, e. g., the classification tasks proposed in this thesis, is required in future work. Thus, aim (3) is seen as sufficiently solved within the scope of this thesis, but with leaving room for more specific, follow-up research.

7.3 Future Work and Concluding Remarks

Based on the deficiencies identified in the previous sections, the following items for the direction of follow-up research work can be named:

Feature sets and noise-robust, real-time, incremental classification methods should be evaluated on even larger, more naturalistic databases (cf. e. g., Ringeval et al. 2013), and semi-supervised learning (e. g., Chapelle et al. 2006; Deng and Schuller 2012) should be used to make use of unlabelled training data. This thesis has focussed on SVMs for static classification and has investigated the effect of various SVM specific parametrisations (balancing of the training sets, and model complexity). Other classifiers should be investigated further, such as deep LSTM-RNNs or graphical models, and SVMs together with the feature normalisation approach should be extended to be able to handle unbalanced training data directly, removing the need for down- or up-sampling.

While it has been shown that the large, brute-forced feature sets are highly effective for many tasks, including complex speech affect recognition and music style identification, minimalistic sets which can be universally used are still to be found. An effort was started in this thesis with the definition of the Geneva Minimalistic Acoustic Parameter Set (GeMAPS). It was shown that this set can compete with the large sets for classification of emotional activation, but fails for more complex tasks such as emotional valence and music style identification. Thus, in future work, experiments in which the features in the large sets are manually analysed wrt. their correlation to the complex tasks, as by, e. g., Weninger et al. (2013b), are necessary. Based on these findings the proposed minimalistic sets can be updated and at the same time, this will lead to a deeper understanding of the acoustic characteristics of each task. The minimalistic feature sets and the algorithms could possible be further improved if more recent auditory models or even models of the auditory neurons and hair cells (cf. e. g., Nicoletti et al. 2010) were used as inspiration to formulate novel acoustic features. Moreover, in Sect. 6.3.3.2 it was discussed that models built with the smaller feature sets potentially can generalise better to unseen data. Thus, the feature sets should be compared in cross-database classification experiments.

A movement contrary to defining minimalistic, hand-crafted feature sets, and which goes beyond large, brute-forced feature sets, is the automatic learning of features: Especially in the field of Automatic Speech Recognition (ASR), deep neural network auto-encoders have become popular to learn a compressed representation of an input (typically spectra) in an unsupervised way for a given task (cf. e. g., Sainath et al. 2012). It was also shown that such methods are suitable for ASR in noisy and reverberated conditions (Feng et al. 2014). Thus, the next step is to adapt such methods to paralinguistic speech and music analysis. While these techniques may perform well for machine learning, they give us little to no information about the underlying acoustic properties of various speech and music phenomena. Therefore it is important, to always consider both automatic learning of features and brute-forcing of descriptors, for building good systems on big data, combined with an in-depth manual analysis of the descriptors, to gain a deeper understanding of the underlying principles.

Concluding, it can be said that, the tools and methods developed for this thesis have set a reference in the field and have expedited the pace at which research is conducted there. This thesis has laid the foundation in acoustic feature extraction on which novel data-mining and machine learning methods can rapidly be built, evaluated, and compared on standardised sets of acoustic descriptors. All in all, we are thus getting closer to building a universal sound analysis system, as it is envisioned for speech by Schuller et al. (2014b) in the iHEARu research project.

References

S. Alghowinem, R. Goecke, M. Wagner, J. Epps, M. Breakspear, G. Parker, Detecting Depression: A Comparison between Spontaneous and Read Speech. In *Proceedings of the ICASSP 2013*, Vancouver, Canada, May 2013. IEEE, pp. 7547–7551

S. Alghowinem, R. Goecke, M. Wagner, J. Epps, M. Breakspear, G. Parker, From Joyous to Clinically Depressed: Mood Detection Using Spontaneous Speech. In *Proceedings of the FLAIRS Conference*, 2012

A. Bhattacharya, W. Wu, Z. Yang, Quality of experience evaluation of voice communication: an affect-based approach. Human-centric Comput. Info. Sci. **2**(1), 1–18 (2012). doi:10.1186/2192-1962-2-7

M.P. Black, P.G. Georgiou, A. Katsamanis, B.R. Baucom, S.S. Narayanan. You made me do it: Classification of Blame in Married Couples' Interactions by Fusing Automatically Derived Speech and Language Information. In *Proceedings of the INTERSPEECH 2011*, Florence, Italy, August 2011. ISCA, pp. 89–92

M.P. Black, A. Katsamanis, B.R. Baucom, C.-C. Lee, A.C. Lammert, A. Christensen, P.G. Georgiou, S.S. Narayanan, Toward automating a human behavioral coding system for married couples' interactions using speech acoustic features. Speech Commun. **55**(1), 1–21 (2013). doi:10.1016/j.specom.2011.12.003

D. Bone, M. P. Black, M. Li, A. Metallinou, S. Lee, S. Narayanan, Intoxicated Speech Detection by Fusion of Speaker Normalized Hierarchical Features and GMM Supervectors. In *Proceedings of the INTERSPEECH 2011*, Florence, Italy, August 2011. ISCA, pp. 3217–3220

D. Bone, M. Li, M.P. Black, S.S. Narayanan, Intoxicated speech detection: A fusion framework with speaker-normalized hierarchical functionals and GMM supervectors. Comput. Speech Lang. **28**(2), 375–391 (2014). doi:10.1016/j.csl.2012.09.004. ISSN 0885-2308

O. Chapelle, B. Schölkopf, A. Zien, *Semi-Supervised Learning* (MIT Press, Cambridge, 2006)

J. Deng, B. Schuller, Confidence Measures in Speech Emotion Recognition Based on Semi-supervised Learning. In *Proceedings of the INTERSPEECH 2012*, Portland, OR, USA, September 2012. ISCA

A. Dhall, R. Goecke, J. Joshi, M. Wagner, T. Gedeon, Emotion recognition in the wild challenge 2013. In *Proceedings of the 15th ACM International conference on multimodal interaction (ICMI) 2013*, Sydney, Australia, December 2013. ACM, pp. 509–516

F. Eyben, S. Petridis, B. Schuller, M. Pantic, Audiovisual Vocal Outburst Classification in Noisy Acoustic Conditions. In *Proceedings of the ICASSP 2012*, Kyoto, Japan, March 2012c. IEEE, pp. 5097–5100

F. Eyben, S. Petridis, B. Schuller, G. Tzimiropoulos, S. Zafeiriou, M. Pantic, Audiovisual Classification of Vocal Outbursts in Human Conversation Using Long-Short-Term Memory Networks. In *Proceedings of the ICASSP 2011*,Prague, Czech Republic, May 2011. IEEE, pp. 5844–5847

F. Eyben, F. Weninger, F. Gross, B. Schuller. Recent developments in openSMILE, the munich open-source multimedia feature extractor. In *Proceedings of the ACM Multimedia 2013*, Barcelona, Spain, 2013a, ACM, pp. 835–838

F. Eyben, F. Weninger, M. Woellmer, B. Schuller. openSMILE version 2.0rc1—source code, open-source research only license, http://opensmile.sourceforge.net. 2013b

F. Eyben, M. Wöllmer, B. Schuller, Open Emotion and Affect Recognition (openEAR), http://sourceforge.net/projects/openart/. September 2009a

F. Eyben, M. Wöllmer, B. Schuller. openEAR—Introducing the Munich Open-Source Emotion and Affect Recognition Toolkit. In *Proceedings of the 3rd International Conference on Affective Computing and Intelligent Interaction (ACII 2009)*, IEEE. vol I, Amsterdam, The Netherlands, pp. 576–581, September 2009b

F. Eyben, M. Wöllmer, B. Schuller. openSMILE—The Munich Versatile and Fast Open-Source Audio Feature Extractor. In *Proceedings of the ACM Multimedia 2010*, Florence, Italy 2010, ACM, pp. 1459–1462

X. Feng, Y. Zhang, J. Glass, Speech feature denoising and dereverberation via deep autoencoders for noisy reverberant speech recognition. In *Proceedings of the ICASSP 2014*, Florence, Italy, May 2014. IEEE, pp. 1778–1782

A.V. Ivanov, G. Riccardi, A.J. Sporka, J. Franc, Recognition of Personality Traits from Human Spoken Conversations. In *Proceedings of the INTERSPEECH 2011*, Florence, Italy, August 2011. ISCA, pp. 1549–1552

T. Jacykiewicz F. Ringeval, *Automatic Recognition of Laughter using Verbal and Non-Verbal Acoustic Features*. Master's thesis, Department of Informatics, University of Fribourg, Switzerland, January 2014

J. H. Jeon, R. Xia, Y. Liu, Level of interest sensing in spoken dialog using multi-level fusion of acoustic and lexical evidence. In *Proceedings of the INTERSPEECH 2010*, Makuhari, Japan, 2010. ISCA, pp. 2802–2805

J. Kim, N. Kumar, A. Tsiartas, M. Li, S. Narayanan, Intelligibility classification of pathological speech using fusion of multiple subsystems. In *Proceedings of the INTERSPEECH 2012*, Portland, OR, USA, 2012. ISCA

C.-C. Lee, E. Mower, C. Busso, S. Lee, S. Narayanan, Emotion recognition using a hierarchical binary decision tree approach. In *Proceedings of the INTERSPEECH 2009*, Brighton, UK, 2009. ISCA, pp. 320–323

C.-C. Lee, A. Katsamanis, M.P. Black, B.R. Baucom, A. Christensen, P.G. Georgiou, S.S. Narayanan, Computing vocal entrainment: A signal-derived pca-based quantification scheme with application to affect analysis in married couple interactions. Comput. Speech Lang. **28**(2), 518–539 (2014)

G. Lukacs, M. Jani, G. Takacs, Acoustic feature mining for mixed speech and music playlist generation. In *Proceedings of the 55th International Symposium ELMAR 2013*, pp. 275–278, Zadar, Croatia, September 2013. IEEE

I. Mporas, T. Ganchev, Estimation of unknown speaker's height from speech. Intern. J. Speech Technol. **12**(4), 149–160 (2009). doi:10.1007/s10772-010-9064-2

A. Muaremi, B. Arnrich, G. Tröster, Towards measuring stress with smartphones and wearable devices during workday and sleep. *BioNanoScience*, pp. 1–12, 2013. doi:10.1007/s12668-013-0089-2

M. Nicoletti, M. Rudnicki, W. Hemmert, A model of the auditory nerve for acoustic- and electric excitation. Frontiers in Computational Neuroscience (September 2010). doi:10.3389/conf.fncom.2010.51.00104

N. Nikolaou, *Music Emotion Classification*. Doctoral dissertation, Technical University of Crete, Crete, Greece, 2011. p. 140

O. Räsänen, J. Pohjalainen, Random subset feature selection in automatic recognition of developmental disorders, affective states, and level of conflict from speech. In *Proceedings of the INTERSPEECH 2013*, Lyon, France, 2013. ISCA, pp. 210–214

D. Reidsma, K. Truong, H. van Welbergen, D. Neiberg, S. Pammi, I. de Kok, B. van Straalen, Continuous interaction with a virtual human. J. Multimod. User Interfaces (JMUI) **4**(2), 97–118 (2011). doi:10.1007/s12193-011-0060-x. ISSN 1783-7677

F. Ringeval, A. Sonderegger, J. Sauer, D. Lalanne, Introducing the RECOLA multimodal corpus of remote collaborative and affective interactions. In *Proceedings of the 2nd International Workshop on Emotion Representation, Analysis and Synthesis in Continuous Time and Space (EmoSPACE), held in conjunction with FG 2013*, Shanghai, China, April 2013. IEEE, pp. 1–8

T. Sainath, B. Kingsbury, B. Ramabhadran, Auto-encoder bottleneck features using deep belief networks. In *Proceedings of the ICASSP 2012*, pp. 4153–4156, Kyoto, Japan, March 2012. IEEE. doi:10.1109/ICASSP.2012.6288833

B. Schuller, A. Batliner, S. Steidl, F. Schiel, J. Krajewski, The INTERSPEECH 2011 Speaker State Challenge. In *Proceedings of the INTERSPEECH 2011*, Florence, Italy, August 2011a, ISCA, pp. 3201–3204

B. Schuller, S. Steidl, A. Batliner, F. Burkhardt, L. Devillers, C. Müller, S. Narayanan, The INTERSPEECH 2010 Paralinguistic Challenge. In *Proceedings of the INTERSPEECH 2010*, Makuhari, Japan, September 2010, ISCA, pp. 2794–2797

B. Schuller, S. Steidl, A. Batliner, J. Epps, F. Eyben, F. Ringeval, E. Marchi, Y. Zhang, The INTERSPEECH 2014 computational paralinguistics challenge: Cognitive and physical load. In *Proceedings of the INTERSPEECH 2014*, Singapore, 2014a. ISCA (to appear)

B. Schuller, S. Steidl, A. Batliner, F. Jurcicek, The INTERSPEECH 2009 Emotion Challenge. In *Proceedings of the INTERSPEECH 2009*, (Brighton, UK, September 2009), pp. 312–315

B. Schuller, S. Steidl, A. Batliner, E. Nöth, A. Vinciarelli, F. Burkhardt, R. van Son, F. Weninger, F. Eyben, T. Bocklet, G. Mohammadi, B. Weiss, The INTERSPEECH 2012 Speaker Trait Challenge. In *Proceedings of the INTERSPEECH 2012*, ISCA, Portland, OR, USA, September 2012a

B. Schuller, S. Steidl, A. Batliner, A. Vinciarelli, K. Scherer, F. Ringeval, M. Chetouani, et al., The INTERSPEECH 2013 Computational Paralinguistics Challenge: Social Signals, Conflict, Emotion, Autism. In *Proceedings of the INTERSPEECH 2013*, (ISCA, Lyon, France, 2013), pp. 148–152

B. Schuller, M. Valstar, R. Cowie, M. Pantic, AVEC 2012: the continuous audio/visual emotion challenge—an introduction, in *Proceedings of the 14th ACM International Conference on Multimodal Interaction (ICMI) 2012*, ed. by L.-P. Morency, D. Bohus, H.K. Aghajan, J. Cassell, A. Nijholt, J. Epps (ACM, Santa Monica, CA, USA, 2012b), pp. 361–362

B. Schuller, M. Valstar, F. Eyben, G. McKeown, R. Cowie, M. Pantic, AVEC 2011—The First International Audio/Visual Emotion Challenge, in *Proceedings of the First International Audio/Visual Emotion Challenge and Workshop*, AVEC 2011, held in conjunction with the International HUMAINE Association Conference on Affective Computing and Intelligent Interaction (ACII) 2011, vol II, ed. by B. Schuller, M. Valstar, R. Cowie, M. Pantic (Springer, Memphis, TN, USA, 2011), pp. 415–424

B. Schuller, Y. Zhang, F. Eyben, F. Weninger, Intelligent systems' Holistic Evolving Analysis of Real-life Universal speaker characteristics. In B. Schuller, P. Buitelaar, L. Devillers, C. Pelachaud, T. Declerck, A. Batliner, P. Rosso, S. Gaines, eds, *Proceedings of the 5th International Workshop on Emotion Social Signals, Sentiment & Linked Open Data (ES^3LOD 2014), satellite of the 9th Language Resources and Evaluation Conference (LREC) 2014*, Reykjavik, Iceland, May 2014b. ELRA. p. 8

M. Suzuki, S. Nakagawa, K. Kita, Emotion recognition method based on normalization of prosodic features. In *Proceedings of the 2013 Asia-Pacific Signal and Information Processing Association Annual Summit and Conference (APSIPA), Kaohsiung*, IEEE, October 2013. doi:10.1109/APSIPA.2013.6694147, pp. 1–5

M. Valstar, B. Schuller, K. Smith, F. Eyben, B. Jiang, S. Bilakhia, S. Schnieder, R. Cowie, M. Pantic. AVEC 2013—The Continuous Audio/Visual Emotion and Depression Recognition Challenge, In *Proceedings of the ACM Multimedia 2013*, Barcelona, Spain, October 2013. ACM

F. Weninger, F. Eyben, B.W. Schuller, M. Mortillaro, K.R. Scherer, On the Acoustics of Emotion in Audio: What Speech, Music and Sound have in Common. *Frontiers in Psychology*, 4(Article ID 292), pp. 1–12, May 2013b. doi:10.3389/fpsyg.2013.00292

Appendix A

A.1 Detailed Descriptions of the Baseline Feature Sets

A.1.1 Details: INTERSPEECH 2009 Set (IS09)

The INTERSPEECH 2009 Emotion Challenge feature set contains 16 Low-level Descriptors (LLDs) and 16 corresponding delta regression coefficients (Sect. 2.3.2). To those 32 descriptors, 12 functionals are applied, which yields a total of **384 features**. Due to the success of Mel-Frequency Cepstral Coefficient (MFCC) for speech emotion recognition in prior work (e.g., Schuller et al. 2003), and the importance of prosodic features (e.g., Batliner and Möbius 2005; Batliner et al. 1999), the 16 LLDs are composed of 12 cepstral descriptors (MFCC) and 4 prosodic descriptors:

		#
MFCC 1–12	computed from a 26 band Mel-scale band spectrum (Sect. 2.2.3.4) covering the frequency range 0–8 kHz. Cepstral liftering with $L = 22$ is performed after the Discrete Cosine Transformation (DCT) (Sect. 2.2.10.1),	12
Root Mean Square (RMS) Energy	as defined in Sect. 2.2.2, Eq. (2.41),	1
Zero-Crossing Rate (ZCR)	expressed in units of crossings per second (see Sect. 2.2.1.1),	1
F_0	fundamental frequency computed via the Autocorrelation function (ACF)/Cepstrum method described in Sect. 2.2.11.3,	1
Prob. of voicing	probability of voicing computed from ACF (Sect. 2.2.11.2, Eq. (2.191)).	1
	Total # descriptors:	16

F. Eyben, *Real-time Speech and Music Classification by Large Audio Feature Space Extraction*, Springer Theses, DOI 10.1007/978-3-319-27299-3

Prior to computation of the prosodic LLDs, the audio samples are scaled to the range $[-1; +1]$. For the MFCC a scaling to the 16-bit sample range $[-32767; 32768]$ is used, in order to be numerically compatible with the Hidden Markov Toolkit (Young et al. 2006) (HTK) toolkit (Young et al. 2006). Short-time analysis (cf. Sect. 2.1.3) with a frame size of 25 ms and a step of 10 ms is performed for all LLDs. A pre-emphasis filter is applied to each frame (cf. Sect. 2.1.4) with $k = 0.97$ and the frame is multiplied by a Hamming window function (Sect. 2.1.3).

All LLD are smoothed with a symmetric moving average window of $W = 3$ frames length (one previous, one current, and one future frame), and delta regression coefficients with a context window size of $W = 2$ (Sect. 2.3.2) are computed. The 12 functionals which are next applied to the LLDs and their deltas are:

		#
Arithmetic mean	Section 2.4.2.1,	1
Moments	Standard deviation, Skewness, Kurtosis (Sect. 2.4.2.2),	3
Maximum value	and position of the maximum value (frame index)	2
Minimum value	and position of the minimum value (frame index)	2
Range	difference of maximum to minimum value	1
Linear regression	coefficients a (slope) and b (offset)	2
Regression error	quadratic error between line and actual signal	1
	Total # descriptors:	12

A.1.2 Details: INTERSPEECH 2010 Set (IS10)

The INTERSPEECH 2010 Paralinguistics Challenge set (IS10) contains 38 LLDs and 38 corresponding delta coefficients. 21 functionals are applied to 34 of these LLDs (group A in the ongoing) and their corresponding delta coefficients, yielding $34 \cdot 2 \cdot 21 = 1{,}428$ features. To the remaining 4 LLDs (group B in the ongoing) and the corresponding delta coefficients, 19 functionals are applied, yielding another $4 \cdot 2 \cdot 19 = 152$ features. The 34 LLDs in group A are:

		#
Loudness	computed with a narrow band approximation from the signal intensity as suggested by Kießling (1997, pp. 156–157) (cf. Sect. 2.2.2, Eq. (2.43))	1
MFCC 0–14	MFCC coefficients 1–14 and 0th coefficient (energy), computed from power Mel-spectrum (26 bands, triangular filter, range 20–8000 Hz—Sect. 2.2.3.4), cepstral liftering with $L = 22$ (Sect. 2.2.10.1),	15

log. MFB 1–8	natural logarithm of Mel-Frequency Band (MFB) power for 8 bands in the range 20–6500 Hz, triangular filters, FFT power spectrum input (Sect. 2.2.3.4)	8
LSP 0–7	8 Line Spectral Pair (LSP) are computed from Linear Prediction (LP) analysis with order $p = 8$ (ACF method). 25 ms frames after Hamming window are resampled to 11 kHz prior to LP analysis.	8
F_0 envelope	computed by holding the last F_0 value (Subharmonic Summation (SHS) algorithm, cf. below) from the previous voiced region during the adjacent unvoiced region.	1
Prob. of voicing	voicing probability from SHS algorithm (Sect. 2.2.11.4)	1
	Total # descriptors:	34

The 4 LLD in group B are:

		#
F_0	via SHS method (Sect. 2.2.11.4), without spectral smoothing, enhancement, and auditory weighting; range 52–620 Hz; 15 harmonics, compression 0.85; no Viterbi post-smoothing.	1
Jitter	relative search range $\alpha = 0.25$, minimum correlation 0.5, see Sect. 2.2.13.1	1
δ Jitter	see Sect. 2.2.13.1	1
Shimmer	see Sect. 2.2.13.2	1
	Total # descriptors:	4

Prior to computation of the LLDs, the audio samples are scaled to the range $[-1; +1]$, with the exception of MFCC for which a scaling to the 16-bit sample range $[-32767; 32768]$ is used, in order to be numerically compatible with the HTK toolkit (Young et al. 2006). Short-time analysis (cf. Sect. 2.1.3) with a frame size of 25 ms and a step of 10 ms is performed for all LLDs, except F_0 related LLDs (which includes Jitter and Shimmer) for which a frame size of 60 ms is used in order to not distort low pitch of male voices down to approx. 50 Hz. A pre-emphasis filter is applied to each 25 ms frame (cf. Sect. 2.1.4) with $k = 0.97$ and the frame is multiplied by a Hamming window function (Sect. 2.1.3). A Gaussian window with $\sigma = 0.25$ is applied to the 60 ms frames.

All LLDs are smoothed with a symmetric moving average window of $W = 3$ frames length (one previous, one current, and one future frame), and delta regression coefficients with a context window size of $W = 2$ (Sect. 2.3.2) are computed. The 21 functionals for group A are (those functionals marked with a * are not applied to group B):

		#
Arithmetic mean	see Sect. 2.4.2.1,	1
Moments	standard deviation, skewness, kurtosis (Sect. 2.4.2.2),	3
Maximum pos.	position of the maximum value (frame index)	1
99 %-tile	as robust substitute for the maximum value	1
Minimum pos.	position of the minimum value (frame index)	1
1 %-tile*	as robust substitute for the minimum value	1
Range 1–99 %*	range of signal expressed as percentile 1 to 99 range	1
Linear regression	coefficients a (slope) and b (offset)	2
Regression error	quadratic error $\overline{e}$ and linear error $\overline{e}_a$ between line and actual signal (Sect. 2.4.2.6)	2
Quartiles	quartiles 1–3 and inter-quartile ranges 1–2, 2–3, 1–3 (Sect. 2.4.2.4)	6
Up-level times	up-level time 75 and 90 % (Sect. 2.4.2.7)	2
	Total # descriptors:	21

Two more descriptors are included to reflect speaking rate and the segment length, respectively: the number of F_0 onsets (N_{onsets}, cf. onset functionals, Sect. 2.4.2.10) and the total duration of the segment in seconds. Overall, the IS10 paralinguistic feature set contains **1,582 features.**[1]

A.1.3 *Details: INTERSPEECH 2011 Set (IS11)*

Similar to the IS10 set, the IS11 set contains two groups of LLDs: group A with 54 energy, spectral, and cepstral LLDs, and group B with 5 source/excitation signal

[1] In the 2.0 openSMILE release three versions of this set are provided: `IS10_paraling.conf`—which is the original configuration file as used in 2010 with SVN revision 281 of openSMILE; `IS10_paraling.conf_oldcompat.conf`—which provides numerically compatible features with the 2010 challenge features with the latest version (2.0) of openSMILE; and `IS10_paraling.conf_fixed.conf` which includes fixes and improvements to the features, which are thus not numerically compatible (further, the number of F_0 onsets has been replaced by the onset rate to be independent of the segment length).

related LLD. To group A, 37 functionals are applied, and to group B, 36 functionals. This results in $54 \cdot 2 \cdot 37 = 3{,}996$ features from group A and $5 \cdot 2 \cdot 36 = 360$ features from group B, giving a total of 4,356 features. In addition to this brute-force combination of LLDs and functionals, the IS11 set contains temporal statistic descriptors based on voiced/unvoiced segments which are computed by applying the following functionals to the F_0 LLD and its first order delta coefficients:

		#
Ratio of non-zero values	ratio of non-zero values, i.e., percentage of voiced frames of total frames	1
Segment lengths	minimum, mean, maximum, and standard deviation of lengths of voiced segments, i.e., $F_0 > 0$ (cf. Sect. 2.4.2.9)	4
	Total # descriptors:	5

Additionally, as in the IS10 set, the total duration of the segment (in seconds) is included as a feature. All together this adds 11 more features, and thus, in total the IS11 set contains **4,367 features**.

The 54 energy, spectral, and cepstral LLDs in group A are:

		#
Loudness	computed as the sum of a simplified auditory spectrum $X_{P,aud}(b)$ (cf. Sect. 2.2.9.3): 26 Mel-frequency bands (triangular filters, power spectrum), range 20–8000 Hz, auditory weighting with equal loudness curve (Eq. (2.177)) and cubic-root compression (exponent 0.33 in Eq. (2.180)),	1
Modulated loudness	computed as the sum of a simplified RelAtive Spectral TrAnsform (RASTA) filtered auditory spectrum $X_{P,aud}(b)$ (cf. Sect. 2.2.9.3): 26 Mel-frequency bands (triangular filters, power spectrum), range 20–8000 Hz, RASTA style bandpass filter applied to Mel-frequency power-spectral bands (Sect. 2.2.9.5), auditory weighting with equal loudness curve (Eq. (2.177)) and cubic-root compression (exponent 0.33 in Eq. (2.180)). This loudness measure reflects the loudness contained in the speech signal while suppressing loudness influence from near stationary or high-frequency noise,	1
RMS Energy	as defined in Sect. 2.2.2, Eq. (2.41),	1
ZCR	expressed in units of crossings per second (see Sect. 2.2.1.1),	1

RASTA auditory bands 1–26	per band raw values of the RASTA filtered auditory band levels $X_{P,aud}(b)$ (cf. modulated loudness, above), replacing the log. MFB of the IS10 set,	26
MFCC 1–12	MFCC coefficients 1–12 without 0th coefficient. This coefficients was removed due to the redundancy with the RMS energy. MFCC are computed from a power Mel-spectrum (26 bands, triangular filter, range 20–8000 Hz—Sect. 2.2.3.4), cepstral liftering with $L =$ 22 (Sect. 2.2.10.1),	12
Band energy	for the linear-scale frequency band from 25–650 Hz and 1–4 kHz (sum of frequency bin magnitudes),	2
Spectral RoP	25, 50, 75, and 90 % spectral Roll-off Points (RoPs) (Sect. 2.2.4.9),	4
Spectral flux	normalised spectral difference between two successive frames (Sect. 2.2.4.12),	1
Spectral entropy	as defined in (Sect. 2.2.4.8),	1
Spec. moments	spectral variance, spectral skewness, and spectral kurtosis (Sect. 2.2.4.7),	3
Spectral slope	as defined in Sect. 2.2.4.2,	1
	Total # descriptors:	54

The 5 source/excitation signal related LLDs in group B are:

		#
$\mathbf{F_0}$	via SHS method (Sect. 2.2.11.4), with spectral smoothing, enhancement, and auditory weighting; $F_0 = 0$ for unvoiced regions; voicing probability threshold $\theta_v =$ 0.7; range: 52–620 Hz; 15 harmonics, compression 0.85; "non-greedy" peak detection (Sect. 2.2.11.4); no Viterbi post-smoothing.	1
Prob. of voicing	raw voicing probability obtained from the SHS algorithm (Sect. 2.2.11.4)	1
Jitter	relative search range $\alpha = 0.25$, minimum correlation 0.5, see Sect. 2.2.13.1	1
δ Jitter	see Sect. 2.2.13.1	1
Shimmer	see Sect. 2.2.13.2	1
	Total # descriptors:	5

Prior to computation of the LLDs, the audio samples are scaled to the range $[-1; +1]$, with the exception of MFCC for which a scaling to the 16-bit sample range $[-32767; 32768]$ is used, in order to be numerically compatible with the HTK toolkit (Young et al. 2006). Short-time analysis (cf. Sect. 2.1.3) with a frame size of 20 ms and a step of 10 ms is performed for all LLDs, except F_0 related LLDs (groupB—which includes Jitter and Shimmer) for which a frame size of 60 ms (step width remains at 10 ms) is used in order to not distort low pitch of male voices down to approx. 50 Hz. In contrast to the IS10 feature set, *no* pre-emphasis filter is applied to the 20 ms frames, as this filter was designed originally to emphasize formant information (Young et al. 2006), however, it may discard or corrupt important prosodic information. The Hamming window function (Sect. 2.1.3) is kept the same as in the IS10 set. A Gaussian window with $\sigma = 0.4$ is applied to the 60 ms frames. All LLDs are smoothed with a symmetric moving average window of $W = 3$ frames length (one previous, one current, and one future frame), and delta regression coefficients with a context window size of $W = 2$ (Sect. 2.3.2) are computed for each LLD.

The functionals which are applied to the LLDs are given in the following table (functionals marked with ^{B} are only applied to LLDs of group B, and functionals marked with ^{A} are only applied to LLDs of group A):

		#
Arithmetic mean	see Sect. 2.4.2.1,	1
Quadratic mean^B	see Sect. 2.4.2.1,	1
Moments	standard deviation, skewness, kurtosis (Sect. 2.4.2.2),	3
Quartiles	quartiles 1–3 and inter-quartile ranges 1–2, 2–3, 1–3 (Sect. 2.4.2.4),	6
1 %-tile	as robust substitute for the minimum value,	1
99 %-tile	as robust substitute for the maximum value,	1
Range 1–99 %	robust range of the signal expressed as percentile 1 to 99 range,	1
Linear regression	coefficient a (slope)—offset b is *not* included (Sect. 2.4.2.6),	1
Regression error	quadratic error $\overline{e}$ between line and actual signal (Sect. 2.4.2.6),	1
Quad. regression	quadratic regression coefficients a and b—offset c is *not* included (Sect. 2.4.2.6),	2
Quad. regr. error	quadratic error $\overline{e}$ between parabola (estimated from quadratic regression) and actual signal (Sect. 2.4.2.6),	1

Centroid	temporal centroid as described in Sect. 2.4.2.5, time scale as relative percentage of segment length ($t(n) = n/N$),	1
Peak mean	mean of peak amplitudes and distance of mean of peak amplitudes to the arithmetic mean (Sect. 2.4.2.8),	2
Peak distance	mean and standard deviation of peak to peak distances Sect. 2.4.2.8,	2
Segment length^A	mean, minimum, maximum, and standard deviation of segment lengths; segmentation is performed by relative thresholds (25 % for all LLD in this group and additionally 75 % only for auditory RASTA bands, spectral features, and MFCC)—see Sect. 2.4.2.9,	4
Up-level time	up-level time 90 % (75 % up-level time from IS10 was removed in favour of the 25 % down-level time) (Sect. 2.4.2.7),	1
Down-level time	down-level time 25 % (Sect. 2.4.2.7),	1
Rise/Fall-time	relative time in which signal is rising/falling (Sect. 2.4.2.7),	2
Curvature-time^B	relative time in which signal has left/right curvature (Sect. 2.4.2.7),	2
LP gain	LP analysis gain (error amplitude) (see Sects. 2.4.3.2 and 2.2.7.1),	1
LP coefficient 1–5	see Sect. 2.4.3.2.	5
	Total # descriptors:	40

Functionals applied to group B are applied only to regions where the LLD is non-zero. The time scale is thereby altered because the regions with zero values are removed from $x(n)$, which makes $x(n)$ shorter, i.e., for the new length N' the following holds true: $N' < N$.

Note: According to Schuller et al. (2011),[2] the IS11 set contains 4,368 features in total. This is the size of the baseline feature vectors provided for the challenge. However, the duration of the segment is counted twice there, due to the way it was implemented in the openSMILE configuration file. Thus, the correct number of unique features in IS11 is 4,367.

[2]And the openSMILE configuration file.

A.1.4 Details: INTERSPEECH 2012 Set (IS12)

For the INTERSPEECH 2012 Speaker Trait Challenge a yet larger set of baseline acoustic features has been proposed by the author of this thesis (Schuller et al. 2012). At the same time, the systematic brute-forcing, i.e., applying all functionals to all LLDs, has been reduced. This means, some functionals, which—when applied to delta LLDs—do not describe meaningful signal statistics, have been removed from the set.

As the IS11 set, the IS12 set contains two group of LLDs: group A with 58 energy, spectral, and cepstral LLDs, and group B with 6 source/excitation signal related LLD. To the 58 LLDs of group A, 58 functionals are applied. To the delta LLDs of group A, 38 functionals are applied. To the LLDs of group B, 56 functionals are applied, and to the delta LLD of group B, 36 functionals are applied. This results in a total of 6,120 features.

In addition to this (semi) brute-force combination of LLDs and functionals, the IS12 set contains a part of the temporal statistic descriptors from the IS11 set. These are based on voiced/unvoiced segments which are computed by applying the following functionals to the F_0 LLD (not to δ as in IS11):

		#
Ratio of non-zero values	ratio of non-zero values, i.e., percentage of voiced frames of total frames	1
Segment lengths	minimum, mean, maximum, and standard deviation of lengths of voiced segments, i.e., $F_0 > 0$ (cf. Sect. 2.4.2.9)	4
	Total # descriptors:	5

In contrast to the IS11 set, the total duration of the segment is *not* included as a feature. The decision to remove the duration has been made because the tasks in the challenge involve speaker trait recognition (Schuller et al. 2012). Speaker trait is a long-term phenomenon, and thus the duration of a single utterance is irrelevant if the data-set contains mixed utterances with random lengths. With these 5 additional descriptors, the IS12 set totals **6,125 features**.

The 58 energy, spectral, and cepstral LLDs in group A are in detail:

		#
Loudness	computed as the sum of a simplified auditory spectrum $X_{P,aud}(b)$ (cf. Sect. 2.2.9.3): 26 Mel-frequency bands (triangular filters, power spectrum), range 20–8000 Hz, auditory weighting with equal loudness curve (Eq. (2.177)) and cubic-root compression (exponent 0.33 in Eq. (2.180)).	1

Modulated loudness	computed as the sum of a simplified RASTA filtered auditory spectrum $X_{P,aud}(b)$ (cf. Sect. 2.2.9.3): 26 Mel-frequency bands (triangular filters, power spectrum), range 20–8000 Hz, RASTA style bandpass filter applied to Mel-frequency power-spectral bands (Sect. 2.2.9.5), auditory weighting with equal loudness curve (Eq. (2.177)) and cubic-root compression (exponent 0.33 in Eq. (2.180)). This loudness measure reflects the loudness contained in the speech signal while suppressing loudness influence from near stationary or high-frequency noise.	1
RMS Energy	as defined in Sect. 2.2.2, Eq. (2.41),	1
ZCR	expressed in units of crossings per second (see Sect. 2.2.1.1),	1
RASTA auditory bands 1–26	per band raw values of the RASTA filtered auditory band levels $X_{P,aud}(b)$ (cf. modulated loudness, above),	26
MFCC 1–14	MFCC coefficients 1–14 without 0th coefficient. MFCC are computed from a power Mel-spectrum (26 bands, triangular filter, range 20–8000 Hz—Sect. 2.2.3.4), cepstral liftering with $L = 22$ (Sect. 2.2.10.1),	14
Band energy	for the linear-scale frequency band from 250–650 Hz and 1–4 kHz (sum of frequency bin magnitudes),	2
Spectral RoP	25, 50, 75, and 90 % spectral RoP (Sect. 2.2.4.9),	4
Spectral flux	normalised spectral difference between two successive frames (Sect. 2.2.4.12),	1
Spectral entropy	as defined in (Sect. 2.2.4.8),	1
Spec. moments	spectral variance, spectral skewness, and spectral kurtosis (Sect. 2.2.4.7),	3
Spectral slope	as defined in (Sect. 2.2.4.2),	1
Harmonicity (spectral)	see Sect. 2.2.4.13,	1 1
Sharpness (spectral)	auditory sharpness (psychoacoustically scaled spectral centroid), see Sect. 2.2.4.10.	1
	Total # descriptors:	59

Compared to IS11, the LLDs harmonicity and sharpness have been added and the linear band energy for the lower band has been corrected from 25–650 Hz to 250–650 Hz as it was previously in the IS10 set and used by Schuller (2006).

The 6 source/excitation signal related LLDs in group B are:

		#
F_0	via SHS method (Sect. 2.2.11.4), with spectral smoothing, enhancement, and auditory weighting; $F_0 = 0$ for unvoiced regions; voicing probability threshold $\theta_v = 0.7$; range: 52–620 Hz; 15 harmonics, compression 0.85; "*non-greedy*" peak detection (Sect. 2.2.11.4); with Viterbi post-smoothing with the following settings: $w_{Tvv} = 14$, $w_{Tvuv} = 25$, $w_{Range} = 2$, and a maximum path buffer length of 0.3 s; all other parameters remain at the defaults given in Sect. 2.2.11.5. An absolute RMS energy threshold of 0.005 is applied to the final F_0 values and F_0 in frames with an energy value below the threshold is set to 0 (the energy is computed from the corresponding 60 ms frame here)—this removes noisy F_0 values which the SHS algorithm produces for low energy frames.	1
Prob. of voicing	raw voicing probability obtained from the SHS algorithm (Sect. 2.2.11.4)	1
Jitter	relative search range $\alpha = 0.25$, minimum correlation 0.5, see Sect. 2.2.13.1	1
δ Jitter	see Sect. 2.2.13.1	1
Shimmer	see Sect. 2.2.13.2	1
log. HNR	logarithmic Harmonics to Noise Ratio (HNR) $HNR_{wf,log}$ computed via wave form matching as defined in Sect. 2.2.13.3.	1
	Total # descriptors:	6

As in the IS11 set, prior to computation of the LLDs, the audio samples are scaled to the range $[-1; +1]$, with the exception of MFCC for which a scaling to the 16-bit sample range $[-32767; 32768]$ is used, in order to be numerically compatible with the HTK toolkit (Young et al. 2006). Short-time analysis (cf. Sect. 2.1.3) with a frame size of 20 ms and a step of 10 ms is performed for all LLDs, except F_0 related LLDs (group B—which includes Jitter and Shimmer) for which a frame size of 60 ms (step width remains at 10 ms) is used in order to not distort low pitch of male voices down to approx. 50 Hz. The 20 ms frames are multiplied by a Hamming window function

(Sect. 2.1.3)—no pre-emphasis filter is used. A Gaussian window with $\sigma = 0.4$ is applied to the 60 ms frames. All LLDs are smoothed with a symmetric moving average window of $W = 3$ frames length (one previous, one current, and one future frame), and delta regression coefficients with a context window size of $W = 2$ (Sect. 2.3.2) are computed for each LLD.

In comparison to the IS11 set, the number of total functionals applied has been significantly increased from 40 to 53 in the IS12 set. The functionals which are applied to the LLDs are given in the following table (functionals marked with ^{B} are only applied to LLDs in group B, functionals marked with ^{A} are only applied to the LLDs in group A, and functionals marked with * are not applied to the delta LLDs, while functionals marked with $^\delta$ are only applied to delta LLDs):

		#
Arithmetic mean*	see Sect. 2.4.2.1,	1
Positive Arithmetic mean$^\delta$	arithmetic mean of positive values only (μ_{a+}, Sect. 2.4.2.1),	1
Root-quadratic mean	see Sect. 2.4.2.1,	1
Flatness	see Sect. 2.4.2.1,	1
Moments	standard deviation, skewness, kurtosis (Sect. 2.4.2.2),	3
Quartiles	quartiles 1–3 and inter-quartile ranges 1–2, 2–3, 1–3 (Sect. 2.4.2.4)	6
1 %-tile	as robust substitute for the minimum value	1
99 %-tile	as robust substitute for the maximum value	1
Range 1–99 %	robust range of the signal expressed as percentile 1 to 99 range	1
Max./Min. position	position of the maximum/minimum value relative to the input length,	2
Full range	full range from single minimum value to single maximum value. This feature has been re-added, since together with the percentile range it adds information on the occurrence of LLD outliers.	1
Linear regression*	coefficient a (slope) *and* offset b (Sect. 2.4.2.6)	2
Regression error*	quadratic error $\overline{e}$ between line and actual signal (Sect. 2.4.2.6)	1
Quad. regression*	quadratic regression coefficients a and b *and* offset c (Sect. 2.4.2.6)	3

Quad. regr. error*	quadratic error $\overline{e}$ between parabola (estimated from quadratic regression) and actual signal (Sect. 2.4.2.6)	1
Centroid*	temporal centroid as described in Sect. 2.4.2.5, time scale as relative percentage of segment length ($t(n) = n/N$)	1
Peak mean*	mean of peak amplitudes, distance of mean of peak amplitudes to the arithmetic mean, *and mean of peak amplitudes relative to the arithmetic mean* (Sect. 2.4.2.8)	3
Peak distance*	mean and standard deviation of peak to peak distances Sect. 2.4.2.8	2
Peak range*	range of peak amplitude values, absolute value and value relative to the arithmetic mean of $x(n)$,	2
Valley range*	range of valley amplitude values relative to the arithmetic mean of $x(n)$,	1
Peak-valley slopes*	mean and standard deviation of rising slopes (valley to peak) and falling slopes (peak to valley),	4
Segment length^{A}	mean, minimum, maximum, and standard deviation of segment lengths; segmentation is performed by relative thresholds (25 % for all LLD in this group and additionally 75 % only for auditory RASTA bands, spectral features, and MFCC)—see Sect. 2.4.2.9	4
Up-level time	up-level time 25, 50, 75, and 90 %; all common values (cf. Schuller 2006) were included in a brute-forcing manner to be able to evaluate their use (Sect. 2.4.2.7)	4
Down-level time	down-level time 25, 50, 75, 90 % (Sect. 2.4.2.7)	4
Rise/Fall-time	relative time in which signal is rising/falling (Sect. 2.4.2.7)	2
Curvature-time^{B}	relative time in which signal has left/right curvature (Sect. 2.4.2.7)	2
LP gain	LP analysis gain (error amplitude) (see Sects. 2.4.3.2 and 2.2.7.1)	1
LP coefficient 1–5	see Sect. 2.4.3.2	5
	Total # descriptors:	61

The peak and regression functionals as well as arithmetic and root-quadratic mean and flatness are applied to all inputs of features in group B. All other functionals applied to LLD group B are applied only to regions where the LLD is non-zero. The time scale is thereby altered because the regions with zero values are removed from $x(n)$, which makes $x(n)$ shorter, i.e., for the new length N' the following holds true: $N' < N$.

A.1.5 Details: INTERSPEECH 2013 ComParE Set (ComParE)

For the ComParE set, the IS12 feature set has been slightly enlarged by approximately 200 features. Irregularities and redundancies have been removed, parameters have been tuned, and updates to some feature extraction algorithms have been made.

As for IS12, two groups of LLDs are defined. To the 59 LLDs in group A, 54 functionals are applied and to the 59 delta LLDs of group A, 46 functionals are applied. To the 6 LLDs in group B, 39 functionals are applied, and to the 6 delta LLDs of set B, 39 functionals are applied. This results in a total of 6,368 features.

In addition to this (semi) brute-force combination of LLDs and functionals, the ComParE set contains the same 5 global temporal statistics as the IS12 set. These are based on voiced/unvoiced segments which are computed by applying the following functionals to the F_0 LLD:

		#
Ratio of non-zero values	ratio of non-zero values, i.e., percentage of voiced frames of total frames	1
Segment lengths	minimum, mean, maximum, and standard deviation of lengths of voiced segments, i.e., $F_0 > 0$ (cf. Sect. 2.4.2.9)	4
	Total # descriptors:	5

In total, the ComParE set contains **6,373 features**.

The 59 energy, spectral, and cepstral LLDs in group A are:

		#
Loudness	computed as the sum of a simplified auditory spectrum $X_{P,aud}(b)$ (cf. Sect. 2.2.9.3): 26 Mel-frequency bands (triangular filters, power spectrum), range 20–8000 Hz, auditory weighting with equal loudness curve (Eq. (2.177)) and cubic-root compression (exponent 0.33 in Eq. (2.180)).	1

Modulated loudness	computed as the sum of a simplified RASTA filtered auditory spectrum $X_{P,aud}(b)$ (cf. Sect. 2.2.9.3): 26 Mel-frequency bands (triangular filters, power spectrum), range 20–8000 Hz, RASTA style bandpass filter applied to Mel-frequency power-spectral bands (Sect. 2.2.9.5), auditory weighting with equal loudness curve (Eq. (2.177)) and cubic-root compression (exponent 0.33 in Eq. (2.180)). This loudness measure reflects the loudness contained in the speech signal while suppressing loudness influence from near stationary or high-frequency noise.	1
RMS Energy	as defined in Sect. 2.2.2, Eq. (2.41),	1
ZCR	expressed in units of crossings per second (see Sect. 2.2.1.1),	1
RASTA auditory bands 1–26	per band raw values of the RASTA filtered auditory band levels $X_{P,aud}(b)$ (cf. modulated loudness, above),	26
MFCC 1–14	MFCC coefficients 1–14 without 0th coefficient. MFCC are computed from a power Mel-spectrum (26 bands, triangular filter, range 20–8000 Hz—Sect. 2.2.3.4), cepstral liftering with $L = 22$ (Sect. 2.2.10.1),	14
Band energy	for the linear-scale frequency band from 250–650 Hz and 1–4 kHz (sum of frequency bin magnitudes),	2
Spectral RoP	25, 50, 75, and 90 % spectral RoPs (Sect. 2.2.4.9),	4
Spectral flux	normalised spectral difference between two successive frames (Sect. 2.2.4.12),	1
Spectral centroid	linear frequency spectral centroid (new in IS12—cf. Sect. 2.2.4.6),	1
Spectral entropy	as defined in Sect. 2.2.4.8,	1
Spec. moments	spectral variance, spectral skewness, and spectral kurtosis (Sect. 2.2.4.7),	3
Spectral slope	computed from linear scale frequency bins, as defined in Sect. 2.2.4.2,	1
Harmonicity (spectral)	see Sect. 2.2.4.13,	1 1

Sharpness (spectral)	auditory sharpness (psychoacoustically scaled spectral centroid), see Sect. 2.2.4.10.	1
	Total # descriptors:	60

Compared to IS12, the LLDs spectral centroid was added in order to provide both the linear version (centroid) and the auditory version (sharpness).

The 6 source/excitation signal related LLDs in group B are the same as in the IS12 set, with the difference that the algorithm for selection of F_0 candidates was improved (greedy version—see Sect. 2.2.11.4) and the RMS energy threshold for F_0 noise removal was reduced from 0.005 to 0.001:

		#
$\mathbf{F_0}$	via SHS method (Sect. 2.2.11.4), with spectral smoothing, enhancement, and auditory weighting; $F_0 = 0$ for unvoiced regions; voicing probability threshold $\theta_v = 0.7$; range 52–620 Hz; 15 harmonics, compression 0.85; "greedy" peak detection (Sect. 2.2.11.4); with Viterbi post-smoothing with the following settings: $w_{Tvv} = 10$, $w_{Tvvd} = 5$, $w_{Thr} = 4$, $w_{Tvuv} = 10$, $w_{Range} = 1$, and a maximum path buffer length of 0.3 s; all other parameters remain at the defaults given in Sect. 2.2.11.5. An absolute RMS energy threshold of **0.001** is applied to the final F_0 values and F_0 in frames with an energy value below the threshold is set to 0 (the energy is computed from the corresponding 60 ms frame here)—this removes noisy F_0 values which the SHS algorithm produces for low energy frames.	1
Prob. of voicing	raw voicing probability obtained from the SHS algorithm (Sect. 2.2.11.4)	1
Jitter	relative search range $\alpha = 0.25$, minimum correlation 0.5, see Sect. 2.2.13.1	1
δ **Jitter**	see Sect. 2.2.13.1	1
Shimmer	see Sect. 2.2.13.2	1
log. HNR	logarithmic HNR $HNR_{wf,log}$ computed via wave form matching as defined in Sect. 2.2.13.3.	1
	Total # descriptors:	6

As in the other sets, prior to computation of the LLDs, the audio samples are scaled to the range $[-1; +1]$, with the exception of MFCC for which a scaling to the 16-bit sample range $[-32767; 32768]$ is used, in order to be numerically compatible with the HTK toolkit (Young et al. 2006). Short-time analysis (cf. Sect. 2.1.3) with a frame size of 20 ms and a step of 10 ms is performed for all LLDs, except F_0 related

LLDs (groupB—which includes Jitter and Shimmer) for which a frame size of 60 ms (step width remains at 10 ms) is used in order to not distort low pitch of male voices down to approx. 50 Hz. No pre-emphasis filter is applied, the frame is multiplied by a Hamming window function (Sect. 2.1.3). A Gaussian window with $\sigma = 0.4$ is applied to the 60 ms frames. All LLDs are smoothed with a symmetric moving average window of $W = 3$ frames length (one previous, one current, and one future frame), and first order delta regression coefficients with a context window size of $W = 2$ (Sect. 2.3.2) are computed for each LLD.

The functionals which are applied to the LLD are given in the following table (functionals marked with ^{B} are only applied to the LLDs in group B, functionals marked with ^{A} are only applied to the LLDs in group A, and functionals marked with * are not applied to the delta LLDs, while functionals marked with $^{\delta}$ are only applied to delta LLDs):

		#
Arithmetic mean$^{A^*,B}$	see Sect. 2.4.2.1,	1
Positive Arithmetic mean$^{A_\delta,B}$	arithmetic mean of positive values only (μ_{a+}, Sect. 2.4.2.1),	1
Root-quadratic mean	see Sect. 2.4.2.1,	1
Flatness	see Sect. 2.4.2.1,	1
Moments	standard deviation, skewness, kurtosis (Sect. 2.4.2.2),	3
Quartiles	quartiles 1–3 and inter-quartile ranges 1–2, 2–3, 1–3 (Sect. 2.4.2.4)	6
1 %-tile	as robust substitute for the minimum value	1
99 %-tile	as robust substitute for the maximum value	1
Range 1–99 %	robust range of the signal expressed as percentile 1 to 99 range	1
Max./Min. position	position of the maximum/minimum value relative to the input length,	2
Full range	full range from single minimum value to single maximum value. This feature has been re-added, since together with the percentile range it adds information on the occurrence of LLD outliers.	1
Linear regression$^{A^*,B}$	coefficient a (slope) *and* offset b (Sect. 2.4.2.6)	2
Regression error$^{A^*,B}$	quadratic error $\overline{e}$ between line and actual signal (Sect. 2.4.2.6)	1

Quad. regression$^{A^*,B}$	quadratic regression coefficients a and b *and* offset c (Sect. 2.4.2.6)	3
Quad. regr. error$^{A^*,B}$	quadratic error $\overline{e}$ between parabola (estimated from quadratic regression) and actual signal (Sect. 2.4.2.6)	1
Centroid$^{A^*,B}$	temporal centroid as described in Sect. 2.4.2.5, time scale as relative percentage of segment length ($t(n) = n/N$)	1
Peak mean^{A}	mean of peak amplitudes, distance of mean of peak amplitudes to the arithmetic mean, *and mean of peak amplitudes relative to the arithmetic mean* (Sect. 2.4.2.8)	3
Peak distance^{A}	mean and standard deviation of peak to peak distances Sect. 2.4.2.8	2
Peak range^{A}	range of peak amplitude values, absolute value and value relative to the arithmetic mean of $x(n)$,	2
Valley range^{A}	range of valley amplitude values relative to the arithmetic mean of $x(n)$,	1
Peak-valley slopes^{A}	mean and standard deviation of rising slopes (valley to peak) and falling slopes (peak to valley),	4
Segment length^{A}	mean, minimum, maximum, and standard deviation of segment lengths; segmentation is performed by relative thresholds (25 % for all LLD in this group and additionally 75 % only for auditory RASTA bands, spectral features, and MFCC)—see Sect. 2.4.2.9	4
Up-level time	up-level time 25, 50, 75, and 90 %; all common values (cf. Schuller 2006) were included in a brute-forcing manner to be able to evaluate their use (Sect. 2.4.2.7); *down-level times were removed because they are redundant with the respective (inverse) up-level times—for the 25, 50 and 75 %*; as 10 % up-level time is not contained, there is no replacement for the 90 % down-level time, however, for consistency it has been removed nonetheless;	4
Rise-time	relative time in which the signal is rising (*fall-time has been removed because it is redundant for signals with more than a few frames*) (Sect. 2.4.2.7),	1
Curvature-time	relative time in which signal has left curvature (Sect. 2.4.2.7)—*right curvature was removed due to redundancy,*	1

LP gain	LP analysis gain (error amplitude) (see Sects. 2.4.3.2 and 2.2.7.1)	1
LP coefficient 1–5	see Sect. 2.4.3.2	5
	Total # descriptors:	55

All functionals applied to group B LLDs are applied only to regions where the LLDs is non-zero. The time scale is thereby altered because the regions with zero values are removed from $x(n)$, which makes $x(n)$ shorter, i.e., for the new length N' the following holds true: $N' < N$.

A.1.6 Details: 2013 Audio-Visual Emotion Challenge Set (AVEC 2013)

As in the IS11, IS12, and ComParE sets, the AVEC 2013 feature sets contain two groups of LLDs: group A with spectral, cepstral, and energy related LLDs, and group B with voicing related LLDs. Group A contains 32 LLDs, and group B contains 6 LLDs. A total of 42 functionals is considered in the set, however, only 41 are applied to the LLDs in group A, 20 to the delta LLDs of group A, 31 to the LLDs in group B, and 20 to the delta LLDs of group B. This combination results in 2,258 features. In addition to these, 10 temporal statistics are included: *Ratio of non-zero values* (ratio of non-zero values, i.e., percentage of voiced frames of total frames), mean, min. max. standard deviation of the voiced segment (i.e., where $F_0 > 0$) lengths (as in the IS12 and ComParE sets), mean, min. max. standard deviation of unvoiced segment (i.e., where $F_0 = 0$) lengths (see Sect. 2.4.2.9), and the duration of the input (in seconds). In total the AVEC 2013 audio baseline feature set contains **2,268 features**.

In group A, the following 32 spectral, cepstral, and energy related LLDs are contained:

		#
Loudness	computed as the sum of a simplified auditory spectrum $X_{P,aud}(b)$ (cf. Sect. 2.2.9.3): 26 Mel-frequency bands (triangular filters, power spectrum), range 20–8000 Hz, auditory weighting with equal loudness curve (Eq. (2.177)) and cubic-root compression (exponent 0.33 in Eq. (2.180)),	1
ZCR	expressed in units of crossings per second (see Sect. 2.2.1.1),	1
MFCC 1–16	MFCC coefficients 1–16 computed from power Mel-spectrum (26 bands, triangular filter, range 20–8000 Hz—Sect. 2.2.3.4), cepstral liftering with $L = 22$ (Sect. 2.2.10.1),	10

Spectral RoP	25, 50, 75, and 90 % spectral RoPs (Sect. 2.2.4.9),	4
Spectral band energy	spectral energy in bands 250–650 Hz, 1–4 kHz (Sect. 2.2.4.1),	2
Spectral flux	normalised spectral difference between two successive frames (Sect. 2.2.4.12),	1
Spectral entropy	as defined in Sect. 2.2.4.8,	1
Spectral flatness	see Sect. 2.2.4.5,	1
Spec. moments	spectral variance, spectral skewness, and spectral kurtosis (Sect. 2.2.4.7),	3
Harmonicity (spectral)	see Sect. 2.2.4.13,	1
Sharpness (spectral)	auditory sharpness (psychoacoustically scaled spectral centroid), see Sect. 2.2.4.10.	1
	Total # descriptors:	26

Additionally the following set of 6 voicing related LLD is considered:

		#
$\mathbf{F_0}$	via SHS method (Sect. 2.2.11.4), with spectral smoothing, enhancement, and auditory weighting; $F_0 = 0$ for unvoiced regions; voicing probability threshold $\theta_v = 0.7$; range 42–620 Hz; 15 harmonics, compression 0.85; "*greedy*" peak detection (Sect. 2.2.11.4); with Viterbi post-smoothing with the following settings: $w_{Tvv} = 10$, $w_{Tvuv} = 10$, $w_{Range} = 1$, $w_{Thr} = 4$, $w_{Tvvd} = 5$, and a maximum path buffer length of 0.9 s; all other parameters remain at the defaults given in Sect. 2.2.11.5. An absolute RMS energy threshold of 0.0008 is applied to the final F_0 values and F_0 in frames with an energy value below the threshold is set to 0 (the energy is computed from the corresponding 60 ms frame here)—this removes noisy F_0 values which the SHS algorithm produces for low energy frames.	1
Prob. of voicing	raw voicing probability obtained from the SHS algorithm (Sect. 2.2.11.4)	1
Jitter	relative search range $\alpha = 0.25$, minimum correlation 0.5, see Sect. 2.2.13.1	1
δ **Jitter**	see Sect. 2.2.13.1	1
Shimmer	see Sect. 2.2.13.2	1

log. HNR	logarithmic HNR $HNR_{wf,log}$ computed via wave form matching as defined in Sect. 2.2.13.3.	1
	Total # descriptors:	6

As in the other sets, prior to computation of the LLDs, the audio samples are scaled to the range $[-1; +1]$, with the exception of MFCC for which a scaling to the 16-bit sample range $[-32767; 32768]$ is used, in order to be numerically compatible with the HTK toolkit (Young et al. 2006). Short-time analysis (cf. Sect. 2.1.3) with a frame size of 25 ms and a step of 10 ms is performed for all LLDs, except F_0 related LLDs (groupB—which includes Jitter and Shimmer) for which a frame size of 60 ms (step width remains at 10 ms) is used in order to not distort low pitch of male voices down to approx. 50 Hz. No pre-emphasis filter is applied in this set, the frame is multiplied by a Hamming window function (Sect. 2.1.3). A Gaussian window with $\sigma = 0.4$ is applied to the 60 ms frames. All LLDs are smoothed with a symmetric moving average window of $W = 3$ frames length (one previous, one current, and one future frame), and first order delta regression coefficients with a context window size of $W = 2$ (Sect. 2.3.2) are computed for each LLD.

To the above LLDs the following 42 functionals are applied (functionals marked with ^{A} are only applied to the spectral and energy related LLDs, *not* the voicing related LLDs, and functionals marked with * are not applied to the delta LLDs, while functionals marked with $^\delta$ are only applied to delta LLDs):

		#
Arithmetic mean*	see Sect. 2.4.2.1,	1
Positive Arithmetic mean$^\delta$	arithmetic mean of positive values only (μ_{a+}, Sect. 2.4.2.1),	1
Root-quadratic mean	see Sect. 2.4.2.1,	1
Flatness	see Sect. 2.4.2.1,	1
Moments	standard deviation, skewness, kurtosis (Sect. 2.4.2.2),	3
Quartiles	quartiles 1–3 and inter-quartile ranges 1–2, 2–3, 1–3 (Sect. 2.4.2.4)	6
1 %-tile	as robust substitute for the minimum value	1
99 %-tile	as robust substitute for the maximum value	1
Range 1–99 %	robust range of the signal expressed as percentile 1 to 99 range	1
Up-level time	up-level time 25, 50, and 90 %; all common values (cf. Schuller 2006) were included in a brute-forcing manner to be able to evaluate their use (Sect. 2.4.2.7)	4

Rise-time	relative time in which signal contour is rising (Sect. 2.4.2.7)	1
Segment length^{A*}	mean, minimum, maximum, and standard deviation of segment lengths; segmentation is performed by relative thresholds (25 % for all LLD)—see Sect. 2.4.2.9	4
Linear regression*	coefficient a (slope) (Sect. 2.4.2.6)	1
Regression error*	linear error $\overline{e}$ between line and actual signal (Sect. 2.4.2.6)	1
Quad. regression*	quadratic regression coefficients a (Sect. 2.4.2.6)	1
Quad. regr. error*	linear error $\overline{e}$ between parabola (estimated from quadratic regression) and actual signal (Sect. 2.4.2.6)	1
Peak-valley slopes*	mean and standard deviation of rising slopes (valley to peak) and falling slopes (peak to valley),	4
Peak distance*	mean and standard deviation of peak to peak distances Sect. 2.4.2.8	2
Peak mean*	mean of peak amplitudes relative to the arithmetic mean (Sect. 2.4.2.8)	1
Peak range*	range of peak amplitude values relative to the arithmetic mean of $x(n)$,	1
Valley range*	range of valley amplitude values relative to the arithmetic mean of $x(n)$,	1
LP gain^{A*}	LP analysis gain (error amplitude) (see Sects. 2.4.3.2 and 2.2.7.1)	1
LP coefficient 1–5^{A*}	see Sect. 2.4.3.2	5
	Total # descriptors:	44

A.1.7 Details: Geneva Minimalistic Acoustic Parameter Set (GeMAPS)–v.o.g (pre-release)

The Geneva Minimalistic Acoustic Parameter Set contains the following compact set of 18 LLDs:

		#
Pitch	logarithmic F_0 on a semitone frequency scale starting at 27.5 Hz (semitone 0); however, as the value of 0 is reserved for unvoiced frames, every value below semitone 1 (29.136 Hz) is clipped to the minimum value of 1. F_0 is computed via the SHS method (Sect. 2.2.11.4), with spectral smoothing, enhancement, and auditory weighting; $F_0 = 0$ for unvoiced regions; voicing probability threshold $\theta_v = 0.7$; range 55–1000 Hz; 15 harmonics, compression 0.85; "greedy" peak detection (Sect. 2.2.11.4); with Viterbi post-smoothing with the following settings: $w_{Tvv} = 10$, $w_{Tvvd} = 5$, $w_{Thr} = 4$, $w_{Tvuv} = 10$, $w_{Range} = 1$, and a maximum path buffer length of 0.4 s; all other parameters remain at the defaults given in Sect. 2.2.11.5. An absolute RMS energy threshold of **0.001** is applied to the final F_0 values and F_0 in frames with an energy value below the threshold is set to 0 (the energy is computed from the corresponding 60 ms frame here)—this removes the noisy F_0 values which the SHS algorithm produces for low energy frames,	1
Jitter	local (period to period) jitter (average over all periods in one 60 ms frame), relative search range $\alpha = 0.1$, minimum correlation 0.5 (see Sect. 2.2.13.1),	1
Shimmer	relative amplitude difference ratios in dB (see Sect. 2.2.13.2),	1
Loudness	computed as the sum of a simplified auditory spectrum $X_{P,aud}(b)$ (cf. Sect. 2.2.9.3): 26 Mel-frequency bands (triangular filters, power spectrum), range 20–8000 Hz, auditory weighting with equal loudness curve (Eq. (2.177)) and cubic-root compression (exponent 0.33 in Eq. (2.180)),	1
HNR	in dB. F_0 is computed via the SHS algorithm with Viterbi smoothing, as described above for *Pitch*; the HNR is then computed according to Eq. (2.234) from the ACF of a 60 ms frame and converted to dB with Eq. (2.236),	1
Spectral Slope	from 0–500 Hz and 500–1500 Hz, computed from a log-power spectrum (see Sect. 2.2.4.2),	2
Alpha Ratio	given in dB, see Sect. 2.2.4.4,	1
Hammarberg Index	see Sect. 2.2.4.3,	1

Formant 1	centre frequency (in Hz) and bandwidth (in Hz) of F_1 via Linear Predictive Coding (LPC) polynomial roots (see Sect. 2.2.8.1); relative level/strength of F_1, expressed as the ratio of the level of the F_0 harmonic closest to F_1 to the level of F_0 itself,	3
Formant 2–3	relative level of F_2 and F_3 (see Formant 1 for description), and centre frequency,	4
Harmonics	harmonic ratios, computed from the relative amplitudes of F_0 harmonics: H1–H2 and H1–A3 (see Sect. 2.2.12).	2
	Total # descriptors:	18

All LLDs signals are smoothed over time with a moving average filter of length 3 frames. In contrast to previous, brute-forced feature sets, no delta regression coefficients are computed from the LLDs in this set. F_0, harmonic features, HNR, jitter, and shimmer are computed from 60 ms frames multiplied with a Gaussian window with $\sigma = 0.4$. Loudness, spectral slope, Formants, Harmonics, Hammarberg Index, and Alpha Ratio are computed from 25 ms windows with a Hamming window applied.

Arithmetic mean and **Coefficient of Variation** (Sect. 2.4.2.2) are applied as functionals to all 18 LLDs, yielding 36 parameters. To pitch and loudness the following functionals are additionally applied: **percentiles 20, 50, and 80**, the **range of percentiles 20—80**, and the **mean and standard deviation of the slope of rising/falling signal parts**. Functionals applied to the LLDs pitch, jitter, shimmer, and all formant related LLDs, are applied to voiced regions only. This leads to 52 parameters. Additionally, 6 temporal features are included: the rate of loudness peaks per second (see Sect. 2.4.2.8), mean length and standard deviation of continuous voiced segments ($F_0 > 0.0$, see *Pitch* above), mean length and standard deviation of unvoiced segments ($F_0 = 0$; approximating pauses), and the rate of voiced segments per time (approximating the pseudo syllable rate). Overall, **58 features** are contained in this minimalistic parameter set.

An *extension* to the minimalistic set is proposed which contains the following 10 extra LLDs:

		#
Spectral Energy Proportions	of energy in 0–500 Hz band and 0–1000 Hz band to total energy of each frame,	2 2
MFCC 1–4	first four cepstral coefficients, computed from a 26-band power Mel-spectrum (20–8000 Hz) Sect. 2.2.3.4), cepstral liftering with $L = 22$ (Sect. 2.2.10.1)	4
Pitch (linear)	additionally to log-Pitch, linearly scaled F_0,	1

Formant 2–3 bandwidth	via roots of LPC polynomial (see Sect. 2.2.8.1),	2
Spectral Flux	unnormalised, reflecting the dynamics of change in the spectral shape *and amplitude* from one frame to the next (see Sect. 2.2.4.12).	1
	Total # descriptors:	12

Arithmetic mean and standard deviation as functionals are applied to all of these 10 additional LLDs resulting in 20 additional features. To the linear scaled pitch the 8 additional functionals, which are also applied to the semitone scale pitch, are applied. Further, the equivalent sound level (LEq) is included. It is computed as the average RMS energy ($\overline{E}_{rms}$), converted to dB by $LEq = 10\log_{10}(\overline{E}_{rms})$. In total this extended Geneva Minimalistic Acoustic Parameter Set (eGeMAPS) consists of **87 features** (v.o.g pre-release).

A.1.8 Details: Music Genre Set (Genre)

This set contains 1,351 acoustic descriptors, optimised for music signal type discrimination. 87 of these features are rhythmic descriptors computed from Mel-frequency band envelopes (Sect. A.1.8.1). The remaining 1,264 features (see Sect. A.1.8.2) are constructed by applying 14 statistical functionals to 40 generic LLDs and 12 functionals to 12 CHROMA Energy-distribution Normalised Statistics (CENS) (cf. Sect. 2.2.14.3 and (Müller et al. 2005) for details) descriptors.

A.1.8.1 Rhythmic Descriptors

The 87 rhythmic features[3] are the same as those originally introduced and evaluated by Schuller et al. (2007):

Overlapping frames of 23.3 ms length (256 samples) are sampled every 10 ms from the audio signal (down-sampled to 11,025 Hz). A Hamming window is applied before the Fast Fourier Transformation (FFT). No auditory weighting of the spectrum is performed. The resulting 128 FFT magnitude bins are combined to 9 non-linear Mel-frequency bands with a triangular filter bank in the frequency domain. The band energies are converted to a logarithmic representation and then smoothed along the time axis by a 15 frame long raised-cosine low-pass filter. A local short-time context sensitive differential as described by Schuller et al. (2007) is computed from each band envelope after the low-pass filtering. The envelope of each band is fed to a comb-filter bank consisting of 57 filters, with filter delays τ ranging from 18 frames to 74 frames. The filtered band envelopes are added up for each comb filter across all 9 band and the total energy output of each filter is computed by summing over the full

[3]The rhythmic features are currently not part of the openSMILE code and thus not publicly available.

input length. Thus, 57 energy values correlated to signal periodicities are obtained, which are referred to as the *Tatum vector*. The Tatum tempo is obtained from this vector by peak picking. In the next step a second comb filter bank with filters at 18 multiples of the Tatum tempo is set up and the output is computed in analogy. This output is the 18 dimensional *meter vector*. From this vector the main tempo is derived, which is also used as a rhythmic descriptor feature. The remaining 11 descriptors are statistics derived from the Tatum vector, the estimated meter (duple and triple), and the two most likely Tatum tempo candidates.

For the rhythmic features meaningful values can only be extracted if the length of the analysed segment is longer than approx. 10 s. This is due to the need to have at least two to three beats for each metrical level present in the meter vector and at least a few beats for the main tempo and its close harmonics in order to have a stable output of these comb filters.

A.1.8.2 Statistical Descriptors

The 1,264 statistical features are composed of two sets: one set with $40 \cdot 2 \cdot 14 =$ 1,120 features (40 LLDs, 40 delta regression coefficients of these 40 LLDs, 14 statistical functionals), the other set with $12 \cdot 12$ features (12 CENS descriptors and 12 functionals). The 40 LLDs are:

		#
Loudness	computed via the narrow band approximation, Sect. 2.2.2, Eq. (2.43),	1
F_0 envelope	computed by holding the last F_0 value (computed via Cepstrum/ACF method Sect. 2.2.11.3) from the previous voiced region during the following unvoiced region; voicing probability derived from ACF peak amplitude,	2
ZCR	expressed in units of crossings per second (see Sect. 2.2.1.1),	1
Auditory spectral bands 1–12	$X_{P,aud}(b)$ (cf. Sect. 2.2.9.3): 12 Bark-frequency bands (triangular filters, power spectrum), range 0–5 kHz, auditory weighting with equal loudness curve (Eq. (2.177)), logarithmic output and cubic-root compression (exponent 0.11 in Eq. (2.180)),	12
LSP 0–7	8 LSP are computed from LP analysis with order $p = 8$ (auto-correlation method),	8
Spectral	roll-off point (95 %), entropy, centroid, and variance (cf. Sect. 2.2.4)	4

MFCC 1–12	first twelve cepstral coefficients, computed from a 16-band power Mel-spectrum (0–5 kHz) Sect. 2.2.3.4), cepstral liftering with $L = 22$ (Sect. 2.2.10.1)	12
	Total # descriptors:	40

For all LLDs the sampling rate of the input signal is reduced to 11 kHz in order to match the sampling rate the rhythmic descriptors are extracted at and to be computationally efficient. A window size of 64 ms (at a rate of 20 ms) is chosen and each window is multiplied with a Gaussian window ($\sigma = 0.25$)—except for ZCR and loudness. An additional 12 CENS descriptors are computed from 12 Chroma features in a standard way as described in Sect. 2.2.14.3. The Chroma descriptors were computed from FFT based magnitude spectra by transforming the spectrum to an octave scale and warping the spectrum over 6 octaves starting at the note A with 55 Hz. The LSP are derived from a LP analysis. More precisely, the LSPs are computed from a partial factorisation of the polynomial of the predictor obtained by the LP analysis. The LSPs are closely related to speech formants and therefore might add beneficial information if vocals are present in a song. All LLDs signals are post-smoothed with a moving average filter of length 3 frames. Delta-regression coefficients with a context window size of $W = 2$ are computed from the above 40 LLDs and the following list of 14 functionals is applied to the resulting set of 80 descriptors:

		#
Arithmetic mean	see Sect. 2.4.2.1,	1
Moments	standard deviation, skewness, kurtosis (Sect. 2.4.2.2),	3
Quartiles	quartiles 1–3 (Sect. 2.4.2.4),	3
1 %-tile	as robust substitute for the minimum value,	1
99 %-tile	as robust substitute for the maximum value,	1
Range 1–99 %	robust range of the signal expressed as percentile 1 to 99 range,	1
Up-/Down-level time	down-level time 25 % and up-level time 75 % (Sect. 2.4.2.7),	2
Peaks	**mean** of peak amplitudes and **number** of peaks (Sect. 2.4.2.8),	2
	Total # descriptors:	14

To the CENS LLDs, 12 functionals are applied, which include the above list without the 3 quartiles, and additionally the distance of the arithmetic mean to the minimum value.

A.2 Detailed Affect Classification Results Per Database

Detailed results per database of the systematic evaluation of speech and singing voice affect classification as introduced in Sect. 6.4 and evaluated in Sect. 6.3 are shown here. Average results over all the settings (13 Support Vector Machine (SVM) complexities, balancing, feature normalisation, feature set), except the one of interest are shown for classification of binary activation, and valence, as well as the original categories for each database.

A.2.1 ABC

Figure A.1 shows the results for binary activation and binary valence classification for the ABC database with various parameters. Figure A.2 (left) shows the results for

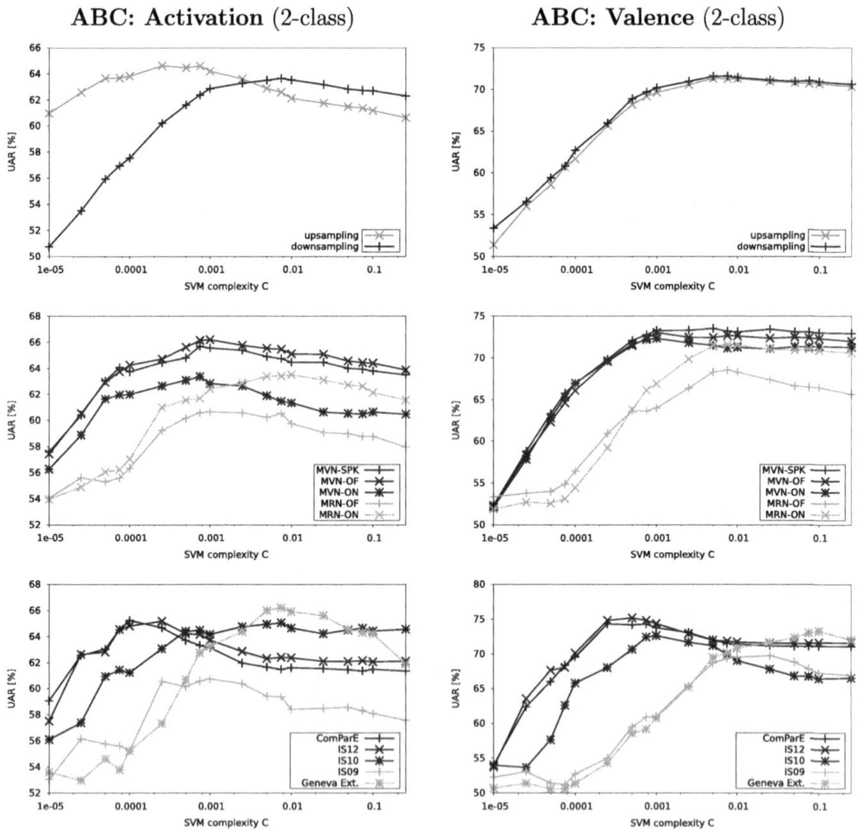

Fig. A.1 UAR performance over a range of 18 SVM complexities, for binary activation and binary valence classification; ABC database; comparison of down- and up-sampling for balancing of the training set, 5 feature normalisation methods, and 5 selected feature sets

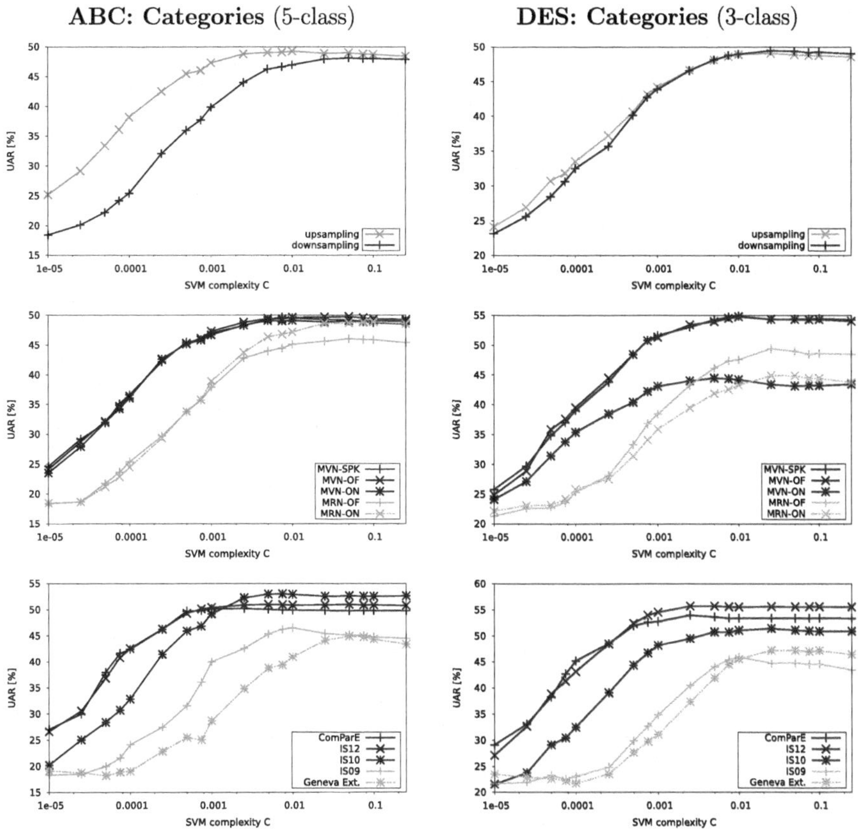

Fig. A.2 UAR multi-class performance over a range of 18 SVM complexities; ABC (*left*) and DES (*right*) databases; comparison of down- and up-sampling for balancing of the training set, 5 feature normalisation methods, and 5 selected feature sets

the classification of the five affect and speaker state categories of ABC. Averaged results for each configuration over 13 SVM complexities are printed for the ABC database in Table A.1.

A.2.2 TUM-AVIC

Figure A.3 (left) shows the results for binary activation classification for the TUM-AVIC database with various parameters. Figure A.3 (right) shows the results for the classification of the three level of interest categories of TUM-AVIC. Averaged results for each configuration over 13 SVM complexities are printed for the TUM-AVIC database in Table A.2. Note, that for AVIC the results for activation and valence

Table A.1 Average UAR results for the ABC database; average over 13 SVM complexities

Database (# categories)	Parameter		UAR (%)		
			Activation	Valence	Categories
ABC (5)	Balancing	Up-sampling	62.7	70.1	**47.8**
		Down-sampling	62.7	**70.4**	43.8
	Ft. norm.	MVN-SPK	64.6	**72.8**	47.8
		MVN-ON	61.7	71.4	47.8
		MVN-OF	**65.2**	72.2	**48.2**
		MRN-ON	62.5	68.9	43.5
		MRN-OF	59.6	65.9	41.7
	Ft. set	eGeMAPS	63.6	67.0	36.8
		IS09	59.3	65.4	41.6
		IS10	**64.5**	69.2	**50.6**
		ComParE	62.2	**72.3**	49.6
		Genre	62.0	71.4	48.1

$\Delta_{p5} = 0.4\,\%$

in Table A.2 are identical due to the chosen mapping of the Level of Interest (LOI) levels to arousal and valence labels (Sect. 6.3.1). Thus, no plots for binary valence classification are shown for TUM-AVIC in Table A.2.

A.2.3 DES

Figure A.4 shows the results for binary activation and binary valence classification for the DES database with various parameters. Figure A.2 (right) shows the results for the classification of the five affect and speaker state categories of DES. Averaged results for each configuration over 13 SVM complexities are printed for the DES database in Table A.3.

A.2.4 EMO-DB

Figure A.5 shows the results for binary activation and binary valence classification for the EMO-DB database with various parameters. Figure A.6 (left) shows the results for the classification of the seven emotion categories of the EMO-DB database. Averaged results for each configuration over 13 SVM complexities are printed for the EMO-DB database in Table A.4.

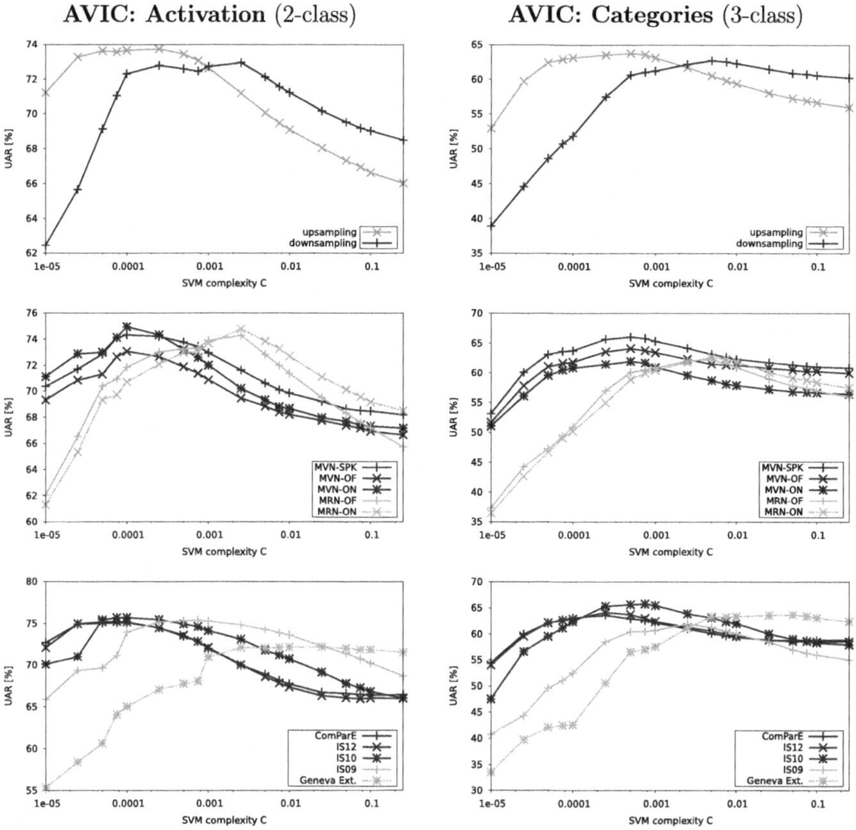

Fig. A.3 UAR performance over a range of 18 SVM complexities, for binary activation and 3-level Level of Interest (LOI) classification; TUM-AVIC database; comparison of down- and up-sampling for balancing of the training set, 5 feature normalisation methods, and 5 selected feature sets

A.2.5 eNTERFACE'05

Figure A.7 shows the results for binary activation and binary valence classification for the eNTERFACE'05 database with various parameters. Figure A.6 (right) shows the results for the classification of the six emotion categories of the eNTERFACE'05 database. Averaged results for each configuration over 13 SVM complexities are printed for this database in Table A.5.

Table A.2 Average UAR results for the TUM-AVIC database; average over 13 SVM complexities

Database (# categories)	Parameter		UAR (%)		
			Activation	Valence	Categories
AVIC (3)	Balancing	Up-sampling	69.82	69.82	60.03
		Down-sampling	**71.14**	**71.14**	**61.10**
	Ft. norm.	MVN-SPK	70.74	70.74	**63.15**
		MVN-ON	69.75	69.75	58.77
		MVN-OF	69.05	69.05	61.75
		MRN-ON	**71.92**	**71.92**	59.75
		MRN-OF	70.96	70.96	59.41
	Ft. set	eGeMAPS	70.91	70.91	60.72
		IS09	**73.16**	**73.16**	58.99
		IS10	71.00	71.00	**62.12**
		ComParE	69.26	69.26	60.45
		Genre	70.79	70.79	61.28

$\Delta_{p5} = 0.2\,\%$

A.2.6 GEMEP

Figure A.8 shows the results for binary activation and binary valence classification for the Geneva Multimodal Emotion Portrayals (GEMEP) database with various parameters. Figure A.9 (left) shows the results for the classification of the 12 emotion categories of the GEMEP database. Averaged results for each configuration over 13 SVM complexities are printed for this database in Table A.6. No significant difference between up-sampling and down-sampling can be seen because the GEMEP database is (almost) balanced by design.

A.2.7 SAL

Figure A.10 shows the results for binary activation and binary valence classification for the Belfast Sensitive Artificial Listener (SAL) database with various parameters. Figure A.9 (right) shows the results for the classification of the four activation/valence space quadrants of the SAL database. Averaged results for each configuration over 13 SVM complexities are printed for this database in Table A.7.

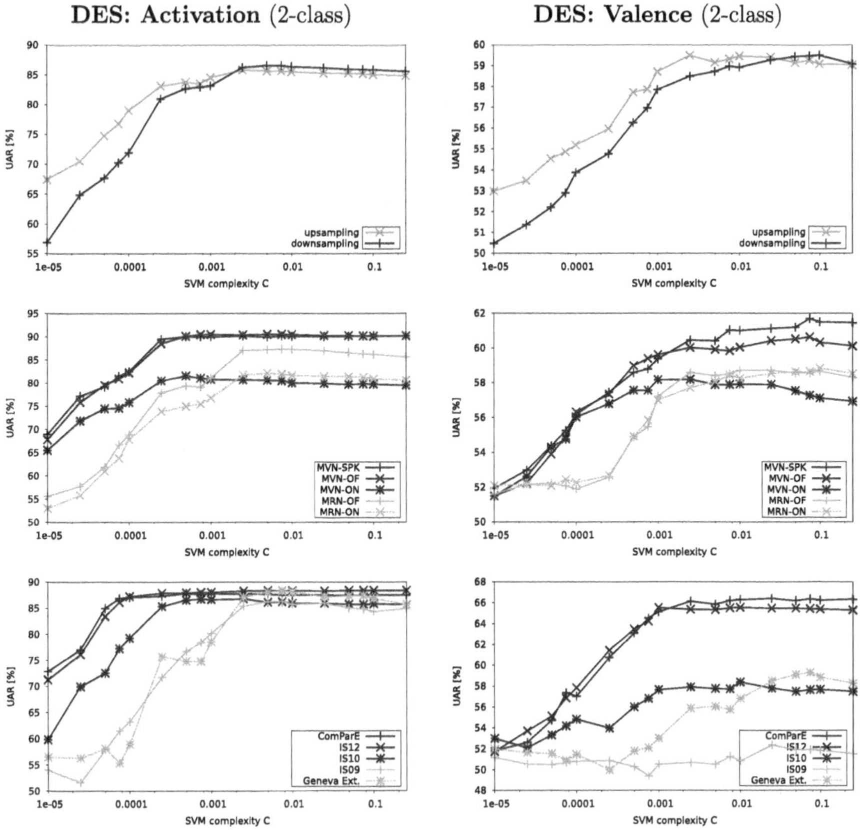

Fig. A.4 UAR performance over a range of 18 SVM complexities, for binary activation and binary valence classification; DES database; comparison of down- and up-sampling for balancing of the training set, 5 feature normalisation methods, and 5 selected feature sets

A.2.8 GeSiE

Figure A.11 shows the results for binary activation and binary valence classification for the Geneva Singing Voice Emotion (GeSiE) database with various parameters. Figure A.12 (left) shows the results for the classification of the 11 emotion categories of the singing voice database. Averaged results for each configuration over 13 SVM complexities are printed for this database in Table A.8.

Table A.3 Average UAR results for the DES database; average over 13 SVM complexities

Database (# categories)	Parameter		UAR (%)		
			Activation	Valence	Categories
DES (5)	Balancing	Up-sampling	84.8	**58.7**	46.3
		Down-sampling	85.0	58.3	46.2
	Ft. norm.	MVN-SPK	**90.0**	**60.3**	**52.6**
		MVN-ON	80.3	57.6	42.9
		MVN-OF	**90.2**	59.8	**52.5**
		MRN-ON	79.5	57.4	39.9
		MRN-OF	84.5	57.5	43.5
	Ft. set	eGeMAPS	83.8	55.8	39.7
		IS09	82.8	51.1	40.0
		IS10	86.1	57.3	48.9
		ComParE	**87.6**	**65.3**	**52.9**
		Genre	83.3	57.8	46.8

$\Delta_{p5} = 0.4\,\%$

A.2.9 SUSAS

Figure A.13 shows the results for binary activation and binary valence classification for the SUSAS database with various parameters. Figure A.12 (right) shows the results for the classification of the 11 emotion categories of the singing voice database. Averaged results for each configuration over 13 SVM complexities are printed for this database in Table A.9.

A.2.10 VAM

Figure A.14 shows the results for binary activation and binary valence classification for the Vera-am-Mittag (VAM) database with various parameters. Figure A.15 (left) shows the results for the classification of the 4 activation/valence-space quadrants of the VAM database. Averaged results for each configuration over 13 SVM complexities are printed for this database in Table A.10.

A.2.11 AIBO

Figure A.15 (right) shows the results for the classification of the 5 affect classes the FAU-AIBO database for all 18 SVM complexities. Averaged results for each configuration over 13 SVM complexities are printed for this database in Table A.11. In contrast to the other databases, due to the size of FAU-AIBO, only downsampling

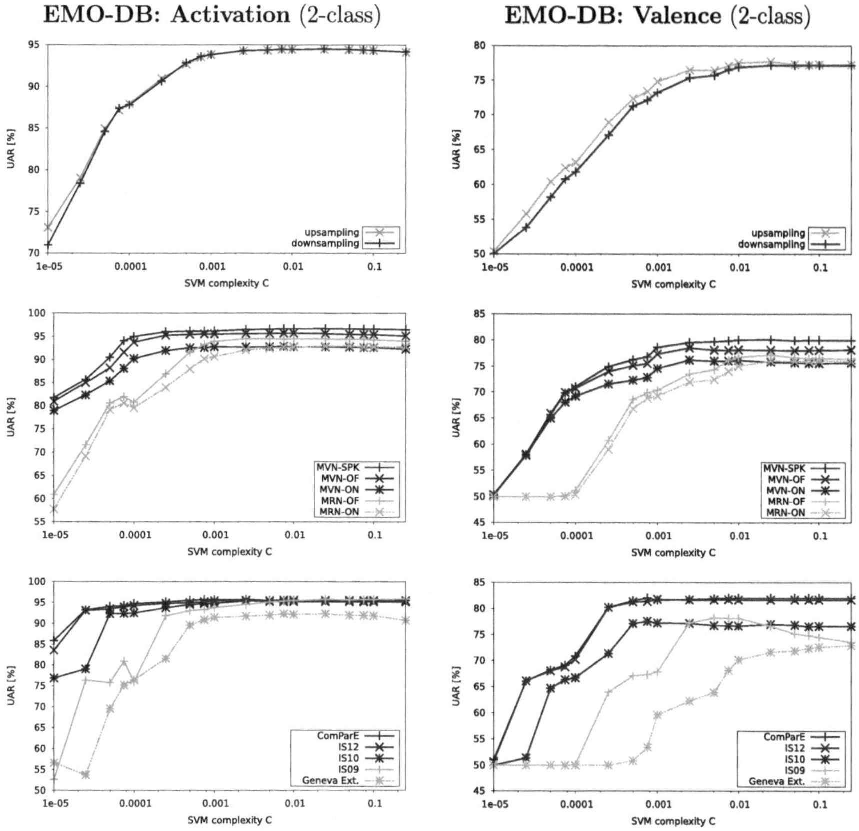

Fig. A.5 UAR performance over a range of 18 SVM complexities, for binary activation and binary valence classification; EMO-DB database; comparison of down- and up-sampling for balancing of the training set, 5 feature normalisation methods, and 5 selected feature sets

was applied as balancing method in all experiments. Also, speaker z-normalisation (MVN-SPK) could not be applied because the speaker IDs in the test of the official INTERSPEECH 2009 Challenge set (Schuller et al. 2009) are not given. The evaluation procedure where a training and an evaluation set are defined by Schuller et al. (2009) is kept and preferred over the 8-fold cross validation due to the size of the corpus and for comparability with other studies on the same data.

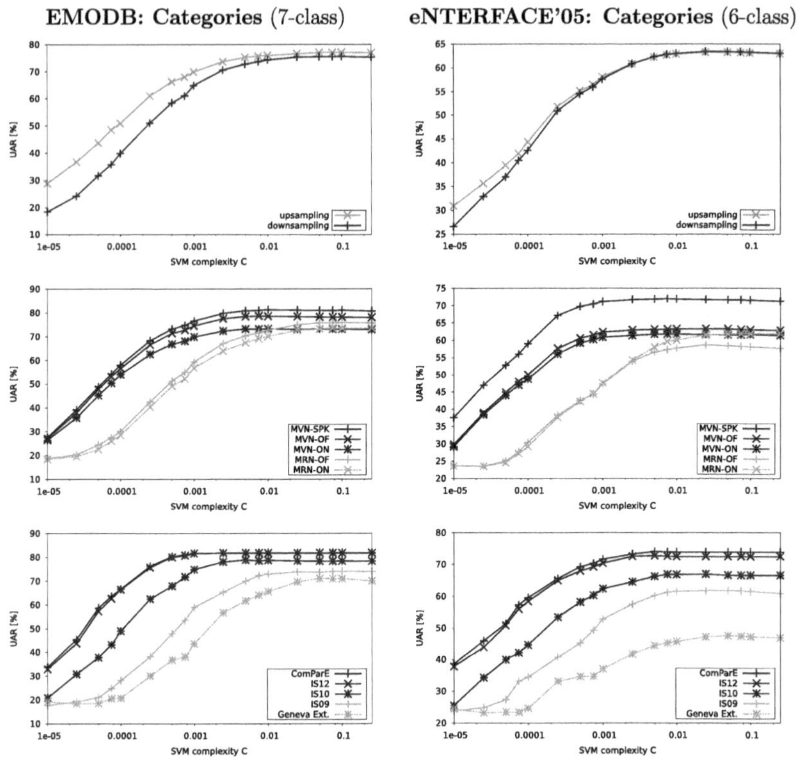

Fig. A.6 UAR multi-class performance over a range of 18 SVM complexities; EMODB (*left*) and ENTERFACE (*right*) databases; comparison of down- and up-sampling for balancing of the training set, 5 feature normalisation methods, and 5 selected feature sets

Table A.4 Average UAR results for EMO-DB; average over 13 SVM complexities

Database (# categories)	Parameter		UAR (%)		
			Activation	Valence	Categories
EMO-DB (7)	Balancing	Up-sampling	93.8	**75.7**	**73.2**
		Down-sampling	93.9	74.9	69.5
	Ft. norm.	MVN-SPK	**96.4**	**78.9**	**78.5**
		MVN-ON	92.6	74.9	71.1
		MVN-OF	95.5	77.3	**76.1**
		MRN-ON	91.3	72.1	64.2
		MRN-OF	93.5	73.3	66.8
	Ft. set	eGeMAPS	90.8	64.6	57.6
		IS09	94.7	73.3	65.3
		IS10	**95.2**	76.5	75.6
		ComParE	**95.5**	**81.8**	**81.2**
		Genre	92.7	75.8	69.2

$\Delta_{p5} = 0.4\,\%$

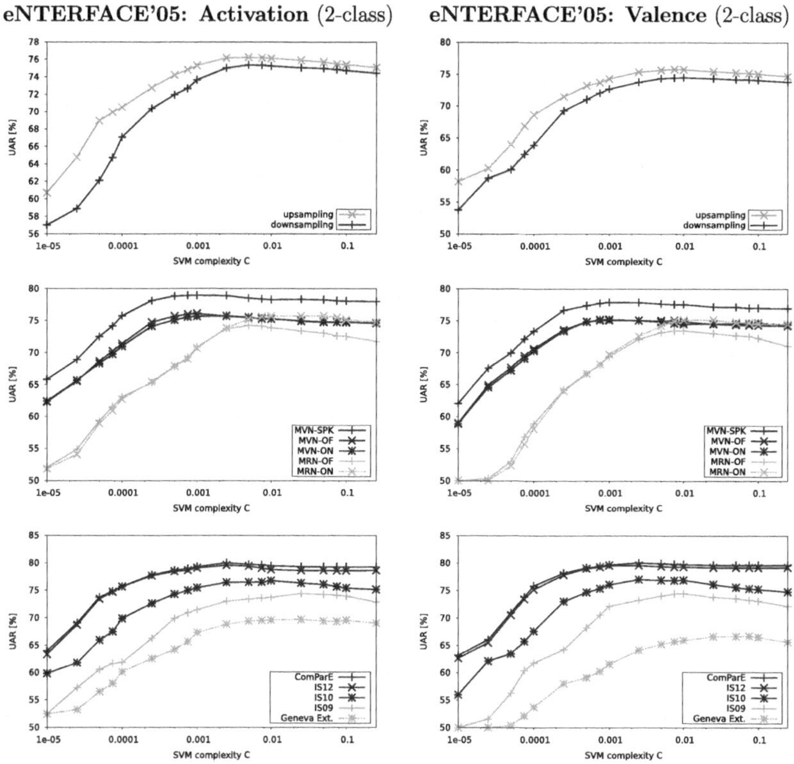

Fig. A.7 UAR performance over a range of 18 SVM complexities, for binary activation and binary valence classification; ENTERFACE database; comparison of down- and up-sampling for balancing of the training set, 5 feature normalisation methods, and 5 selected feature sets

Table A.5 Average UAR results for eNTERFACE'05; average over 13 SVM complexities

Database (# categories)	Parameter		UAR (%)		
			Activation	Valence	Categories
eNTERF. (6)	Balancing	Up-sampling	**75.33**	**74.72**	**60.48**
		Down-sampling	74.12	73.31	60.35
	Ft. norm.	MVN-SPK	**78.44**	**77.39**	**71.07**
		MVN-ON	75.10	74.70	60.84
		MVN-OF	75.25	74.68	**62.32**
		MRN-ON	73.11	72.29	54.84
		MRN-OF	71.74	71.01	53.00
	Ft. set	eGeMAPS	68.02	64.01	42.55
		IS09	72.45	72.11	56.64
		IS10	75.55	75.68	63.99
		ComParE	**79.23**	**79.57**	**72.42**
		Genre	76.25	76.70	61.67

$\Delta_{p5} = 0.3\,\%$

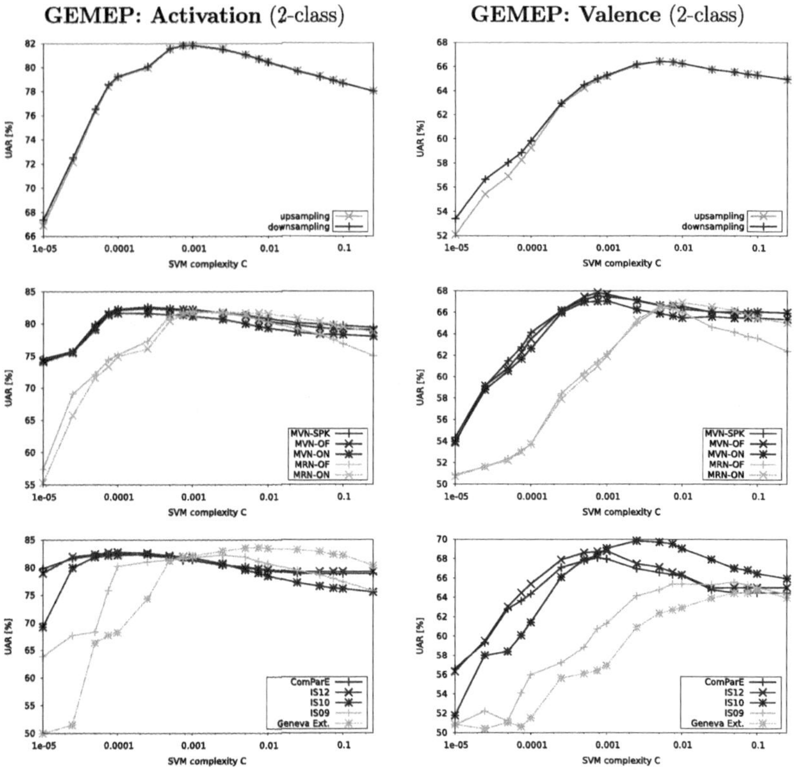

Fig. A.8 UAR performance over a range of 18 SVM complexities, for binary activation and binary valence classification; GEMEP database; comparison of down- and up-sampling for balancing of the training set, 5 feature normalisation methods, and 5 selected feature sets

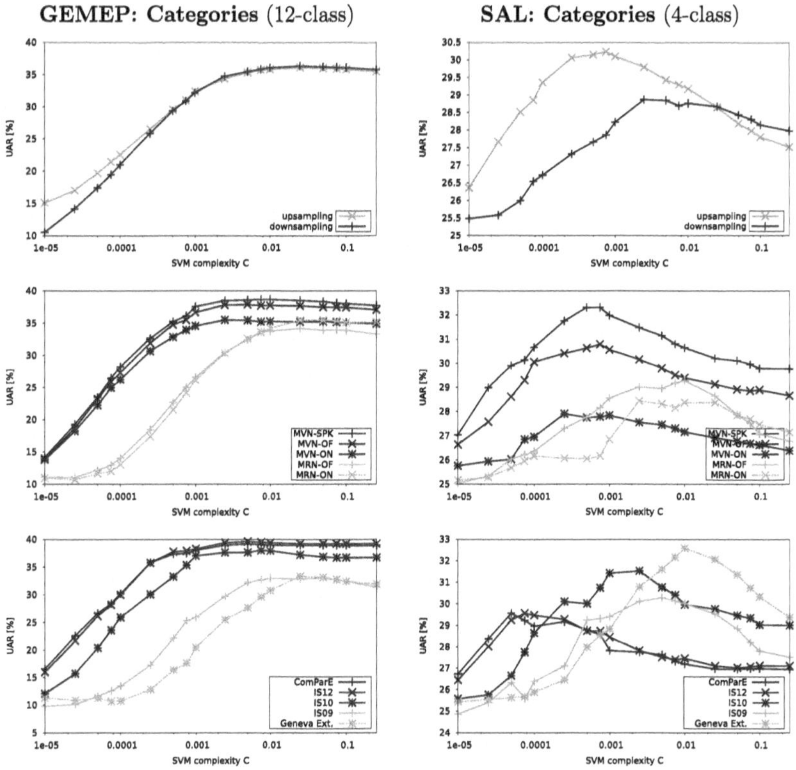

Fig. A.9 UAR multi-class performance over a range of 18 SVM complexities; GEMEP (*left*) and Belfast SAL (*right*) databases; comparison of down- and up-sampling for balancing of the training set, 5 feature normalisation methods, and 5 selected feature sets

Table A.6 Average UAR results for the GEMEP database; average over 13 SVM complexities

Database (# categories)	Parameter		UAR (%)		
			Activation	Valence	Categories
GEMEP (12)	Balancing	Up-sampling	80.28	65.34	33.84
		Down-sampling	80.25	65.38	33.89
	Ft. norm.	MVN-SPK	**81.01**	**66.57**	**37.44**
		MVN-ON	79.74	65.98	34.54
		MVN-OF	80.71	**66.58**	36.69
		MRN-ON	80.42	64.23	30.50
		MRN-OF	79.44	63.42	30.18
	Ft. set	eGeMAPS	**81.83**	61.19	26.48
		IS09	80.06	63.33	29.27
		IS10	79.02	**67.97**	36.24
		ComParE	80.05	66.13	**38.42**
		Genre	79.56	**67.75**	35.49

$\Delta_{p5} = 0.3\,\%$

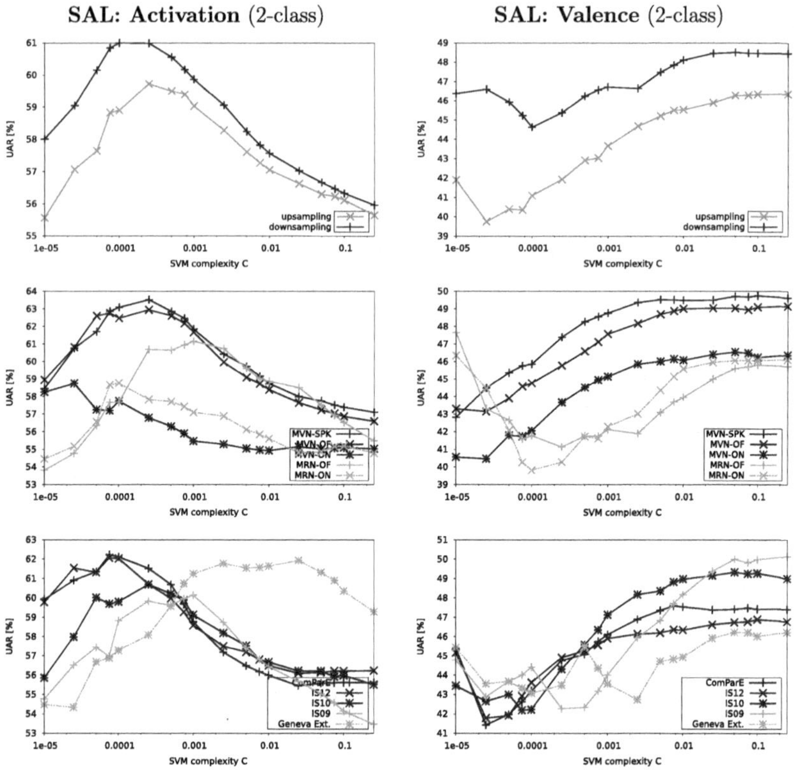

Fig. A.10 UAR performance over a range of 18 SVM complexities, for binary activation and binary valence classification; Belfast SAL database; comparison of down- and up-sampling for balancing of the training set, 5 feature normalisation methods, and 5 selected feature sets

Table A.7 Average UAR results for the Belfast SAL database; average over 13 SVM complexities

Database (# categories)	Parameter		UAR (%)		
			Activation	Valence	Categories
SAL (4q)	Balancing	Up-sampling	57.60	44.90	**29.10**
		Down-sampling	**58.21**	**47.49**	28.29
	Ft. norm.	MVN-SPK	**59.73**	**49.16**	**30.94**
		MVN-ON	55.39	45.73	27.24
		MVN-OF	59.30	48.24	29.67
		MRN-ON	56.11	44.18	27.46
		MRN-OF	58.98	43.64	28.16
	Ft. set	eGeMAPS	**60.77**	45.0	**30.22**
		IS09	57.07	46.9	29.04
		IS10	57.61	**48.0**	**30.12**
		ComParE	57.26	46.8	27.62
		Genre	56.41	45.1	28.61

$\Delta_{p5} = 0.2\,\%$

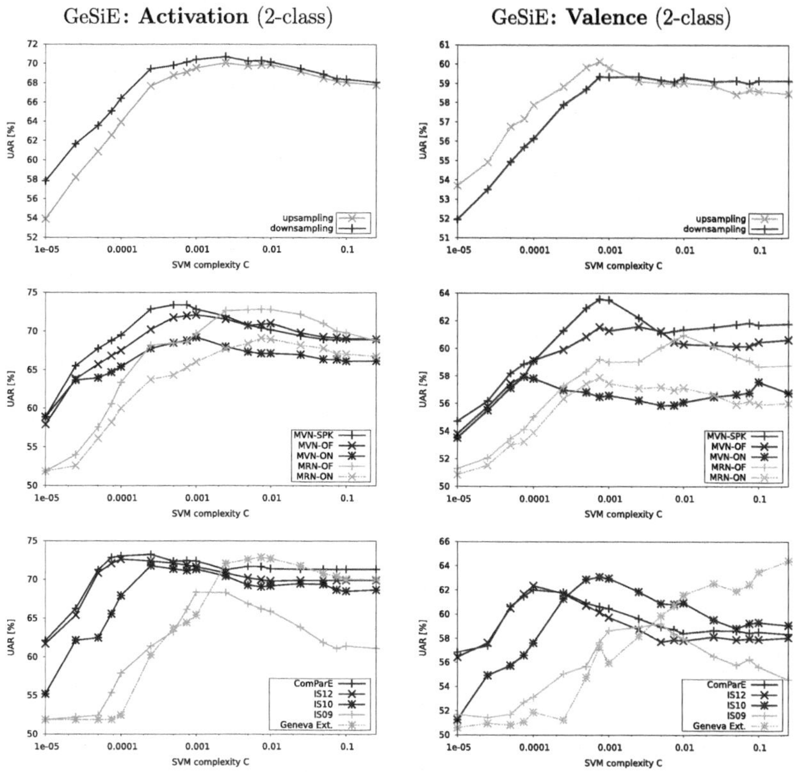

Fig. A.11 UAR performance over a range of 18 SVM complexities, for binary activation and binary valence classification; Geneva Singing Voice Emotion (GeSiE) database; comparison of down- and up-sampling for balancing of the training set, 5 feature normalisation methods, and 5 selected feature sets

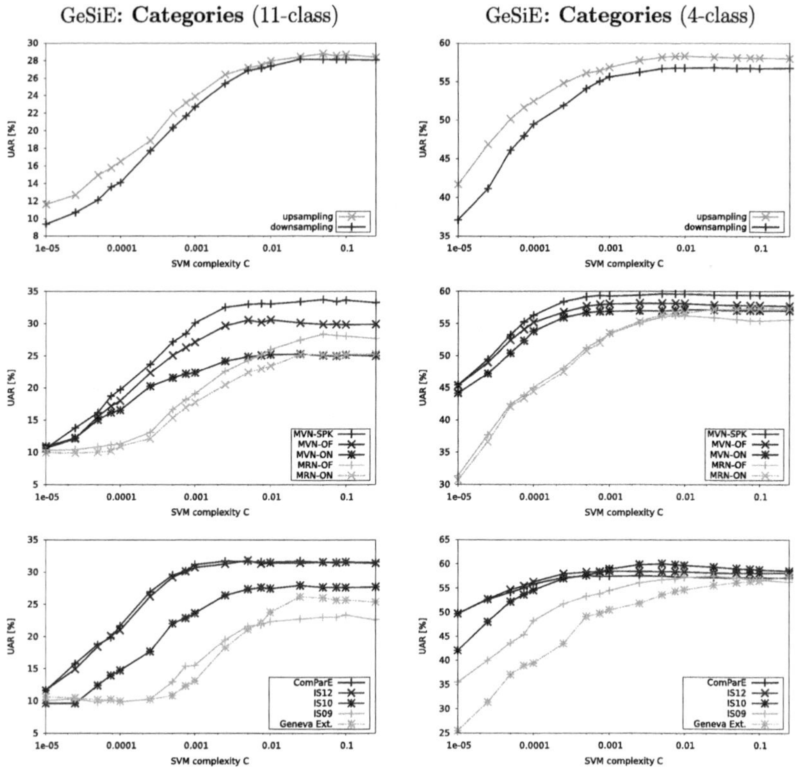

Fig. A.12 UAR multi-class performance over a range of 18 SVM complexities; Geneva Singing Voice Emotion—GeSiE (*left*) and SUSAS (*right*) databases; comparison of down- and up-sampling for balancing of the training set, 5 feature normalisation methods, and 5 selected feature sets.

Table A.8 Average UAR results for the Geneva Singing Voice database; average over 13 SVM complexities

Database (# categories)	Parameter		UAR (%)		
			Activation	Valence	Categories
GeSiE (11)	Balancing	Up-sampling	69.0	59.1	**26.1**
		Down-sampling	**69.6**	59.1	25.4
	Ft. norm.	MVN-SPK	**70.9**	**62.0**	**31.4**
		MVN-ON	67.4	56.6	23.9
		MVN-OF	**70.5**	60.7	28.6
		MRN-ON	67.0	56.8	21.3
		MRN-OF	**70.6**	59.3	23.5
	Ft. set	eGeMAPS	69.0	59.6	20.1
		IS09	64.3	57.0	19.6
		IS10	69.9	**60.8**	25.7
		ComParE	**71.8**	59.4	**30.9**
		Genre	68.7	58.8	26.9

$\Delta_{p5} = 0.5\,\%$

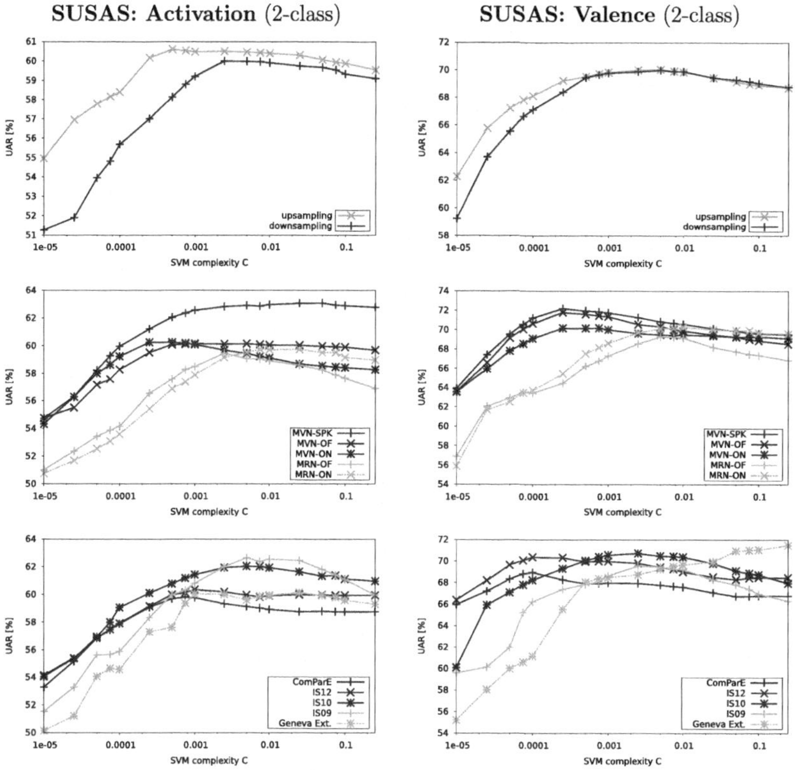

Fig. A.13 UAR performance over a range of 18 SVM complexities, for binary activation and binary valence classification; SUSAS database database; comparison of down- and up-sampling for balancing of the training set, 5 feature normalisation methods, and 5 selected feature sets

Table A.9 Average UAR results for the SUSAS database; average over 13 SVM complexities

Database (# categories)	Parameter		UAR (%)		
			Activation	Valence	Categories
SUSAS (4)	Balancing	Up-sampling	**60.25**	69.49	**57.47**
		Down-sampling	59.26	69.44	55.91
	Ft. norm.	MVN-SPK	**62.66**	**70.77**	**59.33**
		MVN-ON	59.28	69.62	56.90
		MVN-OF	60.00	70.18	57.84
		MRN-ON	58.66	69.18	55.01
		MRN-OF	58.20	67.60	54.37
	Ft. set	eGeMAPS	59.43	69.40	53.04
		IS09	**61.21**	68.18	55.75
		IS10	**61.37**	**69.75**	**58.91**
		ComParE	59.12	67.48	57.30
		Genre	59.06	71.57	58.32

$\Delta_{p5} = 0.2\,\%$

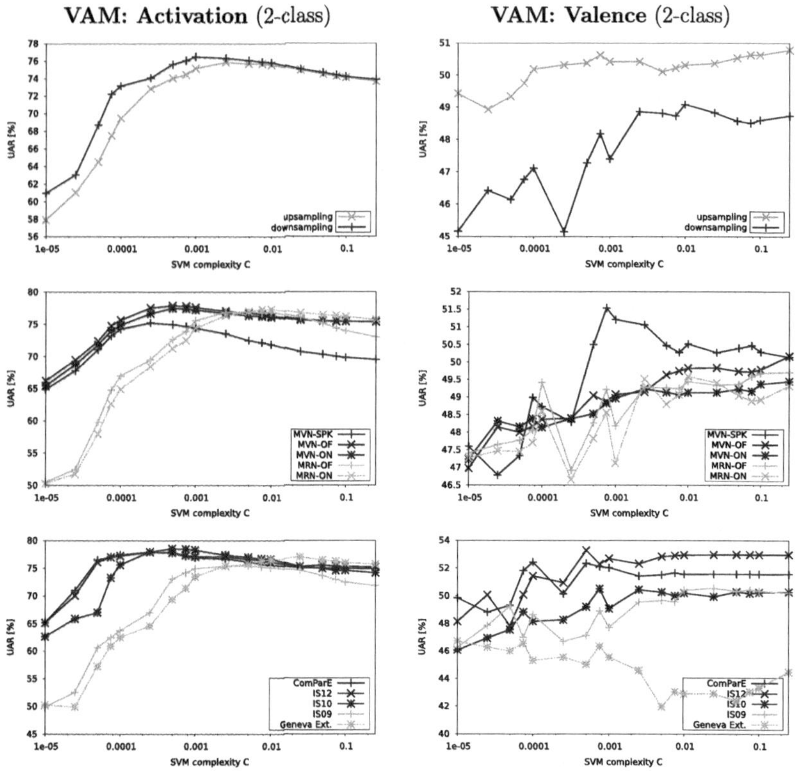

Fig. A.14 UAR performance over a range of 18 SVM complexities, for binary activation and binary valence classification; VAM database database; comparison of down- and up-sampling for balancing of the training set, 5 feature normalisation methods, and 5 selected feature sets

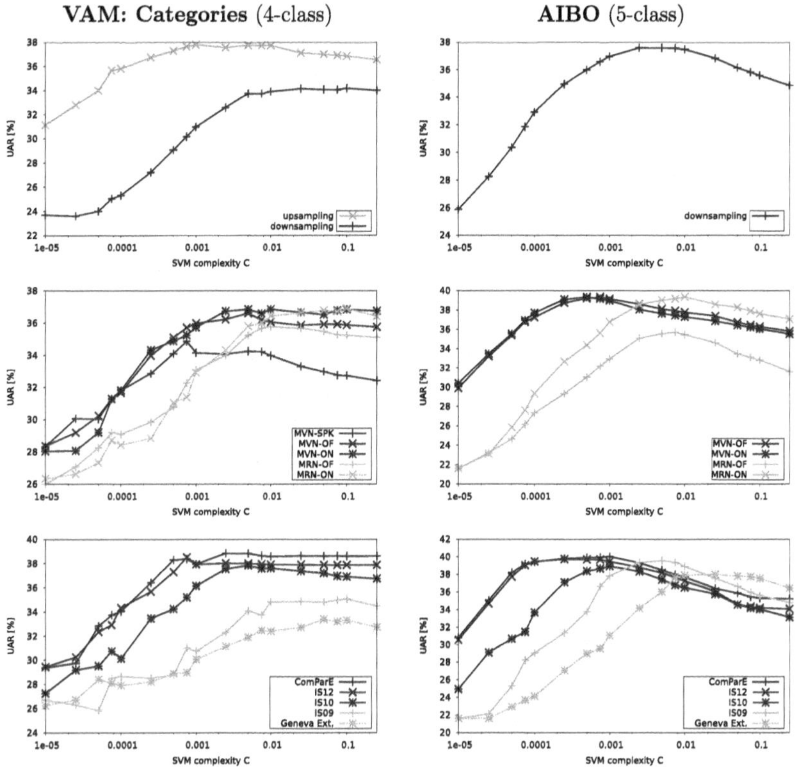

Fig. A.15 UAR multi-class performance over a range of 18 SVM complexities; VAM (*left*) and AIBO (*right*) databases; comparison of down- and up-sampling for balancing of the training set, 5 feature normalisation methods, and 5 selected feature sets

Table A.10 Average UAR results for the Vera-am-Mittag database; average over 13 SVM complexities

Database (# categories)	Parameter		UAR (%)		
			Activation	Valence	Categories
VAM (4q)	Balancing	Up-sampling	74.67	**50.44**	**37.28**
		Down-sampling	**75.27**	48.21	32.45
	Ft. norm.	MVN-SPK	72.27	**50.42**	33.59
		MVN-ON	**76.26**	49.05	**36.21**
		MVN-OF	**76.53**	49.46	35.78
		MRN-ON	75.06	48.67	34.63
		MRN-OF	74.74	49.04	34.11
	Ft. set	eGeMAPS	74.15	43.93	31.49
		IS09	73.51	49.35	32.94
		IS10	**76.49**	49.91	36.52
		ComParE	76.07	**51.58**	**38.38**
		Genre	73.92	50.13	33.82

$\Delta_{p5} = 0.3\,\%$

Table A.11 Average UAR results for the FAU-AIBO database; average over the 13 lowest(!) SVM complexities

Parameters		UAR (%)	
		Valence (2-class)	Categories (5-class)
Ft. norm.	MVN-ON	65.54	36.85
	MVN-OF	**66.61**	**36.93**
	MRN-ON	64.13	34.03
	MRN-OF	59.23	30.88
Ft. set	eGeMAPS	63.46	31.83
	IS09	**65.62**	33.98
	IS10	64.03	35.13
	IS11	64.29	36.83
	ComParE	64.54	**38.19**
	AVEC 2013	65.02	36.56

$\Delta_{p5} \leq 0.1\,\%$

References

A. Batliner, B. Möbius, Prosodic models, automatic speech understanding, and speech synthesis: towards the common ground?, in *The Integration of Phonetic Knowledge in Speech Technology*, ed. by W. Barry, W. Dommelen (Springer, Dordrecht, 2005), pp. 21–44

A. Batliner, J. Buckow, R. Huber, V. Warnke, E. Nöth, H. Niemann, Prosodic feature evaluation: Brute force or well designed?, in *Proceedings of the 14-th ICPhS*, vol. 3, pp. 2315–2318, San Francisco, CA, USA (1999)

A. Kießling, *Extraktion und Klassifikation prosodischer Merkmale in der automatischen Sprachverarbeitung*. Shaker (1997) (ISBN 978-3-8265-2245-1)

M. Müller, F. Kurth, M. Clausen, Audio matching via chroma-based statistical features, in *Proceedings of the 6th International Conference on Music Information Retrieval (ISMIR)*, pp. 288–295, London, UK (2005)

B. Schuller, *Automatische Emotionserkennung aus sprachlicher und manueller Interaktion*. Doctoral thesis, Technische Universität München, Munich, Germany, June 2006

B. Schuller, G. Rigoll, M. Lang, Hidden Markov model-based speech emotion recognition, in *Proceedings of ICASSP 2003*. IEEE, vol. 2, pp. II 1–4, Hong Kong, China, April 2003

B. Schuller, F. Eyben, G. Rigoll, Fast and robust meter and tempo recognition for the automatic discrimination of ballroom dance styles, in *Proceedings of ICASSP 2007*. IEEE, vol. I, pp. 217–220, Honolulu, HI, USA, April 2007

B. Schuller, S. Steidl, A. Batliner, F. Jurcicek, The INTERSPEECH 2009 emotion challenge, in *Proceedings of INTERSPEECH 2009*, pp. 312–315, Brighton, UK, September 2009

B. Schuller, A. Batliner, S. Steidl, F. Schiel, J. Krajewski, The INTERSPEECH 2011 speaker state challenge, in *Proceedings of INTERSPEECH 2011*. ISCA, pp. 3201–3204, Florence, Italy, August 2011

B. Schuller, S. Steidl, A. Batliner, E. Nöth, A. Vinciarelli, F. Burkhardt, R. van Son, F. Weninger, F. Eyben, T. Bocklet, G. Mohammadi, B. Weiss, The INTERSPEECH 2012 speaker trait challenge, in *Proceedings of INTERSPEECH 2012*. ISCA, Portland, OR, USA, September 2012

S. Young, G. Evermann, M. Gales, T. Hain, D. Kershaw, X. Liu, G. Moore, J. Odell, D. Ollason, D. Povey, V. Valtchev, P. Woodland, *The HTK Book*. Cambridge University Engineering Department, for HTK version 3.4 edition (2006)

Appendix B
Mel-Frequency Filterbank Parameters

Table B.1 shows the centre frequencies and bandwidths for 26 Mel-frequency bands from 0–8 kHz as used in Sects. 2.2.3.4 and 2.2.10.1.

Table B.1 Centre frequencies (f_c) and half bandwidths ($f_b/2$) of a non-linear spectral band filterbank with 26 triangular filters spaced equidistantly on the Mel- and Bark-frequency scales from 20 to 8,000 Hz

Band #	$f_{c,mel}$ (Hz)	$f_{bw,mel}/2$ (Hz)	$f_{c,bark}$ (Hz)	$f_{bw,bark}/2$ (Hz)
1	90	73	91	74
2	166	80	168	76
3	250	88	243	75
4	341	96	318	77
5	442	106	398	83
6	553	116	483	89
7	674	127	576	96
8	807	139	675	104
9	952	153	783	112
10	1112	167	899	122
11	1287	184	1027	133
12	1479	201	1166	146
13	1690	221	1318	160
14	1921	242	1486	177
15	2174	266	1672	197
16	2452	291	1880	220
17	2757	320	2133	248
18	3091	350	2376	281
19	3458	384	2676	321
20	3859	421	3019	371
21	4300	462	3418	433
22	4784	507	3886	513
23	5314	556	4443	616
24	5896	610	5118	754
25	6534	669	5952	901
26	7233	733	6921	1024

F. Eyben, *Real-time Speech and Music Classification by Large Audio Feature Space Extraction*, Springer Theses, DOI 10.1007/978-3-319-27299-3

Curriculum Vitae—Florian Eyben

Florian Eyben received his Diploma in information technology in 2008 from TUM (Munich University of Technology) and his doctoral degree (summa cum laude, with honours) in 2015 also from TUM, one of Germany's Excellence Universities. He is currently active in research and development of novel technologies for speech emotion analytics as Chief Technology Officer (CTO) at audEERING GmbH, Gilching, Germany. He focuses on real-time acoustic feature extraction and large-scale machine learning for Affective Computing, Voice Analytics, Computational Paralinguistics, and Music Information Retrieval. Most methods he developed during his PhD at TUM are available as open-source software in his award-winning openSMILE toolkit. The papers introducing the toolkit have been cited more than 700 times. He is an active reviewer for several conferences, including IEEE ICASSP, INTERSPEECH, and ISMIR, as well as several high-profile journals such as Speech Communication and IEEE Transactions on Affective Computing. He has made major contributions to the EU-FP7 SEMAINE project, and has worked for the EU project ASC-Inclusion, and national funded research projects on automatic speech recognition. He has worked as an intern with Toshiba Research, Cambridge, and Google UK on audio and speech analysis and text-to-speech projects. He has authored and co-authored over 110 publications in journals, books, and conference proceedings. His work has been cited over 3400 times and his H-index is 32.[4]

[4] According to Google Scholar.

F. Eyben, *Real-time Speech and Music Classification by Large Audio Feature Space Extraction*, Springer Theses, DOI 10.1007/978-3-319-27299-3

Professional Experience

01/2013–today	**audEERING**, Gilching, Germany. Chief Technology officer, leading development of speech emotion recognition and voice analytics technology.
03/2008–02/2015	**TUM Institute for Human–Machine Communication, Ph. D. candidate and researcher.** Ph. D. topic on acoustic features for automated robust, real-time, real-world emotion and paralinguistics recognition in the Machine Intelligence and Signal Processing (MISP) Group, lead and supervised by Prof. Björn Schuller.
10/2011–03/2011	**Google**, London, UK. Research intern in the speech group, work in Text-To-Speech quality improvement.
01/2011–05/2011	**Toshiba Research**, Cambridge, UK. Worked on using emotion recognition algorithms from the openSMILE toolkit to improve emotional HMM-based Text-To-Speech systems.

Education

03/2008–06/2014	**Doctoral thesis (Ph. D.) in Electrical Engineering**, Technische Universität München (TUM), Munich, Germany; with honours (summa cum laude) grade. Topic: "Real-time Speech and Music Classification by Large Audio Feature Space Extraction".
09/2003–02/2008	**Diploma, Dipl.-Ing. (equiv. to M. Sc.) in Information Technology**, Technische Universität München (TUM), Munich, Germany. Graduated with honours grade.
09/2007–02/2008	**Diploma thesis** with topic "Detection of Non-Verbals and Disfluencies in Conversational Speech"
03/2006–07/2006	**Thesis** (equivalent to Bachelor's thesis) "High Level Rhythmic Audio Features for Robust Music Information Retrieval"
09/1994–05/2003	**German Abitur** (High School graduation). Graduated with honours grade (A+, 1.0).

Academic Awards

2013	**ACM Multimedia Open-Source Software Competition**. Honorable Mention for openSMILE 2.0.
2010	**ACM Multimedia Open-Source Software Competition**. Honorable Mention for openSMILE 1.0.
2008	**Best Lecture Award** for the Tutorial of the Pattern Recognition Lecture at TUM.
1997	**Jugend Forscht**. Winner of the first level of the German young scientist contest with an "easy to use computer program for foreign language vocabulary training".

Printed by Printforce, the Netherlands